AA007549

18th Annual International Astrophysics Conference 2019

The Physics of Energetic Particles: Universal Processes from the Solar Corona to the Very Local Interstellar Medium and the Physics they Enable

Journal of Physics: Conference Series Volume 1332

Pasadena, California, USA
18 - 22 February 2019

Editor:

G.P. Zank

ISBN: 978-1-5108-9988-9
ISSN: 1742-6588

Printed from e-media with permission by:

Curran Associates, Inc.
57 Morehouse Lane
Red Hook, NY 12571

Some format issues inherent in the e-media version may also appear in this print version.

This work is licensed under a Creative Commons Attribution 3.0 International Licence.
Licence details: http://creativecommons.org/licenses/by/3.0/.

No changes have been made to the content of these proceedings. There may be changes to pagination and minor adjustments for aesthetics.

Printed with permission by Curran Associates, Inc. (2020)

For permission requests, please contact the Institute of Physics at the address below.

Institute of Physics
Dirac House, Temple Back
Bristol BS1 6BE UK

Phone: 44 1 17 929 7481
Fax: 44 1 17 920 0979

techtracking@iop.org

Additional copies of this publication are available from:

Curran Associates, Inc.
57 Morehouse Lane
Red Hook, NY 12571 USA
Phone: 845-758-0400
Fax: 845-758-2633
Email: curran@proceedings.com
Web: www.proceedings.com

TABLE OF CONTENTS

Journal of Physics: Conference Series **1332** (2019) 011001 doi:10.1088/1742-6596/1332/1/011001

Preface

G P Zank
Editor, *Proceedings of the 18th Annual International Astrophysics Conference,*
Center for Space Plasma and Aeronomic Research (CSPAR) and Department of Space Physics, University of Alabama in Huntsville, Huntsville AL 35805, USA

E-mail: garyp.zank@gmail.com

The 18th Annual International Astrophysics Conference was held at the Sheraton Pasadena, Pasadena, California, USA from February 18 to February 22, 2019. The theme of the meeting was "The Physics of Energetic Particles: Universal Processes from the Solar Corona to the Very Local Interstellar Medium and the Physics they Enable." Energetic particles are ubiquitous throughout the solar wind, being integral to the solar corona, accelerated by interplanetary shock waves, present in the quiet solar wind, possibly accelerated by magnetic flux ropes or magnetic islands in the vicinity of the heliospheric current sheet, at planetary bow shocks, present in the distant heliosphere and the heliospheric termination shock, in the inner heliosheath, and now observed by Voyager 1 in the very local interstellar medium. Although numerous mechanisms have been identified that accelerate particles, no consensus exists as to which is primary, and indeed several acceleration processes can operate simultaneously in close proximity. The acceleration of particles to high energies represents a fundamental dissipative process for the plasma and can therefore modify the underlying plasma physical processes in important ways. The meeting explored observations, theories, and looked forward to anticipated new missions that will shine a light on this theme, the Parker Solar Probe and IMAP. The 18th AIAC explored these topics via a combination of invited 25 minute talks and a few 40 minute plenary talks.

The 18th AIAC also celebrated Prof. J.R. (Randy) Jokipii's 80th birthday. Randy has been a central figure in the physics of energetic particles and transport, including particle acceleration, particularly at shock waves, and the modulation of cosmic rays in the heliosphere. His seminal early paper, "Cosmic-ray propagation. I. Charged particles in a random magnetic field" defined the field for decades after its publication in 1966. This meeting was an opportunity to celebrate Randy's accomplishments with some selected plenary presentations.

Finally, there are several people to thank for their kindness and considerable help with the 18th AIAC. Adele Corona suffered a horrific car accident on the second day of the meeting. She has thankfully largely recovered. However, we are profoundly grateful to Prof.'s Dave McComas and Nathan Schwadron for selflessly stepping forward to taking over the running and management of the meeting after Adele was rushed to hospital. Their kindness and organizational skills ensured that the meeting continued with virtually no interruption. I should like to thank them both and of course Adele Corona for another outstanding meeting, and her help in providing the logistical support for this volume of papers.

Content from this work may be used under the terms of the Creative Commons Attribution 3.0 licence. Any further distribution of this work must maintain attribution to the author(s) and the title of the work, journal citation and DOI.
Published under licence by IOP Publishing Ltd

Journal of Physics: Conference Series **1332** (2019) 011002 doi:10.1088/1742-6596/1332/1/011002

Peer review statement

All papers published in this volume of *Journal of Physics: Conference Series* have been peer reviewed through processes administered by the proceedings Editors. Reviews were conducted by expert referees to the professional and scientific standards expected of a proceedings journal published by IOP Publishing.

Content from this work may be used under the terms of the Creative Commons Attribution 3.0 licence. Any further distribution of this work must maintain attribution to the author(s) and the title of the work, journal citation and DOI.
Published under licence by IOP Publishing Ltd

A Nearly Incompressible Turbulence-Driven Solar Wind Model

L Adhikari[1], G P Zank[1,2], L.-L Zhao[1]

[1]Center for Space Plasma and Aeronomic Research (CSPAR), University of Alabama in Huntsville, Huntsville,
AL 35899, USA

[2]Department of Space Science, University of Alabama in Huntsville, Huntsville, AL 35899, USA

E-mail: `la0004@uah.edu`

Abstract. We develop a 1D steady-state turbulence driven-solar wind model by coupling recently developed nearly incompressible magnetohydrodynamic turbulence transport model equations and 1D hydrodynamic solar wind equations. The dissipation of majority component quasi-2D turbulence and minority slab turbulence generated by the emergence of the "magnetic carpet" from the photosphere is responsible for heating the coronal plasma. We solve the coupled quasi-2D and slab turbulence transport equations and the solar wind equations from the base of the solar corona until 20 solar radii. We find that i) the coronal temperature increases to $\sim 2.5 \times 10^6$ K near the base of the solar corona; ii) the solar wind accelerates rapidly within a few solar radii; iii) turbulence energy densities decrease and correlation lengths increase with distance; iv) balanced quasi-2D turbulence at the coronal base becomes imbalanced, and imbalanced slab turbulence becomes more imbalanced with distance, and v) the normalized quasi-2D and slab residual energy becomes positive near the coronal base, and both energies become negative with increasing distance.

1. Introduction

The heating of the solar corona to millions of degrees Kelvin and the acceleration of the solar wind are attracting the attention of the science community after the launch of the Parker Solar Probe (PSP). It is thought that the heating of active regions, closed loops in the quiet corona, and open field regions, accelerates the solar wind from subsonic to supersonic speed. The models so far developed to describe the heating of the solar corona have been classified broadly by Cranmer et al [1, 2] into two physical models, i) "wave/turbulence-driven" (W/T) and ii) "reconnection/loop-opening" (RLO) models.

The W/T models assume that the transverse motions of the magnetic footpoints in the photosphere generate Alfvén waves with a speed of $> 10^3$ km s^{-1} that propagate outward through the chromosphere, which partially reflect and dissipate in

Content from this work may be used under the terms of the Creative Commons Attribution 3.0 licence. Any further distribution of this work must maintain attribution to the author(s) and the title of the work, journal citation and DOI.
Published under licence by IOP Publishing Ltd

the chromosphere, and are then transmitted into the lower corona with a smaller flux. The outwardly propagating Alfvén waves produced by the super-Alfvénic transverse photospheric motions are thought to undergo partial non-WKB reflection [3 – 5]. The partially reflected inward Alfvén wave then interacts nonlinearly with an outwardly propagating Alfvén wave [5] resulting in a non-propagating quasi-2D mode. In the W/T models, dissipation of large-scale energy via the generation of quasi-2D nonpropagating fluctuations [6] is thought to be responsible for the heating of coronal plasma. Also, see papers [7 – 12], which employed the ideas of Matthaeus et al [5] to model coronal heating and the origin of fast solar wind in open magnetic field regions. The magnetic field in the photosphere concentrates in intergranular regions or lanes in the quiet Sun ([13, 14]), from which magnetic flux tubes can form. These regions of concentrated and large magnetic field are sometimes called "kilo-Gauss patches" [15]. The flux tubes guide the solar wind and Alfvén waves and expand quickly with increasing height.

The concept of RLO was initially developed to explain plasma heating within closed loops, and was then extended to open magnetic field regions. RLO models require exchange reconnection between an open flux tube and a closed loop [16–18], which can lead to the injection of heated loop plasma into an open flux tube. Of course, the question of how the loop plasma is heated remains to be addressed.

The magnetic flux in the photosphere is mostly concentrated in small tubes of kilogauss strength, and is "stirred" with a characteristic speed of order $\sim 1 - 2$ km s^{-1} [19–20]. A continual flow at the boundaries of photospheric granules of different sizes significantly twist and braid the magnetic fields [21], leading to the release of energy via the generation of current sheets [21]. Rappazzo & Parker [22] used simulations to show that quasi-2D turbulence and current sheets form in a closed magnetic field representing a loop. This is consistent with the nearly incompressible description of MHD [23–25], which shows that the fluctuating fields are dominated by quasi-2D structures in the presence of strong guide field. The variance of the amplitude of the emerging magnetic carpet determines the scale height at which quasi-2D turbulence is driven by the dynamical emergence and interaction of magnetic carpet loops, so that the dissipation of quasi-2D turbulence heats the plasma above the photosphere (Zank et al [26]).

Recently, Zank et al [26] developed coupled nearly incompressible magnetohydrodynamic (NI MHD) quasi-2D and slab turbulence transport model equations for the quiet solar corona in open field regions. The NI MHD turbulence transport equations of Zank et al [26, 27] are appropriate to a low plasma beta β_p. NI MHD in the plasma beta $\beta_p \sim 1$ and $\beta_p \ll 1$ regimes ($\beta_p \equiv 2nk_BT/(B^2/2\mu_0)$, where n is the solar wind density, k_B is Boltzmann's constant, T is the solar wind temperature, B is the large-scale magnetic field, and μ_0 is the magnetic permeability) describes solar wind turbulence as a nonlinearly coupled superposition of a majority quasi-2D and minority slab fluctuations [23, 24, 27]. In Zank et al [26], the coronal turbulence transport model and the background plasma do not interact with each other than through the dissipation of turbulence that heats the background plasma. Zank et al [26] use a prescribed linearly increasing velocity profile, a density profile that conserves mass flux, and a radial average magnetic field. In this paper, we couple the original Parker solar wind model [21] to the Zank et al [26, 27] turbulence transport and heating model so that the turbulence transport model equations couple with the background plasma flow. In Section 2, we present the coupled

Journal of Physics: Conference Series **1332** (2019) 012001 doi:10.1088/1742-6596/1332/1/012001

turbulence transport-driven solar wind model. Section 3 discusses the theoretical results of our solar wind and turbulence transport model equations. Finally, Section 4 presents a discussion and conclusions.

2. The solar wind and turbulence transport model equations

In this section, we present the coupled NI MHD turbulence transport-driven solar wind model equations. The steady-state mass and momentum conservation equations and the solar wind temperature equation in a spherical coordinate system r are given by

$$\frac{d}{dr}(r^2 \rho U) = 0; \tag{1}$$

$$\rho U \frac{dU}{dr} = -\frac{dP}{dr} - \frac{GM_\odot \rho}{r^2}; \tag{2}$$

$$\begin{aligned} U\frac{dT}{dr} &= -(\gamma-1)T\frac{dU}{dr} - (\gamma-1)\frac{2UT}{r} + \frac{1}{3}\frac{m_p}{k_B}\alpha\Bigg[2\frac{\langle z^{*+2}\rangle\langle z^{\infty+2}\rangle\langle z^{\infty-2}\rangle^{1/2}}{L_\infty^+} \\ &+ 2\frac{\langle z^{*-2}\rangle\langle z^{\infty-2}\rangle\langle z^{\infty+2}\rangle^{1/2}}{L_\infty^-} + E_D^*\left(\frac{\langle z^{\infty+2}\rangle\langle z^{\infty-2}\rangle^{1/2}}{L_\infty^+} + \frac{\langle z^{\infty-2}\rangle\langle z^{\infty+2}\rangle^{1/2}}{L_\infty^-}\right) \\ &+ 2\frac{\langle z^{\infty+2}\rangle^2\langle z^{\infty-2}\rangle^{1/2}}{L_\infty^+} + 2\frac{\langle z^{\infty-2}\rangle^2\langle z^{\infty+2}\rangle^{1/2}}{L_\infty^-} + E_D^\infty\Bigg(\frac{\langle z^{\infty+2}\rangle\langle z^{\infty-2}\rangle^{1/2}}{L_\infty^+} \\ &+ \frac{\langle z^{\infty-2}\rangle\langle z^{\infty+2}\rangle^{1/2}}{L_\infty^-}\Bigg)\Bigg], \end{aligned} \tag{3}$$

where $\rho(\equiv m_p n)$ is the mass density, U the solar wind speed, $P(\equiv 2nk_BT)$ the thermal pressure, G the gravitational constant, $M_\odot$ the solar mass, m_p the proton mass, n the solar wind density, and k_B Boltzmann's constant. The parameter α is the von Kármán-Taylor constant. The parameter α determines the nonlinear interaction rate between the forward and backward propagating modes. The dissipation of the turbulence energy heats the coronal/solar wind plasma. Following Matthaeus et al [5] and Zank et al [26], we use $\alpha = 1$. In our future work, we will investigate coronal heating and acceleration of the solar wind as well as the turbulence energy with different values of α $(0 < \alpha < 1)$. The nonlinear dissipative terms become strong or weak depending on whether α is large or small. Therefore, the turbulence energy decays more rapidly when α is large compared to small values of α. Since we are re-considering the Parker problem by including the heating term, we only include the gas pressure gradient and the gravitational force in the momentum equation, and neglect the magnetic field in the momentum equation. In future, we will include the magnetic field and the ponderomotive forces in the momentum equation (see [28–36]). The third term on the right-hand side of Equation (3) is the turbulence heating term, which distinguishes Equation (3) from the original Parker model [21]. Prior heating models used an exponential heating function of the form $h = h_0 \exp(-(r - r_1)/\lambda)$ (Habbal et al [37]), where h_0 is the strength of the heating term, λ is a heating deposition length scale, r heliocentric distance, and $r_1 \sim 2$ solar radii. Habbal et al [37] found that the corona could be heated to millions of degrees Kelvin based on this phenomenological model. However, the form of the exponential heating term is difficult to justify physically.

Flux conservation $r^2B = \text{const} = r_0^2 B_0$ yields the magnetic field as

$$\mathbf{B} = B_0 \left(\frac{r_0}{r}\right)^2 \hat{r},$$

where B_0 is the magnetic field at the reference point r_0, and $\hat{r}$ is the radial direction. From Equations (1), (2), and (3), we derive

$$\frac{C_s^2}{U}\left(M_s^2 - 1\right)\frac{dU}{dr} = \frac{4\gamma k_B}{m_p}\frac{T}{r^2}\left[r - \frac{GM_\odot m_p}{4\gamma k_B}\frac{1}{T} - \frac{\alpha m_p}{6\gamma k_B}\frac{r^2}{UT}\left\{2\frac{\langle z^{*+2}\rangle\langle z^{\infty+2}\rangle\langle z^{\infty-2}\rangle^{1/2}}{L_\infty^+}\right.\right.$$
$$+ 2\frac{\langle z^{*-2}\rangle\langle z^{\infty-2}\rangle\langle z^{\infty+2}\rangle^{1/2}}{L_\infty^-} + E_D^*\left(\frac{\langle z^{\infty+2}\rangle\langle z^{\infty-2}\rangle^{1/2}}{L_\infty^+} + \frac{\langle z^{\infty-2}\rangle\langle z^{\infty+2}\rangle^{1/2}}{L_\infty^-}\right)$$
$$+ 2\frac{\langle z^{\infty+2}\rangle^2\langle z^{\infty-2}\rangle^{1/2}}{L_\infty^+} + 2\frac{\langle z^{\infty-2}\rangle^2\langle z^{\infty+2}\rangle^{1/2}}{L_\infty^-} + E_D^\infty\left(\frac{\langle z^{\infty+2}\rangle\langle z^{\infty-2}\rangle^{1/2}}{L_\infty^+}\right.$$
$$\left.\left.\left. + \frac{\langle z^{\infty-2}\rangle\langle z^{\infty+2}\rangle^{1/2}}{L_\infty^-}\right)\right\}\right], \tag{4}$$

where $M_s = U/C_s$ is the sonic Mach number and $C_s^2 = 2\gamma k_B T/m_p$ is the square of the sound speed. Equation (4) exhibits a sonic point at the location where the terms inside (...) on the left-hand side and the terms inside [...] on the right-hand side become zero simultaneously, i.e., at $M_s^2 = 1$. We use L'Hôspital's rule to solve Equation (4) in the vicinity of the sonic point.

The solar wind equations couple with the quasi-2D and slab turbulence transport equations through the third term on the right hand side of Equation (3). Before writing the turbulence transport equations, we define the quasi-2D and slab Elsässer variables as [26]

$$\mathbf{z}^{\infty\pm} = \mathbf{u}^\infty \pm \frac{\mathbf{B}^\infty}{\sqrt{\mu_0\rho}} \text{ and } \mathbf{z}^{*\pm} = \mathbf{u}_1 \pm \frac{\mathbf{B}^*}{\sqrt{\mu_0\rho}}, \tag{5}$$

respectively. Here, the superscript "∞" refers to core incompressible variables, i.e., quasi-2D turbulence, and the superscript "*" or subscript "1"denotes the higher order incompressible corrections, i.e., slab turbulence. The parameters $\mathbf{u}$ and $\mathbf{B}$ are the fluctuating velocity and magnetic field, ρ is the solar wind density, and μ_0 is the magnetic permeability. Similarly, the quasi-2D and slab variances of the Elsässer variables and the residual energy E_D can be formed from Equation (5) as [26]

$$\langle z^{\infty,*\pm 2}\rangle = \langle \mathbf{z}^{\infty,*\pm}\cdot\mathbf{z}^{\infty,*\pm}\rangle;\ E_D^{\infty,*} = \langle \mathbf{z}^{\infty,*+}\cdot\mathbf{z}^{\infty,*-}\rangle, \tag{6}$$

where $\langle z^{\infty,*\pm 2}\rangle$ denote the ensemble-averaged quasi-2D/slab energy in backward and forward propagating modes with respect to the inward orientation of the magnetic field. Similarly, the correlation functions corresponding to backward/forward propagating modes, and the residual energy can be written as

$$\begin{aligned} L_{\infty,*}^{\pm} &= \int \langle \mathbf{z}^{\infty,*\pm}\cdot\mathbf{z}^{\infty,*\mp}\rangle dy \equiv \langle z^{\infty,*\pm 2}\rangle\lambda_{\perp,*}^{\pm}; \\ L_D^{\infty,*} &= \int \langle \mathbf{z}^{\infty,*+}\cdot\mathbf{z}^{\infty,*-\prime} + \mathbf{z}^{\infty,*+\prime}\cdot\mathbf{z}^{\infty,*-}\rangle dy \equiv E_D^{\infty,*}\lambda_D^{\infty,*}, \end{aligned} \tag{7}$$

where $y = |\mathbf{y}|$ is the spatial lag between fluctuations, and $\mathbf{z}^{\infty,*-'}$ denotes the lagged Elsässer variables. The quantities $\lambda^{\pm}_{\perp,*}$ and $\lambda^{\infty,*}_{D}$ are the quasi-2D/slab correlation lengths corresponding to backward/forward propagating modes, and the residual energy.

The 1D steady-state quasi-2D turbulence transport model equations in a spherical coordinate system r can be expressed as [26, 27, 38]

$$U\frac{d\langle z^{\infty\pm 2}\rangle}{dr} + \frac{1}{2}\left(\langle z^{\infty\pm 2}\rangle + E_D^{\infty}\right)\left(\frac{dU}{dr} + \frac{2U}{r}\right) = -2\frac{\langle z^{\infty\pm 2}\rangle^2\langle z^{\infty\mp 2}\rangle^{1/2}}{L_{\infty}^{\pm}} + 2\frac{\langle z^{*\pm 2}\rangle^2\langle z^{*\mp 2}\rangle^{1/2}}{L_*}; \tag{8}$$

$$U\frac{dE_D^{\infty}}{dr} + \frac{1}{2}(E_D^{\infty} + E_T^{\infty})\left(\frac{dU}{dr} + \frac{2U}{r}\right) = -E_D^{\infty}\left(\frac{\langle z^{\infty+2}\rangle\langle z^{\infty-2}\rangle^{1/2}}{L_{\infty}^{+}} + \frac{\langle z^{\infty-2}\rangle\langle z^{\infty+2}\rangle^{1/2}}{L_{\infty}^{-}}\right) + E_D^{*}\left(\frac{\langle z^{*+2}\rangle\langle z^{*-2}\rangle^{1/2}}{L_*} + \frac{\langle z^{*-2}\rangle\langle z^{*+2}\rangle^{1/2}}{L_*}\right); \tag{9}$$

$$U\frac{dL_{\infty}^{\pm}}{dr} + \frac{1}{2}(L_{\infty}^{\pm} + \frac{L_D^{\infty}}{2})\left(\frac{dU}{dr} + \frac{2U}{r}\right) = 0; \tag{10}$$

$$U\frac{dL_D^{\infty}}{dr} + \frac{1}{2}\left(\frac{dU}{dr} + \frac{2U}{r}\right)\left(L_D^{\infty} + L_{\infty}^{+} + L_{\infty}^{-}\right) = 0, \tag{11}$$

where $E_T^{\infty} = (\langle z^{\infty+2}\rangle + \langle z^{\infty-2}\rangle)/2$ is the total quasi-2D turbulent energy. We assume $L_+^* = L_-^* = L^*$ in the second terms on the right-hand sides of Equations (5) and (6). The assumption $L_+^* = L_-^* = L^*$ reduces the complexity of the equations. Similarly, the 1D steady-state slab model equations can be expressed as ([26, 27, 38])

$$(U \mp V_A)\frac{d\langle z^{*\pm 2}\rangle}{dr} + \frac{1}{2}\frac{dU}{dr}\left(\langle z^{*\pm 2}\rangle - E_D^*\right) - (2b-1)\frac{U}{r}\langle z^{*\pm 2}\rangle + (6b-1)\frac{U}{r}E_D^* \pm 4b\frac{V_A}{r}E_D^* \mp \frac{1}{2}\frac{V_A}{\rho}\frac{d\rho}{dr}\left(\langle z^{*\pm 2}\rangle - E_D^*\right) = -2\frac{\langle z^{*\pm 2}\rangle\langle z^{\infty\pm 2}\rangle\langle z^{\infty\mp 2}\rangle^{1/2}}{L_{\infty}^{\pm}} - 2\frac{\langle z^{*\pm 2}\rangle^2\langle z^{*\mp 2}\rangle^{1/2}}{L_*}; \tag{12}$$

$$U\frac{dE_D^*}{dr} + \frac{1}{2}\frac{dU}{dr}(E_D^* - E_T^*) - (2b-1)\frac{U}{r}E_D^* + (6b-1)\frac{U}{r}E_T^* - 4b\frac{V_A}{r}E_C^* - \frac{1}{2}\frac{V_A}{\rho}\frac{d\rho}{dr}E_C^* = -E_D^*\left(\frac{\langle z^{\infty+2}\rangle\langle z^{\infty-2}\rangle^{1/2}}{L_{\infty}^{+}} + \frac{\langle z^{\infty-2}\rangle\langle z^{\infty+2}\rangle^{1/2}}{L_{\infty}^{-}}\right) - E_D^*\left(\frac{\langle z^{*+2}\rangle\langle z^{*-2}\rangle^{1/2}}{L_*} + \frac{\langle z^{*-2}\rangle\langle z^{*+2}\rangle^{1/2}}{L_*}\right); \tag{13}$$

$$U\frac{dL_*}{dr} + \frac{1}{2}\frac{dU}{dr}\left(L_* - \frac{L_D^*}{2}\right) - (2b-1)\frac{U}{r}L_* + \left(3b - \frac{1}{2}\right)\frac{U}{r}L_D^* = 0; \tag{14}$$

$$U\frac{dL_D^*}{dr} + \frac{dU}{dr}\left(\frac{L_D^*}{2} - L_*\right) - (2b-1)\frac{U}{r}L_D^* + 2(6b-1)\frac{U}{r}L_* = 0, \tag{15}$$

where $E_T^* = (\langle z^{*+2}\rangle + \langle z^{*-2}\rangle)/2$ is the total slab turbulent energy, $E_C^* = (\langle z^{*+2}\rangle - \langle z^{*-2}\rangle)/2$ the slab cross-helicity, and $V_A = B/\sqrt{\mu_0\rho}$ the Alfvén velocity. The parameter b reflects the assumed geometry of the slab turbulence and is related to the covariance of the slab form of the Elsässer variables [26, 39]. We use $b = 0.3$.

Similarly, the 1D steady-state transport equation for the variance in the density fluctuations is [26, 27, 38]

$$U\frac{d\langle\rho^{\infty 2}\rangle}{dr} + 2\langle\rho^{\infty 2}\rangle\frac{dU}{dr} + 4\frac{U}{r}\langle\rho^{\infty 2}\rangle = -\frac{\langle u^{\infty 2}\rangle^{1/2}\langle\rho^{\infty 2}\rangle}{l_u^\infty}, \tag{16}$$

where $\langle u^{\infty 2}\rangle$ and l_u^∞ are the fluctuating quasi-2D kinetic energy and the corresponding correlation length, and are given by

$$\langle u^{\infty 2}\rangle = \frac{\langle z^{\infty+2}\rangle + \langle z^{\infty-2}\rangle + 2E_D^\infty}{4} \text{ and } l_u^\infty = \frac{(E_T^\infty + E_C^\infty)\lambda_\perp^+ + (E_T^\infty - E_C^\infty)\lambda_\perp^- + E_D^\infty\lambda_D^\infty}{2(E_T^\infty + E_D^\infty)}. \tag{17}$$

Notice that the transport equation for backward propagating modes, Equation (12), exhibits a singular point at $U = V_A$. We replace the mixing dissipative term in Equation (12) by an advective term $\langle z^{\infty-2}\rangle^{1/2}\hat{\mathbf{s}}\cdot\nabla\langle z^{*+2}\rangle$ to avoid the Alfvénic singularity [Zank et al 26]. In the vicinity of the Alfvén surface, we solve the modified transport equation

$$\begin{aligned}\left(U - V_A + \langle z^{\infty-2}\rangle^{1/2}\right)\frac{d\langle z^{*+2}\rangle}{dr} &= -\frac{1}{2}\frac{dU}{dr}(\langle z^{*+2}\rangle - E_D^*) + (2b-1)\frac{U}{r}\langle z^{*+2}\rangle \\ -(6b-1)\frac{U}{r}E_D^* - 4b\frac{V_A}{r}E_D^* &+ \frac{1}{2}\frac{V_A}{\rho}\frac{d\rho}{dr}((\langle z^{*+2}\rangle - E_D^*)) - 2\alpha\frac{\langle z^{*+2}\rangle^2\langle z^{*-2}\rangle^{1/2}}{L_*};\end{aligned} \tag{18}$$

$$\begin{aligned}U\frac{dT}{dr} &= -(\gamma-1)T\frac{dU}{dr} - (\gamma-1)\frac{2UT}{r} + \frac{1}{3}\frac{m_p}{k_B}\alpha\Bigg[2\frac{\langle z^{*-2}\rangle\langle z^{\infty-2}\rangle\langle z^{\infty+2}\rangle^{1/2}}{L_\infty^-} + \\ &E_D^*\left(\frac{\langle z^{\infty+2}\rangle\langle z^{\infty-2}\rangle^{1/2}}{L_\infty^+} + \frac{\langle z^{\infty-2}\rangle\langle z^{\infty+2}\rangle^{1/2}}{L_\infty^-}\right) + 2\frac{\langle z^{\infty+2}\rangle^2\langle z^{\infty-2}\rangle^{1/2}}{L_\infty^+} \\ &+ 2\frac{\langle z^{\infty-2}\rangle^2\langle z^{\infty+2}\rangle^{1/2}}{L_\infty^-} + E_D^\infty\left(\frac{\langle z^{\infty+2}\rangle\langle z^{\infty-2}\rangle^{1/2}}{L_\infty^+} + \frac{\langle z^{\infty-2}\rangle\langle z^{\infty+2}\rangle^{1/2}}{L_\infty^-}\right)\Bigg].\end{aligned} \tag{19}$$

We solve the coupled solar wind equations and turbulence transport equations using a 4th order Runge-Kutta method from 1 to 20 $R_\odot$. Since these equations are coupled, the background solar wind speed, density and temperature are influenced by the evolving MHD turbulence and vice versa.

3. Results

We solve the self-consistent 1D steady-state NI MHD turbulence transport model equations and background solar wind equations for the quiet solar corona from 1 to 20 $R_\odot$ using the boundary conditions shown in Table 1. Here we assume balanced quasi-2D turbulence, and imbalanced slab turbulence at the base of the solar corona. The boundary conditions for the quasi-2D Elsässer energies are obtained by assuming 2.7 G magnetic field fluctuations and a solar coronal density of 4×10^8 cm^{-3}, which yields the turbulent Alfvén speed as $V_A \sim 295$ kms^{-1}, or $V_A^2 \sim 8.7 \times 10^4$ km^2 s^{-2} [26]. To obtain the boundary condition for the energy in forward propagating modes (slab), we assume an 80:20 ratio between quasi-2D and slab turbulence. The boundary condition for the correlation function is obtained by multiplying the energy density and the corresponding correlation length. The correlation length of the quasi-2D turbulence is taken from case 2 of Zank et al [26], where $\lambda_\perp = 5 \times 10^4$ km was assumed. This correlation length is 10 times larger than the correlation length suggested by Abramenko et al [40]. For the residual energy, we assume approximate equipartition between the fluctuating kinetic and magnetic energy, slightly dominated by fluctuating magnetic energy at the coronal base.

2D Core Model Equations		Slab Model Equations	
$\langle z^{\infty+2} \rangle$	20000 km^{2}s^{-2}	$\langle z^{*+2} \rangle$	2222.22 km^{2}s^{-2}
$\langle z^{\infty-2} \rangle$	20000 km^{2}s^{-2}	$\langle z^{*-2} \rangle$	5000 km^{2}s^{-2}
E_D^∞	-2200 km^{2}s^{-2}	E_D^*	-115.79 km^{2}s^{-2}
L_∞^+	1.0 $\times 10^9$ km^{3}s^{-2}	L_*	1.92 $\times 10^6$ km^{3}s^{-2}
L_∞^-	1.0 $\times 10^9$ km^{3}s^{-2}	L_D^*	-2.89 $\times 10^6$ km^{3}s^{-2}
L_D^∞	-1.1 $\times 10^8$ km^{3}s^{-2}		
U	29.26 kms^{-1}		
n	4×10^8 cm^{-3}		
T	5×10^4 K		
$\langle \rho^{\infty 2} \rangle$	1.6×10^{15} cm^{-6}		

Table 1. Assumed base values at 1 $R_\odot$ used to solve the turbulence transport and the background (large-scale) solar wind equations.

The numerical solutions of the 1D hydrodynamic solar wind equations when coupled to the turbulence transport equations are shown in Figure 1. The top left panel of Figure 1 shows the solar wind speed and the Alfvén velocity as a function of normalized heliocentric distance. The solar wind plasma accelerates rapidly near the base of the solar corona, and reaches $\sim$ 500 kms^{-1} at 20 $R_\odot$. Similarly, the Alfvén velocity increases rapidly to a peak value of $\sim$ 2600 kms^{-1} and then decreases with increasing heliocentric distance. Here the acceleration of the solar wind plasma and the increase in the Alfvén velocity are associated with the decay rate of turbulence, which can be adjusted by the von Kármán-Taylor constant α. The numerical solutions of the solar wind speed and the Alfvén velocity show that the solar wind speed and the Alfvén velocity intersect at $\sim$ 7.5 $R_\odot$, which identifies the Alfvén critical point. The Alfvén surface is the location at which the solar wind and the Alfvén speed become equal, i.e., $|\mathbf{U}| = |\mathbf{V}_A|$. The Alfvén

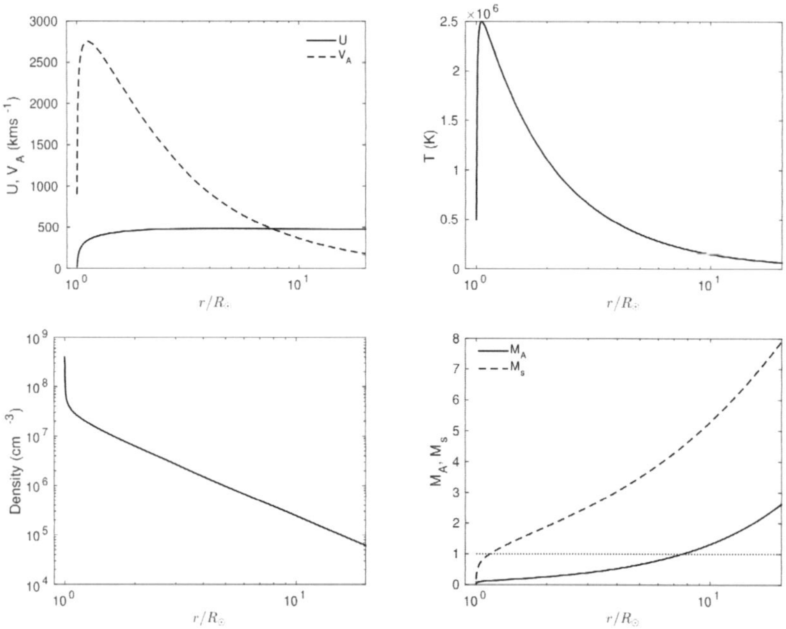

Figure 1. Solar wind parameters as a function of heliocentric distance normalized to the solar radius. Top left: The solar wind speed, and the Alfvén velocity. Top right: The solar wind temperature. Bottom left: The solar wind density. Bottom right: The Alfvén and sonic Mach number. The horizontal blue dotted line shows $M_A = M_S = 1$.

surface distinguishes the sub-Alfvénic region $|\mathbf{U}| \ll |\mathbf{V}_A|$ and the super-Alfvénic region $|\mathbf{U}| \gg |\mathbf{V}_A|$. Recently, Adhikari et al [41] used the low plasma beta turbulence transport model equations to show that the solar wind turbulence at the Alfvén surface does not turn off.

The top right panel of Figure 1 shows the coronal plasma temperature as a function of normalized heliocentric distance. It shows that the coronal temperature increases to $\sim 2.5 \times 10^6$ K near the coronal base, and then decreases with increasing heliocentric distance. The bottom left panel of Figure 1 illustrates the solar wind density as a function of normalized heliocentric distance. The solar wind density decreases rapidly near the coronal base, and then slowly with distance. The rapid decrease in the solar wind density corresponds to a rapid increase in the solar wind speed to conserve mass flux.

The bottom right panel of Figure 1 shows the Alfvén Mach number and the sonic Mach number as a function of normalized heliocentric distance. Both Mach numbers are less than 1 at 1 $R_\odot$, become equal to 1 at a certain distance, and then become larger than 1 with increasing distance. The sonic Mach number (dashed curve) intersects the horizontal blue line at ~ 1.14 $R_\odot$, which is the sonic point. The Alfvén Mach number (solid curve) intersects the horizontal line at ~ 7.5 $R_\odot$, which is the Alfvén surface, also shown in the top panel of Figure 1 (intersection point of the solar wind speed and the

Alfvén velocity). The locations of the sonic point and the Alfvén surface depend on the heating rate due to the dissipation of turbulence.

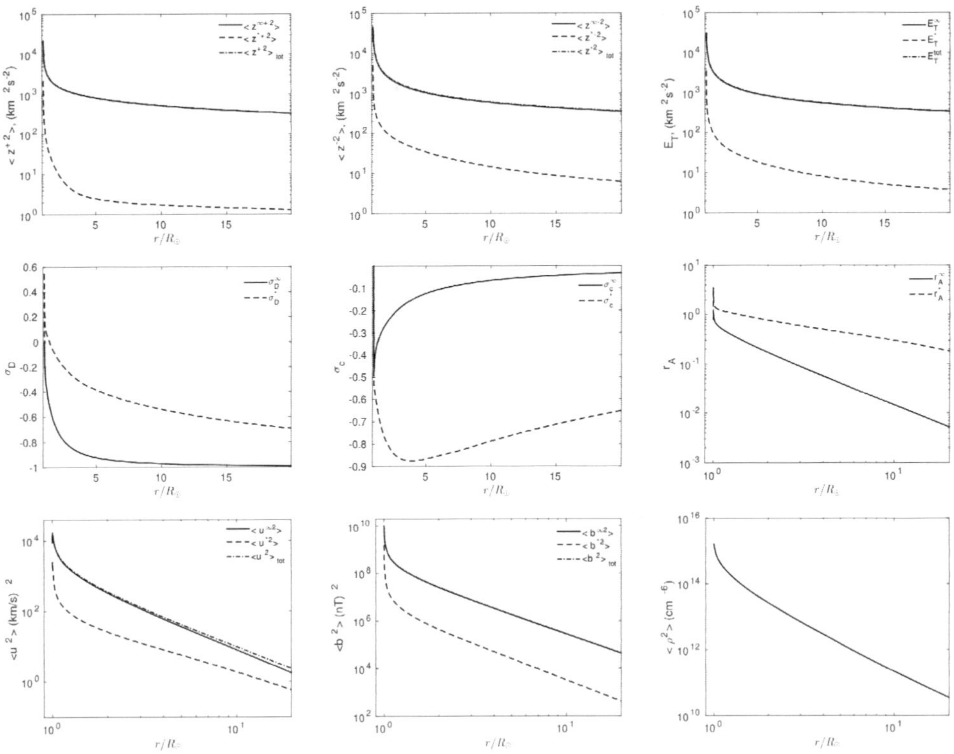

Figure 2. Turbulence quantities plotted as a function of normalized heliocentric distance. First row: Left, middle and right panels are the energy in backward and forward propagating modes, and the total turbulent energy. Second row: Left, middle and right panels are the normalized residual energy, the normalized cross-helicity and the Alfvén ratio. Third row: Left, middle and right panels are the fluctuating kinetic energy, fluctuating magnetic energy, and the variance of the density fluctuations. In all panels, the solid curves depict quasi-2D turbulence quantities, the dashed curves slab turbulence quantities, and the dashed-dotted-dashed curves the sum of quasi-2D and slab turbulence quantities.

Unlike the Zank et al [26] model, the radial profiles of the turbulence quantities shown in Figure 2 and the background plasma shown in Figure 1 influence each other. In Figure 2, the solid curves depict majority quasi-2D turbulence, the dashed curves minority slab turbulence, and the dashed-dotted-dashed the total turbulence, i.e., the sum of quasi-2D and slab turbulence. In Figure 2, the left, middle, and right panels of the first row illustrate the energy in backward and forward propagating modes, and the total turbulent energy as a function of normalized heliocentric distance. The energy in quasi-2D turbulence is larger than that in slab turbulence between 1 and 20 $R_\odot$, and

Journal of Physics: Conference Series **1332** (2019) 012001 doi:10.1088/1742-6596/1332/1/012001

both energies decrease with increasing heliocentric distance.

The left, middle, and right panels of the second row of Figure 2 show the normalized residual energy, the normalized cross-helicity, and the Alfvén ratio as a function of normalized heliocentric distance. Here the left panel of the second row shows that the quasi-2D and slab normalized residual energies become positive near the coronal base and then become negative with increasing heliocentric distance. The figure also shows that the quasi-2D fluctuations are more magnetically dominated compared to slab fluctuations. As illustrated in the normalized cross-helicity plot (middle panel of the second row), the balanced quasi-2D turbulence at 1 $R_\odot$ evolves to an imbalanced state beyond 1 $R_\odot$, while the imbalanced slab turbulence becomes more imbalanced with increasing heliocentric distance. The slab normalized cross-helicity is larger than the quasi-2D normalized cross-helicity, suggesting that it is primarily outward propagating waves that are present. The right panel of the second row displays the quasi-2D and slab Alfvén ratios as a function of heliocentric distance. It shows that both Alfvén ratios decrease with increasing distance, indicating that both quasi-2D and slab turbulence are dominated by the fluctuating magnetic energy, consistent with the result of the normalized residual energy.

The left panel of the third row of Figure 2 is the fluctuating kinetic energy. Both the quasi-2D and slab fluctuating kinetic energies decrease with increasing heliocentric distance. Here the fluctuating quasi-2D energy is larger than the fluctuating slab kinetic energy between 1 and 20 $R_\odot$. The middle panel of the third row of Figure 2 shows that the quasi-2D fluctuating magnetic energy is the dominant component in the region between 1 and 20 solar radii. The quasi-2D and slab fluctuating magnetic energies decrease with increasing heliocentric distance. Similarly, the right panel of the third row describes the variance of density fluctuations as a function of heliocentric distance. Similar to the turbulence energy, the variance of density fluctuations decreases with increasing distance.

Numerical solutions of the transport equations for the quasi-2D and slab correlation functions are shown in top left and middle panels of Figure 3. Here the quasi-2D correlation functions corresponding to forward and backward modes $L_\infty^\pm$ (black and red curves) and the residual energy L_D^∞ decrease initially and then become approximately constant with increasing distance. Notice that the $L_\infty^\pm$ are positive and L_D^∞ negative throughout the heliosphere, and that L_∞^+ and L_∞^- overlap with each other. Similarly, the correlation function associated with slab turbulence L^* decreases rapidly and then increases slightly with distance, whereas L_D^* gradually decreases with heliocentric distance.

The top right and bottom left panels of Figure 3 describe the quasi-2D and slab correlation lengths (solid and dashed curves, respectively) corresponding to backward and forward propagating modes as a function of normalized heliocentric distance. The correlation lengths of the quasi-2D and slab forward and backward propagating modes increase with increasing distance. Similarly, the bottom middle and right panels of Figure 3 show the correlation lengths corresponding to quasi-2D and slab (solid and dashed curves, respectively) velocity and magnetic field fluctuations, which also increase with increasing heliocentric distance.

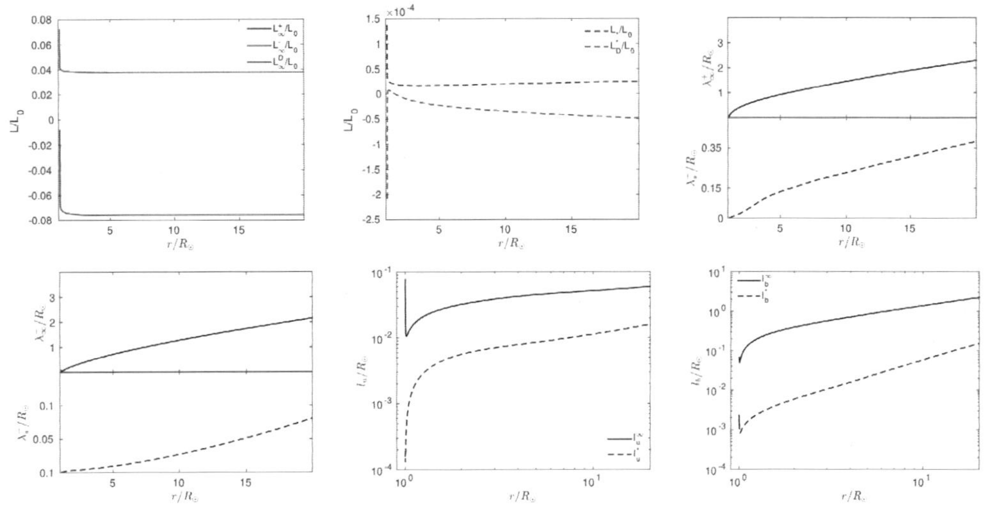

Figure 3. Normalized quasi-2D and slab correlation functions and correlation lengths as a function of normalized heliocentric distance. Top left: Quasi-2D correlation functions. Top middle: Slab correlation functions. Top right: Correlation lengths corresponding to backward propagating modes. Bottom left: Correlation lengths corresponding to forward propagating modes. Bottom middle: Correlation length of velocity fluctuations. Bottom right: Correlation length of magnetic field fluctuations.

4. Discussion and Conclusions

We have developed a turbulence driven-solar wind model by coupling the nearly incompressible magnetohydrodynamic (NI MHD) turbulence transport model equations and the solar wind equations. This work is an extension of Zank et al [26]. Unlike the Zank et al [26] model, in which turbulence and plasma flow do not interact with each other, we couple the NI MHD turbulence model to the background plasma flow. The NI MHD turbulence transport model in the plasma beta regime of order $\beta_p \ll 1$ describes majority quasi-2D (nonpropagating) and minority slab turbulence. Since quasi-2D turbulence is the dominant component in the NI MHD model, it plays a major role in heating and driving the coronal/solar wind plasma. The NI MHD model is therefore different from the wave/turbulence-driven (W/T) models [7 – 12], in which Alfvén waves heat and drive the solar wind. The NI MHD model assumes that the small-scale "magnetic carpet" loops in the photosphere continuously pump quasi-2D fluctuations into the region above the photosphere, and the subsequent dissipation of which heats the plasma above the photosphere. We solved the coupled NI MHD quasi-2D and slab turbulence transport model equations and plasma flow equations from the coronal base to 20 solar radii using a Runge Kutta 4th order method. We summarize our findings as follows.

(i) The solar wind flow speed increases rapidly over a height of $\sim 2 - 4$ $R_\odot$, and then more gradually until ~ 20 solar radii. The sonic point is located at ~ 1.14 $R_\odot$, and the Alfvén surface is located at ~ 7.5 $R_\odot$.

(ii) The coronal plasma temperature rises to $\sim 2.5 \times 10^6$ K within a few solar radii.
(iii) The solar wind density decreases quickly initially, and then more slowly with increasing distance.
(iv) The quasi-2D and slab turbulence energies decrease with distance, and the correlation lengths increase with distance.
(v) The fully balanced quasi-2D turbulence state at 1 $R_\odot$ becomes imbalanced with increasing heliocentric distance. The imbalanced minority slab turbulence at coronal base remains imbalanced between 1 and 20 $R_\odot$, eventually being comprised predominantly of outwardly propagating modes.
(vi) The normalized quasi-2D and slab residual energy becomes positive near the coronal base, and both energies become negative with increasing distance.
(vii) The variance of the density fluctuations decreases with increasing heliocentric distance.

In this paper, we presented a turbulence driven-solar wind model and numerical solutions for a particular set of boundary conditions. In our turbulence driven-solar wind model, the turbulence energy densities decay rapidly with the prescribed boundary conditions shown in Table 1 leaving only a modest turbulence energy at 20 $R_\odot$. The rapid decay of turbulence energy is due to a strong nonlinear dissipation term, and no sources of turbulence have been included in the turbulence transport model equations. In future work, we will investigate the effects of strong and weak dissipation terms and consider sources of turbulence. However, our findings suggest that the dissipation of quasi-2D turbulence is primarily responsible for heating the coronal plasma to several million of degrees Kelvin, and drives the solar wind from a subsonic to a supersonic state. The comparison of the results of our turbulence-driven solar wind model equations with Parker Solar Probe (PSP) measurements will provide a true test of our theory.

Acknowledgments
We acknowledge the partial support of a Parker Solar Probe contract SV4-84017, an NSF-DOE grant PHY-1707247, and an NSF EPSCoR RII-Track-1 cooperative agreement OIA-1655280.

References
[1] Cranmer, S R., Kohl, J L., Miralles, M P., & van Ballegooijen, A A., 2010, *in Astronomical Society of the Pacific Conference Series*, Vol. 428, SOHO-23: Understanding a Peculiar Solar Minimum, ed. S. R. Cranmer, J. T. Hoeksema, & J. L. Kohl, 209
[2] Cranmer, S R., & van Ballegooijen, A A., 2010, *ApJ*, 720, 824
[3] Hollweg, J. V., 1986, *J. Geophys. Res.*, 91, 4111
[4] Marsch, E. & Tu, C.-Y., 1990, *J. Geophys. Res.*, 95, 8211
[5] Matthaeus, W. H., Zank, G. P., Oughton, S., Mullan, D. J., & Dmitruk, P., 1999, *ApJ*, 523, L93
[6] Shebalin, J. V., Matthaeus, W. H., & Montgomery, D., 1983, *Journal of Plasma Physics*, 29, 525
[7] Dmitruk, P., Milano, L. J., & Matthaeus, W. H., 2001, *ApJ*, 548, 482
[8] Dmitruk, P., Matthaeus, W. H., Milano, L. J., 2002, *ApJ*, 575, 571
[9] Oughton, S., Matthaeus, W. H., Dmitruk, P., 2001, *ApJ*, 551, 565
[10] Cranmer, S. R., van Ballegooijen, A. A., & Edgar, R. J., *ApJS*, 171, 520
[11] Cranmer, S. R., van Ballegooijen, A. A., & Woolsey, L. N., *ApJ*, 767, 125
[12] Verdini, A., Velli, M., Matthaeus, W. H., Oughton, S., & Dmitruk, P., 2010, *ApJ*, 708, L116

[13] Fisk, L A., Schwadron, N A., & Zurbuchen, T H., 1999, *J. Geophys. Res.*, 104, 19765
[14] Fisk, L A., 2003, *Journal of Geophysical Research (Space Physics)*, 108, 1157
[15] Tsuneta, S., Ichimoto, K., Katsukawa, Y., 2008, *ApJ*, 688, 1374
[16] Einaudi, G., Boncinelli, P., Dahlburg, R. B., & Karpen, J. T., 1999, *J. Geophys. Res.*, 104, 521
[17] Moore, R. L., Falconer, D. A., Porter, J. G., & Suess, S. T., 1999, *ApJ*, 526, 505
[18] Rappazzo, A. F., Matthaeus, W. H., Ruffolo, D., Servidio, S., & Velli, M., 2012, *ApJ*, 758, L14
[19] Muller, R., Roudier, T., Vigneau, J., & Auffret, H., 1994, *A&A*, 283, 232
[20] Berger, T. E. & Title, A. M., 1996, *ApJ*, 463, 365
[21] Parker, E. N. 1988, *ApJ*, 330, 474
[22] Rappazzo, A. F. & Parker, E. N., 2013, *ApJ*, 773, L2
[23] Zank, G. P. & Matthaeus, W. H., 1992, *J. Geophys. Res.*, 97, 17189
[24] Zank, G. P. & Matthaeus, W. H., 1993, *Physics of Fluids*, 5, 257
[25] Hunana, P. & Zank, G. P. 2010, *ApJ*, 718, 148
[26] Zank, G. P., Adhikari, L., Hunana, P., Tiwari, S. K., Moore, R., Shiota, D., Bruno, R., & Telloni, D., 2018, *ApJ*, 854, 32
[27] Zank, G. P., Adhikari, L., Hunana, P., Shiota, D., Bruno, R., & Telloni, D., 2017, *ApJ*, 835, 147
[28] Holzer, T. E. & Axford, W. I., 1970, *ARA&A*, 8, 31
[29] Leer, E., Holzer, T. E., & Fla, T., 1982, *Space Sci. Rev.*, 33, 161
[30] Hollweg, J. V., 1986, *J. Geophys. Res.*, 91, 4111
[31] Cranmer, S. R., van Ballegooijen, A. A., & Edgar, R. J. 2007, *ApJS*, 171, 520
[32] Cranmer, S. R. & van Ballegooijen, A. A., 2010, *ApJ*, 720, 824
[33] Cranmer, S. R., van Ballegooijen, A. A., & Woolsey, L. N., 2013, *ApJ*, 767, 125
[34] Verdini, A., Velli, M., Matthaeus, W. H., Oughton, S., & Dmitruk, P., 2010, *ApJ*, 708, L116
[35] van Ballegooijen, A. A., Asgari-Targhi, M., Cranmer, S. R., & DeLuca, E. E. 2011, *ApJ*, 736, 3
[36] van Ballegooijen, A. A. & Asgari-Targhi, M. 2016, *ApJ*, 821, 106
[37] Habbal, S. R., Esser, R., Guhathakurta, M., & Fisher, R. R., 1995, *Geophys. Res. Lett.*, 22, 1465
[38] Adhikari, L., Zank, G. P., Hunana, P., Shiota, D., Bruno, R., Hu, Q., & Telloni, D., 2017, *ApJ*, 841, 85
[39] Zank, G. P., Dosch, A., Hunana, P., Florinski, V., Matthaeus, W. H., Webb, & G. M., 2012, *ApJ*, 745, 35
[40] Abramenko, V. I., Zank, G. P., Dosch, A., 2013, *ApJ*, 773, 167
[41] Adhikari, L., Zank, G. P. , and Zhao, L.-L., 2019, *The Astrophysical Journal*, 876, 26

Journal of Physics: Conference Series **1332** (2019) 012002 doi:10.1088/1742-6596/1332/1/012002

Global Energetics of Solar Flares and Coronal Mass Ejections

Markus J. Aschwanden[1], Amir Caspi[2], Christina M.S. Cohen[3], Gordon Holman[4], Ju Jing[5], Matthieu Kretzschmar[6], Eduard P. Kontar[7], James M. McTiernan[8], Richard A. Mewaldt[3], Aidan O'Flannagain[9], Ian G. Richardson[10], Daniel Ryan[11], Harry P. Warren[12], and Yan Xu[5]

[1] Lockheed Martin, Solar and Astrophysics Laboratory, Org. A021S, Building 252, 3251 Hanover Street, Palo Alto, CA 94304, USA; aschwanden@lmsal.com
[2] Planetary Science Directorate, Southwest Research Institute, Boulder, CO 80302, USA; amir.caspi@swri.org
[3] California Institute of Technology, Mail Code 290-17, Pasadena, CA 91125, USA; rmewaldt@srl.caltech.edu; cohen@srl.caltech.edu
[4] Code 671, NASA Goddard Space Flight Center, Greenbelt, MD 20771, USA; gordon.d.holman@nasa.gov
[5] Institute for Space Weather Sciences, New Jersey Institute of Technology, University Heights, Newark NJ 07102-1982, USA; ju.jing@njit.edu, yan.xu@njit.edu
[6] LPC2E, UMR 6115 CNRS and University of Orléans, 3a Avenue de la recherche scientifique, F-45071 Orléans, France; matthieu.kretzschmar@cnrs-orleans.fr
[7] School of Physics and Astronomy, University of Glasgow, Glasgow G12 8QQ, UK; eduard.kontar@astro.gla.ac.uk
[8] Space Sciences Laboratory, University of California, Berkeley, CA 94720, USA; jimm@ssl.berkeley.edu
[9] Astrophysics Research Group, School of Physics, Trinity College Dublin, Dublin 2, Ireland; aidanoflann@gmail.com
[10] GPHI and Dept. of Astronomy, University of Maryland; Code 672, NASA Goddard Space Flight Center, Greenbelt, MD 20770, USA; ian.g.richardson@nasa.gov
[11] NASA Goddard Space Flight Center, 8800 Greenbelt Road, Greenbelt, MD 20770, USA; ryand5@tcd.ie
[12] Space Science Division, Naval Research Laboratory, Washington, DC 20375, USA; harry.warren@nrl.navy.mil

Abstract. We investigate the global energetics and energy closure of various physical processes that are energetically important in solar flares and coronal mass ejections (CMEs), which includes: magnetic energies, thermal energies, nonthermal energies (particle acceleration), direct and indirect plasma heating processes, kinetic CME energies, gravitational CME energies, aerodynamic drag of CMEs, solar energetic particle events, EUV and soft X-ray radiation, white-light, and bolometric energies. Statistics on these forms of energies is obtained from 400 GOES M- and X-class events during the first 3.5 years of the Solar Dynamics Observatory (SDO) mission. A

Content from this work may be used under the terms of the Creative Commons Attribution 3.0 licence. Any further distribution of this work must maintain attribution to the author(s) and the title of the work, journal citation and DOI.
Published under licence by IOP Publishing Ltd

primary test addressed in this study is the closure of the various energies, such as the equivalence of the dissipated magnetic energies and the primary dissipated flare energies (accelerated particles, direct heating, CME acceleration), which faciliate the energy of secondary processes (plasma heating, shock acceleration) and interactions with the solar wind (aerodynamic drag). Our study demonstrates energy closure in the statistical average, while individual events may have considerable uncertainties, requiring improved nonlinear force-free field models, and particle acceleration models with observationally constrained low-energy cutoffs.

1. THEORETICAL CONCEPTS

The energy flow in solar flares and coronal mass ejections (CMEs) passes through several processes, which are depicted in the diagram of Fig. 1. Initially, a stable nonflaring active region exists with a near-potential magnetic field with energy E_p, which then becomes twisted and sheared, building up nonpotential energy E_{np} and the free energy, $E_{free} = E_{np} - E_p$, of which a fraction $E_{mag} \le E_{free}$ is dissipated during a flare (e.g., [1, 2]). There are three primary energy dissipation processes that follow after a magnetic instability, typically a magnetic reconnection process, spawning (i) the acceleration of nonthermal particles (e.g., reviews by [3–7], with electron energy $E_{nt,e}$ and ion energy $E_{nt,i}$, providing (ii) direct heating in the magnetic reconnection region, E_{dir} (e.g., [8–10]); these are often accompanied by (iii) an eruptive process, which can be a complete eruption of a CME or filament, or a semi-eruptive energy release, also known as "failed eruption", in the case of a confined flare (e.g., Török and Kliem [11]). The CME process carries an energy of $E_{cme} = E_{kin} + E_{grav}$, consisting of the kinetic energy E_{kin} and the gravitational potential energy E_{grav}, to lift a CME from the solar surface into the heliosphere. These primary energy dissipation processes allow us to test the primary energy closure equation,

$$E_{mag} = E_{nt} + E_{dir} + E_{cme} = (E_{nt,e} + E_{nt,i}) + E_{dir} + (E_{cme}^{kin} + E_{cme}^{grav} - E_{drag}) \ , \quad (1)$$

where the left side of the equation contains the total (magnetic) energy input (or storage) and the right side contains the total energy output (or dissipation). There is additional energy supply by the aerodynamic drag E_{drag}, which appears as a negative quantity on the right-hand side of Eq. (1) because it is supplied by the slow solar wind and reduces the magnetic energy demand in the flare region. After this primary step in the initiation of a flare and CME, secondary energy dissipation processes kick in. Nonthermal particles are accelerated along bi-directional trajectories that lead out of the magnetic reconnection region, where most particles precipitate down to the chromosphere, then heat the chromospheric plasma and drive evaporation of the heated plasma up into the corona (e.g., Antonucci and Dennis [12]), while other particles escape into interplanetary space (see reviews by [13–15]). The flare arcade that becomes filled with heated chromospheric plasma, radiates, and looses its energy by conduction and radiation in soft X-rays (SXR) and extreme ultraviolet (EUV). The thermal energy content E_{th} can be calculated from the total emission measure observed in SXR and EUV and should not exceed the nonthermal energy, $E_{nt} = E_{nt,e} + E_{nt,i}$, unless there are other heating processes besides the electron beam-driven heating observed in hard

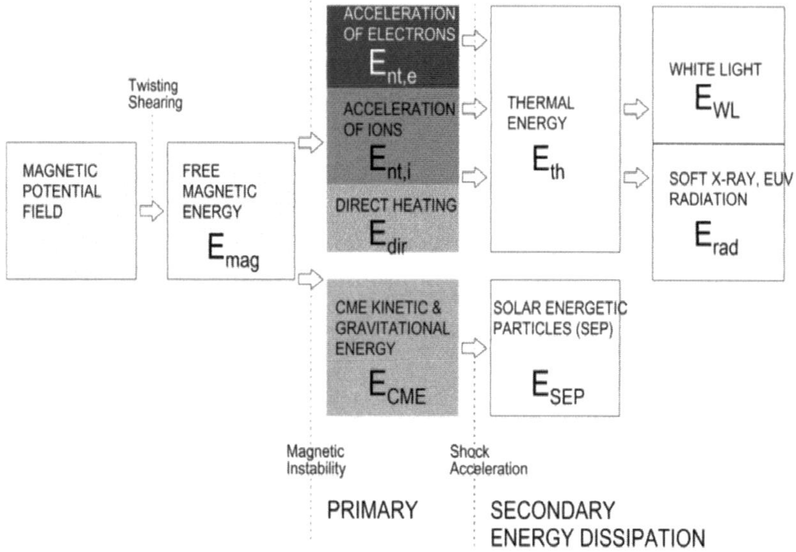

Figure 1. Schematic diagram of energy input (free magnetic energy E_{mag}), primary energy dissipation processes (electron acceleration $E_{nt,e}$, ion acceleration $E_{nt,i}$, direct heating E_{dir}, and launching of CME (E_{cme}), and secondary energy dissipation processes (thermal energy E_{th}, solar energetic particles E_{sep}, bolometric luminosity, and radiative energies observed in white light E_{WL}, soft X-rays, and EUV E_{rad} (Aschwanden et al [27]).

X-rays (according to the thick-target bremsstrahlung model of Brown 1971). Thus we can test the following energy inequality between thermal and nonthermal energies (if we neglect direct heating),

$$E_{th} \leq E_{nt} = E_{nt,e} + E_{nt,i} \ . \tag{2}$$

Radiation is produced not only at SXR and EUV wavelengths (E_{th}), but also in visible and near-ultraviolet wavelengths, recorded as white-light flare emission; this is the largest contributor to the bolometric energy or luminosity E_{bol}, which contains vastly more radiative energy than observed in SXR [16–18]. Using a superimposed epoch analysis of 2100 C, M, and X-class flares, Kretzschmar et al [19] and Kretzschmar ([18] and Table 1 therein) calculated the total solar irradiance for five synthesized flare time profiles. The continuum emission produced by white-light flares determined in this way allows us to compare another pair of energies – the total thermal energy E_{th} and the bolometric luminosity, produced by the flare impact of precipitating particles, radiative backwarming, and locally enhanced ionization, enhancing bound—free and free—continuum emission (e.g., [20–25]),

$$E_{bol} \approx E_{th} \ . \tag{3}$$

Another secondary process is the acceleration of nonthermal particles by the CME, which is produced by shock acceleration in very fast CMEs, observed in the form of *solar*

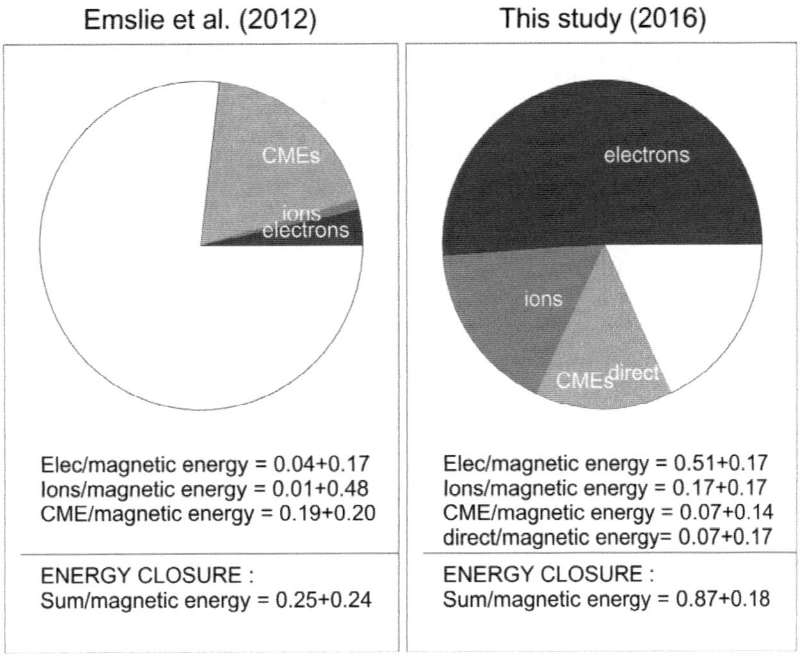

Figure 2. Pie chart of energy closure, obtained from previous work of Emslie et al [28] (left panel) and from the study of Aschwanden et al [27]. The full area of each pie chart is normalized to the total dissipated magnetic energies in flares. Testing the closure of energies, there is a discrepancy of over-estimating the magnetic energy by a factor of 4 in [28], which is reduced to a factor of 1.15 in the study of [27].

energetic particle (SEP) events (e.g., see review by Reames 2013 [26]), which allows us to test another energy inequality,

$$E_{sep} \leq E_{cme} = E_{kin}^{cme} + E_{grav}^{cme} \ . \tag{4}$$

The energy closure studied here depends, of course, on specific physical models of flares and CMEs. In this discussion we should be aware of the caveat that alternative flare models may deviate from the energy closure relationships and the inequalities discussed here. Another important issue in any energy closure relationship concerns the double-counting of energies if there are multiple energy conversion processes acting at the same time or nearly simultaneously. We attempt to distinguish between primary and secondary energy dissipation mechanisms, as shown in Figure 1.

2. OBSERVATIONAL RESULTS

2.1. Magnetic Energies

We measured the magnetic field in a large number (in about 200 cases out of the 400 flaring active regions) (Aschwanden et al [2]) by using the *vertical-current approximation nonlinear force-free field code (VCA-NLFFF)* [29, 30]. This deduces the magnetic potential field $\mathbf{B}_p$ from photospheric magnetograms $B_z(x,y)$ observed with the *Helioseismic and Magnetic Imager (HMI)* onboard the *Solar Dynamic Observatory (SDO)*, and measures the non-potential field $\mathbf{B}_{np}$ by forward-fitting of nonpotential field lines to coronal loop structures observed in EUV with the *Atmospheric Imager Assembly (AIA)* onboard SDO, which are automatically traced with the *Oriented Coronal Curved Loop Tracing (OCCULT)* code (Aschwanden et al [31]). The obtained magnetic parameters exhibit scaling laws for the nonpotential energy, $E_{np} \propto E_p^{1.02}$, for the free energy, $E_{free} \propto E_p^{1.7}$ and $E_{free} \propto B_\varphi^{1.0} L^{1.5}$ (where B_φ is the azimuthally twisted field component), for the dissipated energy, $E_{diss} \propto E_p^{1.6}$ and $E_{diss} \propto E_{free}^{0.9}$, and the energy dissipation volume, $V \propto E^{1.2}$. The potential energies vary in the range of $E_p = 1 \times 10^{31}$ — 4×10^{33} erg, while the free energy has a ratio of $E_{free}/E_p \approx 1\%$ — 25%. The Poynting flux amounts to $F_{flare} \approx 5 \times 10^8 - 10^{10}$ erg cm^{-2} s^{-1} during flares, which averages to $F_{AR} \approx 6 \times 10^6$ erg cm^{-2} s^{-1} during the entire observation period, and is comparable with the coronal heating rate requirement in active regions. Caveats of the calculation of magnetic energies includes the accuracy of non-potential energies obtained from the vertical-current approximation, and the detectability of the step-wise decrease of the free energy during the flare duration. It appears that a coronal illumination effect (e.g., chromospheric evaporation) is necessary to detect the full free energy in the preflare phase.

2.2. Thermal Energies

The thermal energies of the same data set of $\approx$ 400 M and X-class flares have been inferred from the multi-wavelength images observed with the *Atmospheric Imaging Assembly (AIA)* onboard the SDO (Aschwanden et al [32]). We compute the *differential emission measure (DEM)* distribution functions and associated multi-thermal energies, using a spatially-synthesized Gaussian DEM forward-fitting method. The multi-thermal DEM function yields a significantly higher energy (by an average factor of $\approx$ 14) than an isothermal energy estimate at the peak of the DEM. We find a statistical energy ratio of $E_{th}/E_{diss} \approx 2\%$ — 40% between the multi-thermal energy E_{th} and the magnetically dissipated energy E_{diss}, which is an order of magnitude higher than the estimates of Emslie et al [28]. For the analyzed set of M and X-class flares we find the following physical parameter ranges: $L = 10^{8.2} - 10^{9.7}$ cm for the length scale of the flare areas, $T_p = 10^{5.7} - 10^{7.4}$ K for the DEM peak temperature, $T_w = 10^{6.8} - 10^{7.6}$ K for the emission measure-weighted temperature, $n_p = 10^{10.3} - 10^{11.8}$ cm^{-3} for the average electron density, $EM_p = 10^{47.3}$–$10^{50.3}$ cm^{-3} for the DEM peak emission measure, and $E_{th} = 10^{26.8} - 10^{32.0}$ erg for the multi-thermal energies. The deduced multi-thermal energies are consistent with the *Rosner-Tucker-Vaiana 1978 (RTV)* [33] scaling law, $E_{th,RTV} = 7.3 \times 10^{-10} T_p^3 L_p^2$, which predicts extremal values of $E_{th,max} \approx 1.5 \times 10^{33}$ erg for the largest flare, and $E_{th,min} \approx 1 \times 10^{24}$ erg for the smallest coronal nanoflare. The size distributions of the spatial parameters exhibit powerlaw tails that are consistent with the predictions of the fractal-diffusive self-organized criticality model, combined with the RTV scaling law (Aschwanden 2012 [14]). Note that the thermal energies quantify a secondary energy dissipation process (in the thick-target model of Brown 1971), produced after the primary magnetic energy dissipation by accelerated electrons, ions, and direct plasma heating in the magnetic reconnection region. Nevertheless, the multi-thermal energies can be used as an upper limit for the precipitating electron flux that spawns chromospheric evaporation (Eq. 2).

2.3. Non-Thermal Energies

Using the *Ramaty High-Energy Solar Spectroscopic Imager (RHESSI)* data of 191 M and X-class flare events from the same data set discussed before, we fit a thermal and a nonthermal component to RHESSI spectra, which yields the temperature of the differential emission measure (DEM) tail, the nonthermal power law slope and flux, and the thermal/nonthermal cross-over energy e_{co}. From these parameters we calculate the total nonthermal energy E_{nt} in electrons with two different methods: (i) using the observed cross-over energy e_{co} as low-energy cutoff, and (ii) using the low-energy cutoff e_{wt} predicted by the warm thick-target bremsstrahlung model of Kontar et al [35]. Based on a mean temperature of $T_e = 8.6$ MK in active regions we find low-energy cutoff energies of $e_{wt} = 6.2 \pm 1.6$ keV for the warm-target model, which is significantly lower than the cross-over energies $e_{co} = 21 \pm 6$ keV. Comparing with the statistics of magnetically dissipated energies E_{mag} and thermal energies E_{th} from the two previous studies, we find the following mean (logarithmic) energy ratios with the warm-target model: $E_{nt} = 0.41 E_{mag}$, $E_{th} = 0.08 E_{mag}$, and $E_{th} = 0.15 E_{nt}$. The total dissipated magnetic energy exceeds the thermal energy in 95% and the nonthermal energy in 71% of the flare events, which confirms that magnetic reconnection processes are sufficient to explain flare energies. The nonthermal energy exceeds the thermal energy in 85% of the events, which largely confirms the warm thick-target model. In future work we may use the number of accelerated electrons in the flare volume (as estimated from the measured flare area and vertical scale height) as an independent additional constraint for the low-energy cutoff of the nonthermal electron injection spectrum.

2.4. CME Energies

We determine the mass m_{cme}, the kinetic energy E_{kin}, and the gravitational potential energy E_{grav} of CMEs in the same AIA/SDO dataset of 400 M and X-class flare events mentioned before, using a new method based on the EUV dimming effect [36]. EUV dimming is modeled in terms of a 3-D geometric volume that encompasses a CME footprint area in the lower corona and expands adiabatically into the upper corona and heliosphere. Such a dynamic CME model is fitted to the observed evolution of the total emission measure observed in EUV. The model derives the evolution of the CME volume $V(t)$, the mean electron density $n_e(t)$, the emission measure $EM(t)$, the bulk plasma expansion velocity $v(t)$, the mass m_{cme} (being constant in time), and the kinetic energy $E_{kin}(t)$ in the CME source region. The EUV dimming method is truly complementary to the Thomson scattering method in white light, which probes the CME evolution in the heliosphere at distances of $r \gtrsim 2R_\odot$, while the EUV dimming method tracks the CME launch in the corona. CME energies are found to be systematically lower than the dissipated magnetic energies, which is consistent with a magnetic origin of CMEs.

We refine the CME model and apply it to (860) GOES M- and X-class flare events observed during the first seven years (2010—2016) of the SDO mission [37]. The model refinements include: the 3-D geometry of CMEs, the gravitational deceleration, the relationship between the center-of-mass motion and the leading-edge motion observed in white-light coronagraphs, and the application of the Rosner—Tucker—Vaiana scaling law. The refined CME model based on EUV-dimming observed in AIA/SDO data complements the traditional white-light scattering model, using the Large-Angle and Spectrometric Coronagraph Experiment (LASCO) onboard the *Solar and Heliospheric Observatory*. The CME masses, speeds, and kinetic energies determined from both AIA and LASCO are found to agree on average. Empirical scaling laws were found, i.e., $v \propto T_e^{1/2}$, $v \propto (m_{cme})^{1/4}$, and $m_{cme} \propto L^2$, which indicate scaling laws for CME parameters.

The most recent work on CME energetics includes the aerodynamic drag force, which represents an external energy source that accelerates low—speed CME's to the slow solar wind speed of $w \approx 400$ km, and slows down initially fast CMEs to the slow solar wind speed. Preliminary estimates indicate that the CME energy budget reduces from 7% of the total dissipated magnetic energy to $\approx 5\%$, thanks to the acceleration conveyed by the aerodynamic drag. Moreover, the aerodynamic drag model [38, 39] predicts arrival times of CMEs at Earth with an accuracy of $\approx 23\%$ [40].

2.5. Solar Energetic Particles (SEPs)

SEP energies were calculated from observations by the STEREO A and B spacecraft and near-Earth spacecraft. Since most flares are not associated with SEP events, and many SEP events overlap with, and are difficult to separate from other events, estimates of SEP energies were only made for a small number of the 400 M and X-class flares in [27]. Estimates for 10 SEP events observed at both STEREO spacecraft and the Earth are given in Table 1 of [27], where further details of the analysis are also available. These events give a ratio $E_{sep}/E_{cme} = 0.03 \pm 0.45$, comparable with the [28] result of ≈ 0.04. The low ratio is consistent with our notion of CME-driven acceleration leading to SEP events being a secondary energy conversion process The ratio of the magnetic and SEP energies is only available for four events, which give $E_{sep}/E_{mag} = 0.1 \pm 1.64$. Notwithstanding the large error, the low ratio confirms that the magnetic free energy in the flare region is sufficient to explain the energetics of SEP particles, regardless of whether they are accelerated in the coronal flare region or in interplanetary shocks.

3. CONCLUSIONS

The results of the energy budget or energy partition of different forms of energies during a solar flare/CME event is summarized in Fig. 2. In the pioneering work of Emslie et al [28], the major constituents are accelerated electrons ($4\% \pm 17\%$), accelerated ions ($1\% \pm 48\%$), and CME kinetic energies in the slow solar wind frame ($19\% \pm 20\%$), which add up to $25\% \pm 24\%$ of the dissipated magnetic energy. The lack of closure is most likely caused by the (ad hoc) four-fold over-estimate of the magnetic free energy. A series of recent studies dedicated to the global energetics of flares and CMEs finds the following overall average ratios of $51\% \pm 17\%$ for electron acceleration, $17\% \pm 17\%$ for ion acceleration, $7\% \pm 14\%$ for CME kinetic energies, and $7\% \pm 17\%$ for direct heating, which add up to closure ($87\% \pm 18\%$) [2, 32, 41, 27, 40, 36]. *Solar energetic particle (SEP)* events draw a relatively small fraction of a few percent from the energy of a CME shock front. The largest uncertainties, mostly due to a lack of suitable energy and magnetic field models, are attributed to the measurement of changes in the free (magnetic) energy, direct heating during flares, the low-energy cutoff of nonthermal particles, CME halo geometries, and bolometric flare luminosities.

We acknowledge useful discussions with John Raymond, Nat Gopalswamy, Edmond Roelof, Richard Mewaldt, HaiHoing Che, and other participants of the 18th meeting on Energetic Particles in Pasadena, February 2019. This work was partially supported by NASA contracts NNX11A099G, NNG04EA00C (SDO/AIA), and NNG09FA40C (IRIS).

References

[1] Schrijver, C.J., DeRosa, M.L., Metcalf, T., 2008, *ApJ* 675, 1637
[2] Aschwanden, M.J., Xu, Y., Jing, J., 2014, *ApJ* 797, 50
[3] Miller, J.A., Cargill, P.J., Emslie, A.G., 1997, *JGR* 102/A7, 14631
[4] Aschwanden, M.J., 2002, *Space Science Reviews* 101, 1
[5] Aschwanden, M.J.. 2004, *Physics of the Solar Corona. An Introduction*, Berlin: Springer and Praxis, p.216
[6] Benz, A.O., 2008, *Living Reviews in Solar Physics* 5, 1
[7] Holman, G.D., Aschwanden, M.J., Aurass, H., 2011, *Space Sci Rev* 159, 107
[8] Sui, L. and Holman, G.D., 2003, *ApJL* 596, L251
[9] Caspi, A. and Lin, R.P. 2010, *ApJ* 725, L161
[10] Caspi, A., Shih, A.Y., McTiernan, J.M., and Krucker, S. 2015, *ApJ* 811, L1
[11] Török, T. and Kliem, B. 2005, *ApJ* 630, L97
[12] Antonucci, E. and Dennis, B.R. 1983, *Solar Phys.* 86, 67
[13] Hudson, H.S. and Ryan, J. 1995, *ARAA* 33, 239
[14] Aschwanden, M.J. 2012, *A&A* 539, A2.
[15] Lin, R.P. 2007, *Space Sci Rev* 124, 233
[16] Woods, T.N., Eparvier, F.G., Fontenla, J., et al. 2004, *GRL* 31, L10802
[17] Woods, T.N., Kopp, G., and Chamberlin, P.C. 2006, *JGR (Space Physics)*, 111, 10
[18] Kretzschmar, M. 2011, *A&A*, 530, A84
[19] Kretzschmar, M., Dudok de Wit, T., Schmutz, W., et al. 2010, *Nature Physics*, 6(9), 690
[20] Najita, K. and Orrall, F.Q. 1970, *Solar Phys*, 15, 17
[21] Hudson, H.S. 1972, *Solar Phys*, 24, 414
[22] Ding, M.D., Liu, Y., Yeh, C.-T., and Li, J.P. 2003, *A&A*, 403, 1151
[23] Battaglia, M. and Kontar, E.P. 2011, *ApJ*, 735, 42
[24] Battaglia, M., Kontar, E.P., and Hannah,I.G. 2011, *A&A* 526, A3
[25] Xu,Y., Jing, J., Wang, S., and Wang, H. 2014, *ApJ*, 787, 7
[26] Reames, D.V. 2013, *Space Science Reviews*, 175, 53
[27] Aschwanden, M.J., Caspi, A., Cohen, C.M.S., et al. 2017, *ApJ*, 836, 17
[28] Emslie, A.G., Dennis, B.R., Shih, A. Y., et al. 2012, *ApJ*, 759, 71
[29] Aschwanden, M.J. 2013, *Solar Phys.*, 287, 323
[30] Aschwanden, M.J. 2016b, *ApJSS* 224, 25
[31] Aschwanden, M.J., DePontieu, B., and Katrukha, E. 2013, Entropy, 15(8), 3007
[32] Aschwanden, M.J., Boerner, P., Ryan, D., et al. 2015, *ApJ* 802, 53
[33] Rosner, R., Tucker, W.H., and Vaiana, G.S. 1978, *ApJ*, 220, 643
[34] Brown, J.C. 1971, *Solar Phys.* 18, 489
[35] Kontar, E.P., Jeffrey, N.L.S., Emslie, A.G., and Bian, N.H. 2015, *ApJ* 809, 35
[36] Aschwanden, M.J. 2016a, *ApJ* 831, 105
[37] Aschwanden, M.J. 2017, ApJ 847, 27
[38] Cargill, P.J. 2004, *Solar Phys.*, 221, 135
[39] Vrsnak, B., Zic, T., Vrbanec, D., et al. 2013, *Solar Phys.*, 285, 295
[40] Aschwanden, M.J. 2019, *New Millennium Solar Physics*, Section 11.8, New York: Springer, (in press), http://www.lmsal.com/~aschwand/bookmarks_books2.html.
[41] Aschwanden, M.J., Holman, G., O'Flannagain, A., et al. 2016, ApJ 832, 27.

A Brief Review on Particle Acceleration in Multi-island Magnetic Reconnection

H. Che
University of Maryland, College Park, MD, 20742, USA
NASA Goddard Space Flight Center, Greenbelt, MD, 20771, USA

G. P. Zank
Center for Space Plasma and Aeronomic Research (CSPAR), University of Alabama in Huntsville, Huntsville, AL 35805, USA
Department of Space Science, University of Alabama in Huntsville, Huntsville, AL 35899, USA

Abstract. The basic physics and recent progresses in theoretical and particle-in-cell (PIC) simulation studies of particle acceleration in multi-island magnetic reconnection are briefly reviewed. Particle acceleration in multi-island magnetic reconnection is considered a plausible mechanism for the acceleration of energetic particles in solar flares and the solar wind. Theoretical studies have demonstrated that such a mechanism can produce the observed power-law energy distribution of energetic particles if the particle motion is sufficiently randomized in the reconnection event. However, PIC simulations seem to suggest that the first-order Fermi acceleration mechanism is unable to produce a power-law particle energy distribution function in mildly relativistic multi-island magnetic reconnections. On the other hand, while simulations of highly relativistic reconnections appear to be able to produce a power-law energy spectra, the spectral indices obtained are generally harder than the soft power-law spectra with indices ~ -5 commonly observed in the solar wind and solar flare events. In addition, the plasma heating due to kinetic instabilities in 3D magnetic reconnection may "thermalize" the power-law particles, making it even more difficult for multi-island reconnections to generate a power-law spectrum. We discuss the possible reasons that may lead to these problems.

1. Introduction

Magnetic reconnection (MR) is believed to be an essential process for particle energization in magnetospheric substorms [1, 2, 3], solar wind [4, 5], and solar flares [6, 7, 4, 8], and is drawing increasing interests in its possible roles in astrophysical phenomena such as the launching of the solar wind [9, 10], γ-ray flares in the Crab Nebula [11, 12] or pulsar nebulae in general, and γ-ray bursts [13, 12].

The mechanisms of particle acceleration in space and astrophysical plasma is still not fully understood [14]. We need to understand how particles gain energy and how the observed power-law energy distribution $f(W) = W^{-\alpha}$ $(\alpha > 0)$ is produced [4, 15]. Two types of acceleration mechanisms, namely, the first and second order Fermi acceleration (Fig. 1), form the foundation of most of the current models. In 1949, Fermi proposed the first acceleration mechanism for the acceleration of cosmic rays [16]. In essence fast particles move among wandering magnetic

Content from this work may be used under the terms of the Creative Commons Attribution 3.0 licence. Any further distribution of this work must maintain attribution to the author(s) and the title of the work, journal citation and DOI.
Published under licence by IOP Publishing Ltd

fields in space and spiral around the lines of force until they collide with an irregular moving magnetic cloud and are reflected. On average the particles have an energy gain in each reflection, proportional to the second order of the velocity ratio, i.e., $\delta W = (V/c)^2 W$, where V is the average velocity of magnetic clouds, and hence this mechanism is called the second order Fermi acceleration. The randomness of moving magnetic clouds leads to a power-law energy distribution. The second acceleration mechanism concerns the diffusive shock acceleration [17, 18, 19, 20, 21]. The particles are accelerated as they are bounced back and forth between the downstream and upstream of shock waves and energy gain in each crossing is proportional to the first order of V/c, where V is the velocity of shock waves, and hence the mechanism is called first order Fermi-acceleration. The particles are scattered during crossings by plasma turbulence generated by the shock, resulting in a power-law energy distribution.

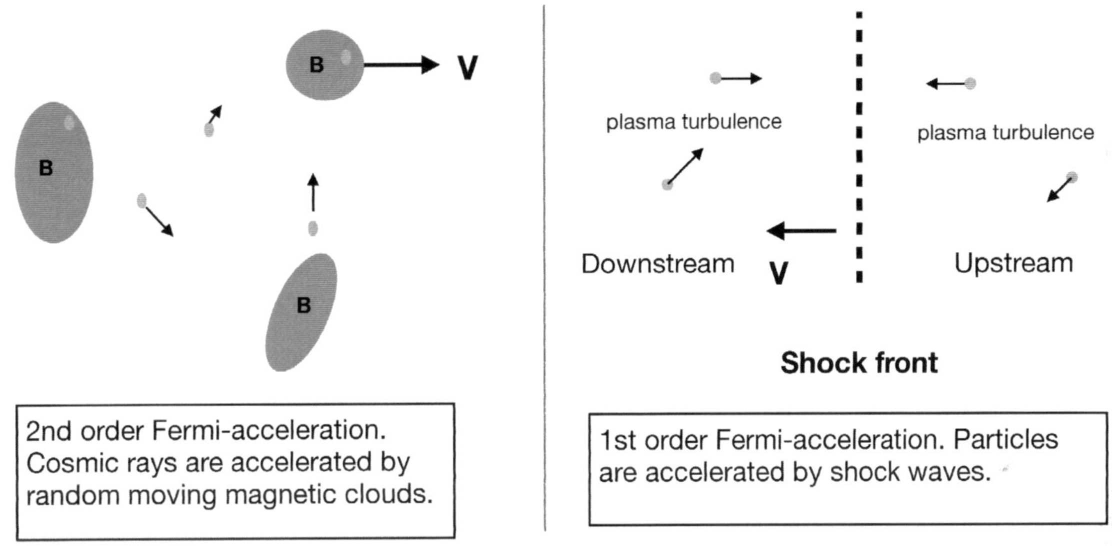

Figure 1: Illustration of 2nd and 1st order Fermi-acceleration in astrophysical media and shock waves.

Observations of substorms and solar flares show that particles are commonly accelerated during MR [22, 4, 15], but the acceleration process involved is complex and diverse [23, 24]. In MR the released magnetic energy predominantly transfers to the electromagnetic energy flux, the thermal energy flow, and the kinetic energy of plasma outflow or the bulk motion [25, 26] (Fig. 2). It is suggested that plasma outflow is accelerated either by the inductive reconnection electric field, such as the perpendicular electric field $\mathbf{V} = -c\mathbf{E} \times \mathbf{B}/B^2$, the parallel electric field [27, 23, 28, 29, 30], or by the gradient/curvature of magnetic field induced drifts [31, 32, 33]. The ohmic heating and wave-particle interactions generated by instabilities and turbulence lead to the conversion to thermal energy [34, 35, 36, 37, 38, 39]. The bulk acceleration from MR can not produce a power-law energy distribution. Turbulent heating generated by instabilities does commonly produce a non-Maxwellian velocity distribution function through Landau damping or nonlinear wave-particle interactions, such as phase mixing by electron holes, but a power-law energy distribution is not a common feature [41, 31, 42, 43, 44, 45, 46]. To solve this problem, a first-order Fermi-acceleration in multi-island MR was proposed [47]. This model assumes that particles are repeatedly accelerated by the randomly distributed magnetic islands generated by

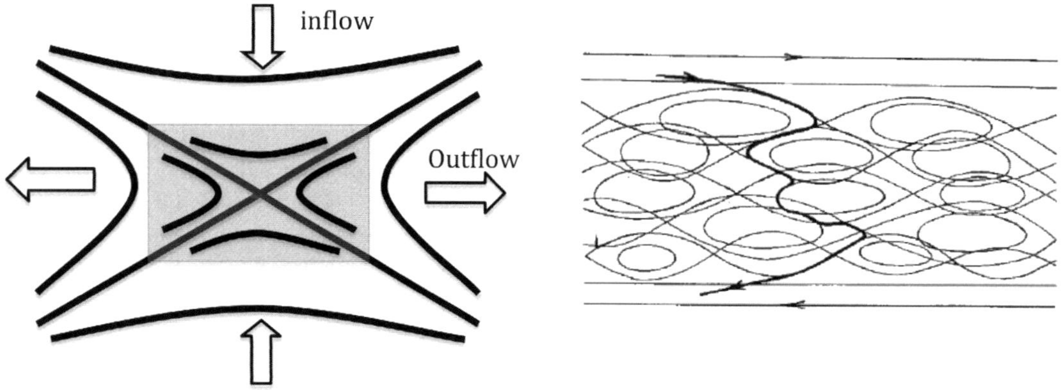

Figure 2: Left: Illustration of magnetic reconnection x-line. The shaded blue area is the diffusion region, where charged particles and the magnetic field decouple. Right: Illustration of patchy magnetic reconnection. Solid line: the stochastic wandering of magnetic field lines resulting in particle penetration of magnetic islands (from Galeev, 1986 [40]).

tearing instability [48, 49].

Tearing instability is driven by the magnetic shear in current sheets and can be triggered when the width of current sheet becomes close to ion inertial scale. MR driven by tearing instability is thus called spontaneous MR [40]. Tearing instability produces multiple magnetic islands and the coalescence of islands lead to stochastic wandering of magnetic field lines and percolation of islands [50, 51, 52], and spontaneous MR is therefore also called patchy MR (Fig. 2). The stochastic motion of particles caused by the coalescence of islands may result in the power-law energy distribution in principle, which is the essential element in the multi-islands MR acceleration model [47, 53, 54]. However, particle-in-cell (PIC) simulations carried by several groups show that similar to shock acceleration, multi-island MR also faces an injection problem. It is found that for extreme relativistic cases with Lorentz factor $\gamma \gg 1$ and extremely magnetized plasma with $B^2/4\pi mc^2 \gg 1$, a power-law energy distribution function can be obtained [55, 56, 57, 58]; but for mildly relativistic cases with $\gamma \sim 1$ and $B^2/4\pi mc^2 \sim 1$, a power-law energy distribution is difficult to achieve, particularly for energetic electrons [44, 45, 46], although in some simulations electrons with lower energy at a few specific regions develop a power-law energy distribution [31, 59, 33].

Since energetic particles accelerated in heliophysics and solar physics are mildly relativistic, the injection problem in multi-island MR acceleration is a serious issue for the mechanism to be applied to particle acceleration in solar flares. This paper discusses progresses and problems in the study of multi-island MR acceleration in the past decade. The basic physics of the multi-island MR acceleration is introduced, the progress and unsolved problems are described, and then we present a summary of the main points and conclusions.

2. Particle Acceleration in Multi-island Magnetic reconnection

The basic picture of multi-island MR is schematically shown in Fig. 3. The size of the islands varies from electron to ion-scales. Opposite directed magnetic fields between neighbouring islands develop localized MR. These small scale MR events can be triggered by turbulence due to the twists and braids of magnetic flux tubes. In large-scale MRs such as those associated with solar flares, plasmoids generated by current-driven instabilities such as the tearing instability

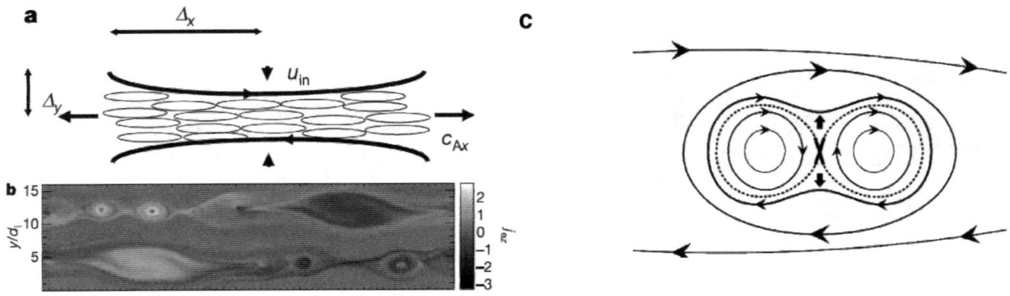

Figure 3: Left Panel: a) Diagram showing volume-filing islands driven by a tearing instability around the magnetic field reversal region. b) PIC simulation of island formation during magnetic reconnection. j_{ez} is the out-of-plane current sheet. (From Drake, Swisdak, Che & Shay, Nature, 2006 [47].) Right Panel: c) a cartoon of a locally oppositely directed magnetic field (lines with arrows) in neighbouring islands experiencing magnetic reconnection (Zank et. al. 2014 [53]).

fill the magnetic field reversal region. In multi-island MR, beside the acceleration produced by reconnection induced electric field, the contraction and merging of the islands also lead to the acceleration of particles [47, 59, 60, 53, 61, 62, 63].

2.1. Acceleration via Contraction of Magnetic Islands in Magnetic Reconnection

To illustrate how island contraction accelerates electrons, let's consider a 2D anti-parallel MR in the xy plane. Without external forces, an elongated magnetic island will gradually become round due to its internal magnetic tension. This effect is illustrated in a 2D simulation in Fig. 4 (a, b). Initially the magnetic island is a very narrow ellipse, and later it becomes a circle with radius r. In this process, the magnetic flux ψ is approximately conserved such that $B_{x,a}a \approx B_{x,b}r$, where $B_{x,a}$ and $B_{x,b}$ are the magnetic field as shown in panels (a) and (b), respectively. In an incompressible plasma, the volume of the magnetic island is conserved, so that $\pi ab = \pi r^2$. The perimeter of magnetic island decreases during this process. The perimeter l of an eclipse approximately is $l \approx \pi(a+b)(1+3h/(10+\sqrt{4-3h})) > \pi(a+b) \geq 2\pi\sqrt{ab} = 2\pi r$, thus the perimeter decrease as the island evolves from eclipse to round — a reason we call the contraction of island.

The loss of magnetic energy $\triangle\epsilon$ can be estimated by [53]:

$$\triangle\epsilon = \frac{b-a}{b}\epsilon_a, \tag{1}$$

where $\epsilon_a = \pi ab B_{x,a}^2$ is the initial magnetic energy stored in the island.

In the simulation in Fig. 4 (a, b), the temperature of the trapped electrons has increased as the island contracts, indicating the magnetic energy is converted to the kinetic and thermal energy of particles. It should be noted that the volume of the magnetic island is generally not conserved and magnetic energy can accumulate during contraction. During the contraction, electrons are accelerated in directions both parallel and perpendicular to the magnetic field. The detailed derivation of the acceleration process can be found in [47] and [53]. Here we describe the basic picture.

The trapped electrons circulate along the magnetic field lines of the contracting islands and gain energy in each circle (Fig. 4 (c, d)). Since the island contraction is relatively slow, the

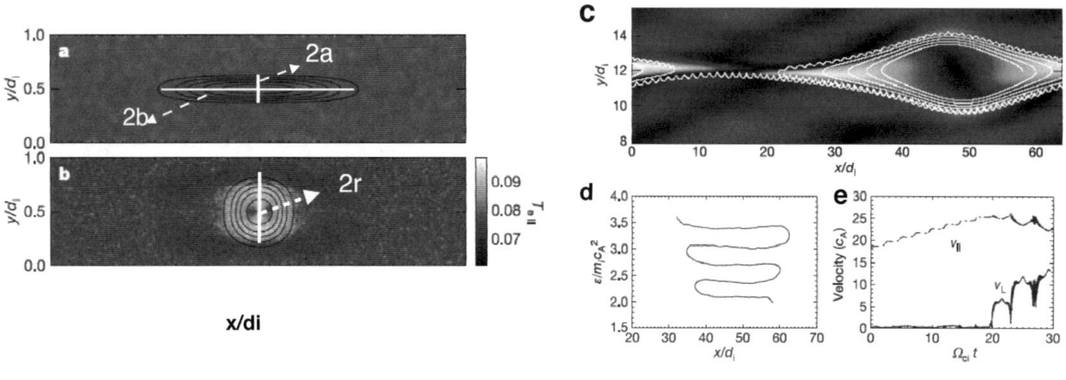

Figure 4: Left Panel: a 2D simulation of a squashed magnetic flux bubble. Right Panel: Test particle orbit and the sudden energy gain of electrons at the ends of island due to the curvature drift. (From Drake, Swisdak, Che & Shay, Nature, 2006 [47]).

second adiabatic invariant $J = m_e \oint v_\parallel ds \sim m_e v_\parallel l$ implies

$$\frac{dv_\parallel}{dt} = -v_\parallel \frac{1}{l}\frac{dl}{dt}. \tag{2}$$

Here $v_\parallel$ is the parallel velocity of a single electron, ds is the displacement along magnetic field lines, and l is the perimeter of the magnetic island. Since $dl/dt < 0$ due to contraction, $dv_\parallel/dt > 0$, and the trapped electrons are accelerated as they bounce inside the island.

The first adiabatic invariant $\mu = m_e v_\perp^2/2B = constant$ and the conservation of magnetic flux $\psi \sim Bl$ describe the electron acceleration in the perpendicular direction:

$$\frac{dv_\perp^2}{dt} = \frac{v_\perp^2}{B}\frac{dB}{dt}, \qquad \frac{dB}{dt} = -B\frac{dl}{dt}. \tag{3}$$

Again since $dl/dt < 0$, $dv_\perp^2/dt > 0$. This energy gain process is shown in a test particle simulation (Fig. 4 (d, e)). In panel (d) we see that the electron energy jumps when the electron arrives at each endpoint of the magnetic island. The perpendicular velocity, i.e. the gyration velocity abruptly increases as $v_\parallel$ slightly decreases, implying that the energy transfer is from the parallel to perpendicular.

Assume the velocity of contraction of an island with perimeter l is $V_l = dl/dt$, both Eq. (3) & (2) can be rewritten in the form of

$$\frac{dW}{dt} \propto \frac{V_l}{c}\frac{1}{\tau_c}W, \tag{4}$$

where $W = m_e v^2/2$ and $\tau_c = l/c$. Thus the contraction of island is in essence a 1st order Fermi acceleration.

The changing of curvature of magnetic islands can accelerate particles in the perpendicular direction, especially, at the ends of islands (Fig.4), as the particles trapped within magnetic islands and slowly spiral along magnetic field lines, i.e. $v_\parallel \gg v_\perp$. The curvature drift is

$$\mathbf{v}_c = \frac{mv_\parallel^2}{eB}\mathbf{b} \times (\mathbf{b} \cdot \nabla \mathbf{b}), \tag{5}$$

where $\mathbf{b} = \mathbf{B}/B$. The curvature drift contributes to the perpendicular velocity. Once $v_\perp$ is sufficiently large, the particle can escape the trap of magnetic island and migrate to another magnetic island.

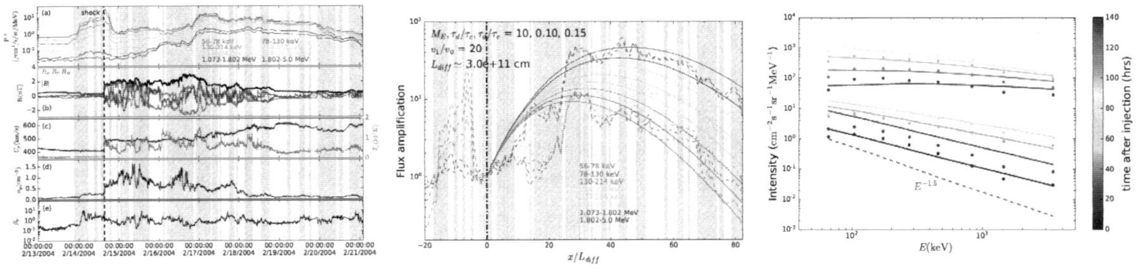

Figure 5: **Left** Ulysses observations. The vertical gray lines show the regions where we have observed (using a Grad-Sharafnov reconstruction technique) the presence of magnetic islands. **Middle:** The comparison of theoretical (dashed) and observational (solid) particle flux. **Right:** The comparison of theoretical (solid) and observational (dotted) differential intensity spectrum. (Zhao, Zank et al. 2018 [65])

2.2. Acceleration via Merging of Magnetic Islands

The merging of islands produces localized MR (Fig. 3). For simplicity, we first neglect the reconnection electric field so that the first and second adiabatic invariants hold. Following Fermo et al. [64], let's assume two round islands with initial areas A_1 and A_2, which are merged into a single island with area $A_f = A_1 + A_2$. The circumference of the merged island satisfies $l = (l_1^2 + l_2^2)^{1/2} < l_1 + l_2$, which implies that the $dl/dt < 0$ in Eq. (2). Consequently, the merging leads to the electron acceleration in the parallel direction.

During merging, the magnetic flux from the two islands is not additive but is equal to the larger of the two initial fluxes, $\max(\psi_1, \psi_2)$ [64, 53]. Before merging, the total magnetic energy of the two islands is $W_i = B_1^2 \pi l_1^2 + B_2^2 \pi l_2^2 = \pi(\psi_1^2 + \psi_2^2)$. The magnetic energy of the merged island $W_f = \pi \max(\psi_1, \psi_2)^2 < W_i$. Thus merging of islands is an alternative path to efficiently dissipate magnetic energy and accelerate electrons in the parallel direction. If we consider the localized electric field caused by the reconnections, part of the magnetic energy is transferred to the kinetic energy of electrons via electric field acceleration.

2.3. First-order Fermi Acceleration and Power-Law Energy Distribution

The MR electric field accelerate particles via $d\mathbf{v}/dt = q/m\delta\mathbf{E}$. In the contraction and merging of islands, first and second adiabatic invariants and magnetic flux are approximately conserved and causes parallel and perpendicular acceleration. To obtain a power-law particle energy distribution, particles must experience sufficient stochastic motion. If the contraction of multi-islands acceleration dominates, particles are scattered by randomly distributed magnetic island or turbulence, then the acceleration is a first-order Fermi-acceleration. Zank et al. studied how various acceleration mechanisms in multi-island MR driven by tearing instability contribute to the index of power-law energy distribution.

By expanding the 6D Liouville equation, Zank et al. [53] obtain:

$$\frac{\partial f}{\partial t} + \mathbf{v} \cdot \frac{\partial f}{\partial \mathbf{x}} + \frac{\mathbf{F}_L}{m} \cdot \nabla_v f = \frac{\delta f}{\delta t}\Big)_s - \frac{(\mathbf{F} - \mathbf{F}_L)}{m} \cdot \nabla_v f - f \nabla_v \cdot (\frac{\mathbf{F}}{m}), \tag{6}$$

where $\mathbf{F} \equiv \mathbf{F_L} + \mathbf{F_c} + \mathbf{F_m} + \mathbf{F_E}$, $\mathbf{F_L} = \mathbf{v} \times \mathbf{B}/c$ is the Lorentz force, $\mathbf{F_c}$ is the force caused by the contraction of islands, $\mathbf{F_m}$ is the force caused by the merging of islands, and $\mathbf{F_E}$ is the force caused by localized reconnection inductive electric field $\delta\mathbf{E}$. While the Lorentz force is a function of v, it can be shown that $\nabla_v(\mathbf{v} \times \mathbf{B}) = 0$. On the other hand, $\nabla_v \cdot \mathbf{F}_c/m = 2\eta_c$, $\nabla_v \cdot \mathbf{F}_m/m = \eta_m - \eta_m = 0$. The pitch-angle scattering term is assumed to have the simplest possible form $\delta f/\delta t)_s = \partial((1/\tau_s)(1-\mu^2)(\partial f/\partial\mu))/\partial\mu$, where $\mu = \cos\theta$ is the cosine of the particle pitch angle. $\tau_s \sim \Omega = eB/m$ is the characteristic time-scale of pitch angle scattering.

Zank et al. [53] considered the MR acceleration in a super-Alfvénic flow, e.g., the solar wind. In this case, the convective flow velocity term typically is much larger than the corresponding advective Alfvénic velocity, allowing the latter term to be safely neglected in the transport formulation. To eliminate the electric field $\mathbf{E} = -\mathbf{U}\times\mathbf{B}$ due to the flow bulk motion, we rewrite the equation in the plasma flow frame using $\mathbf{v} \equiv \mathbf{c} + \mathbf{U}$. Averaging Eq. (6) over the gyro-angle, expanding the gyrophase-averaged distribution function f by Legendre polynomials, then the Laplace transformation yields a general solution for a nearly isotropic particle distribution in the steady state of Eq. (6):

$$f(x, c/c_0) = \frac{n_0}{8\pi c_0^2 v_E}\left(\frac{c}{c_0}\right)^{-(3+M_E+2\tau_{diff}/(3\tau_c)M_E^2)/2}\exp\left(-\frac{2\tau_{diff}}{3\tau_c}M_E\frac{x}{L_{diff}}\right)I_0(\Phi)\times$$
$$H[\ln(\frac{c}{c_0})]\times H[\ln(\frac{c}{c_0})+\frac{2}{M_E}\frac{x}{L_{diff}}], \quad (7)$$

$$\Phi = \left[\frac{\tau_{diff}}{3\tau_c}M_E^2\left(M_A - 3 + \frac{\tau_{diff}}{3\tau_c}M_E^2\right)\right]^{1/2}\times\left[(\ln(c/c_0)^2 + \frac{4}{M_E}\ln(c/c_0)\frac{x}{L_{diff}}\right]^{1/2},$$

where $\tau_c \equiv 1/\eta_c$ is the contraction time, $\tau_{diff} = \kappa/U^2$ is the diffusion time, $L_{diff} = \kappa/U$ is the diffusion length, $M_E = U/v_E$ and v_E is the reconnection electric field accelerated velocity; I_0 is the modified Bessel function of order 0 and $H(x)$ is the Heaviside step function. We can see that the velocity power-law index depends on M_E and τ_{diff}/τ_c. M_E is a unique quantity in MR due to the formation of particle beams. Eq. (7) has successfully explained the enhancement of energetic particle flux produced at the downstream of a shock at 5 AU by multi-island model [65] (Fig.5). If we consider only magnetic island contraction, the steady-state solution yield a velocity power-law with index $\alpha = 3[1+\tau_c/(8\tau_{diff})]$ [53], implying the longer diffusion time produces a harder power-law.

3. Outstanding Issues in the Current Studies of Multi-island Magnetic Reconnection

Analytical studies [66, 53, 61, 62, 63, 54] have shown that a power-law particle energy distribution can be achieved in multi-island MR if the motion of particles can be well randomized. However, in PIC simulations how to produce a power-law particle energy distribution remains a main issue, especially, how to reproduce the observed power-law spectra found in both solar flares and solar wind. PIC simulations [67, 55, 68, 57, 69] that have successfully produced power-law energy spectra have assumed extreme relativistic conditions with typical Lorentz factors $\gamma \sim 100$ to 10^3, and magnetic field $\sim 10^{12}$ G (the corresponding ratio of electron gyro-frequency and plasma frequency is $\Omega_{ce}/\omega_{pe} \sim 100$). These simulations may not be relevant to the bulk of energetic electrons in the solar corona and solar wind whose velocities are typically $0.2c - 0.5c$ ($\gamma \sim 1.02 - 1.3$) and magnetization $\Omega_{ce}/\omega_{pe} < 1$. PIC simulations with similar system sizes but essentially non-relativistic were so far unable to generate the power-law energy spectrum [44, 45, 46]. As an example, we show the electron energy distribution functions obtained from PIC simulations carried by two independent groups in Fig. 6.

It is unclear why mildly relativistic PIC simulations fail to produce the power-law energy distribution. One possible reason is that it is easier for highly relativistic particles to escape from the trapping of a single magnetic island and hence the particles can interact with many different islands. The repetitive acceleration of the particles in different islands makes the motion of particles more stochastic. In mildly relativistic PIC simulations, particles are confined in the

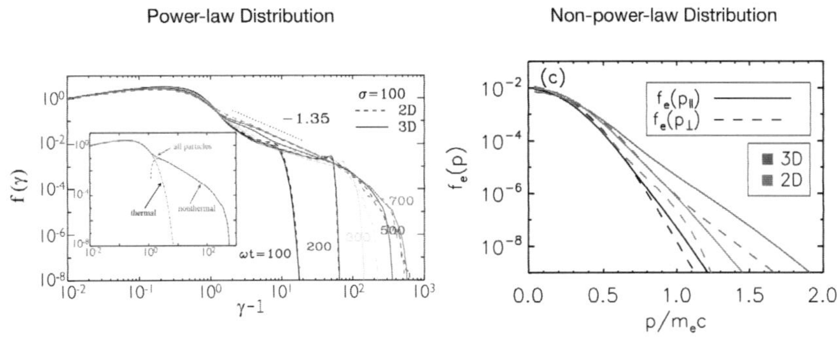

Figure 6: **Left** Power-law electron energy distribution from highly relativistic PIC simulation; Guo et al.[55]. **Right:** Non-power law electron energy distribution from mildly relativistic PIC simulation; Dahlin et al.[46].

magnetic islands for a relatively long period of time and particle transport among magnetic islands is slow. For example, in the PIC simulation by Drake et al.[47], the average time for an electron to be confined in the magnetic island is about 30 Ω_i^{-1} (see Fig. 4, right panel). For a simulation domain of about 200 d_i, the maximum length of the simulation free from complications caused by boundary condition is about 200 Ω^{-1}. The mean number of islands an electron can interact with during the whole simulation is only about 7, which may be too small to randomize the electron population.

Another problem with current PIC simulations is that none can produce the observed steep power-law above the spectral break in flare energetic electrons (Fig. 7) [70, 71, 72, 4, 73]. The typical energy spectral index obtained in the aforementioned relativistic simulations is ~ -1.3, which is close to the observed index below the spectral break ($\sim 50-100$ keV), but far from the index -3.9 ± 0.9 above the break. This index can not match the observed energy spectrum of superhalo electrons in the solar wind with a index ~ -3. According to the steady-state solution from Zank et al. [53], the hard power-law is likely a result of long diffusion time. The softer power-law spectra in solar flares and solar wind indicates particle transport is likely very fast.

Most of the aforementioned simulations are conducted in 2D. In 3D MR, it is almost unavoidable that various kinetic instabilities are triggered. This results in plasma heating that may suppress the generation of a power-law energy distribution function [59, 45, 46, 74]. How to reconcile this issue with the observed power-law spectra is probably the key to understand particle acceleration in solar flares and solar wind.

4. Summary and Conclusion

Contraction of multi-island during MR is an emerging mechanism for particle acceleration in space and astrophysics [47, 4, 53]. In the past decade extensive analytical and numerical studies have been carried out. The followings are the main points of this paper:

- The acceleration from contraction of magnetic island during MR is proportional to the first-order of V_l/c, where V_l is the contraction velocity of the island [47].
- Particles attain a power-law energy distribution $f(W) \propto W^{-\alpha}$ through repetitive interactions with multiple islands. Therefore, this mechanism is a first-order Fermi-acceleration. Analytical studies show that the contribution from contraction to the index of power-law α is related to the ratio of contraction time and particle diffusion time τ_c/τ_{diff}, implying longer diffusion time leads to a harder power-law [53].

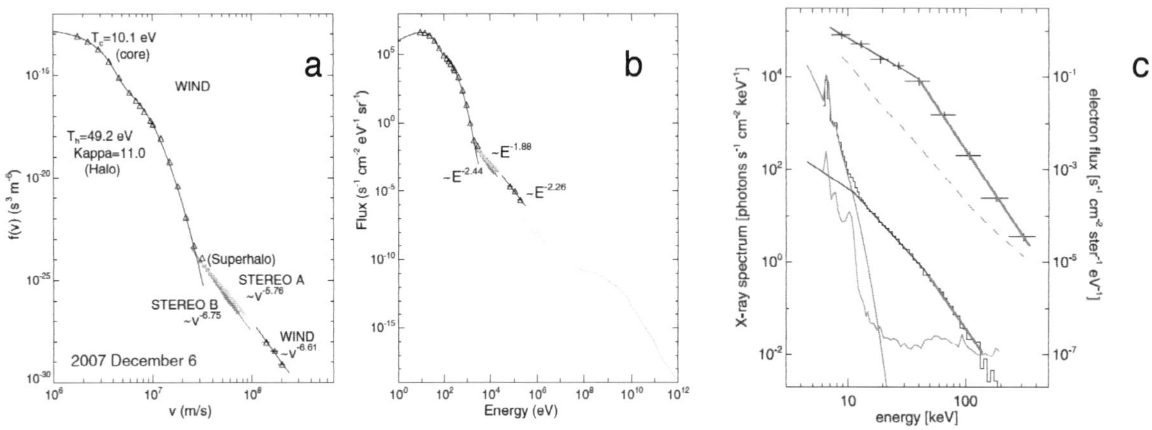

Figure 7: **a)** The quiet-time electron velocity distribution function f(v) at 1AU; **b)** the flux spectrum. a) and b) are obtained from the solar wind observation of Wang et. al.[75]. The electron energy ranges from $\sim$ 9 eV to $\sim$ 200 eV as measured by *WIND* (black) and *STEREO* (green) A (green) and B (red) on 2006 December 6. The solid lines represent the fit by a sum of a Maxwellian and a Kappa distribution to the solar wind core and halo distribution and a power-law fit to the superhalo. In panel (b), the blue diamonds/curves represent the electron flux spectra of the interplanetary electrons at 0.2-2 MeV and galactic cosmic ray at $\sim$ 30 MeV-1 TeV. Panel **c)** is a broken power-law energy spectrum fit to the data (blue line) from the paper by S. Krucker, 2007[71].

- PIC simulations can produce a power-law in highly relativistic MR events [55], but fail to produce a power-law energy distribution in mildly relativistic MR, which is more relevant to particle acceleration in solar wind and solar flares [46].
- In addition, the power-law energy distribution function obtained by highly relativistic PIC simulations is so far too hard ($\alpha \sim 1-2$) to explain the energetic particle energy distribution observed in the solar wind ($\alpha \sim 5$) and solar flares.
- The apparent contradiction between analytical calculations and the PIC simulations implies that the assumption of the particle diffusion time in multi-island MR is long, which may explain the hard power-law in relativistic MR acceleration simulations, but in mildly relativistic MR accelerations in solar flares and solar wind, the diffusion time is likely to be much shorter. What can cause such short diffusion time is the subject of future studies.

5. Acknowledgement
HC is partly supported by NASA Grant No.NNX17AI19G. The simulations and analysis were supported by the NASA High-End Computing (HEC) Program through the NASA Advanced Supercomputing (NAS) Division at Ames Research Center.

References
[1] Dungey J W 1961 *Physical Review Letters* **6** 47–48
[2] Baker D N, Pulkkinen T I, Angelopoulos V, Baumjohann W and McPherron R L 1996 *Journal of Geophysical Research: Space Physics* **101** 12975–13010 URL `https://agupubs.onlinelibrary.wiley.com/doi/abs/10.1029/95JA03753`
[3] Zelenyi L M, Artemyev A V, Malova K V, Petrukovich A A and Nakamura R 2010 *Physics Uspekhi* **53** 933–941
[4] Lin R P 2011 Space Sci. Rev. **159** 421–445 (*Preprint* `1110.1805`)
[5] Khabarova O V, Zank G P, Malandraki O E, Li G, le Roux J A and Webb G M 2017 *Sun and Geosphere* **12** 23–30

[6] Priest E and Forbes T 2000 *Magnetic Reconnection* (Magnetic Reconnection, by Eric Priest and Terry Forbes, pp. 612. ISBN 0521481791. Cambridge, UK: Cambridge University Press, June 2000.)
[7] Dennis B R, Emslie A G and Hudson H S 2011 Space Sci. Rev. **159** 3–17
[8] Benz A O 2017 *Living Reviews in Solar Physics* **14** 2
[9] Gloeckler G, Zurbuchen T H and Geiss J 2003 J. Geophys. Res. **108** 1158
[10] Fisk L A 2003 J. Geophys. Res. **108** 1157
[11] Bühler R and Blandford R 2014 *Reports on Progress in Physics* **77** 066901
[12] Blandford R, Yuan Y, Hoshino M and Sironi L 2017 Space Sci. Rev. **207** 291–317
[13] Kouveliotou C, Wijers R A M J and Woosley S 2012 *Gamma-ray Bursts*
[14] Treumann R A and Baumjohann W 2015 *The Astronomy and Astrophysics Review* **23** 4
[15] Oka M, Birn J, Battaglia M, Chaston C C, Hatch S M, Livadiotis G, Imada S, Miyoshi Y, Kuhar M, Effenberger F, Eriksson E, Khotyaintsev Y V and Retinò A 2018 Space Sci. Rev. **214** 82 (*Preprint* 1805.09278)
[16] Fermi E 1949 *Physical Review* **75** 1169–1174
[17] Axford W I 1977 The three-dimensional structure of the interplanetary medium *Study of Travelling Interplanetary Phenomena* (*Astrophysics and Space Science Library* vol 71) ed Shea M A, Smart D F and Wu S T pp 145–163
[18] Bell A R 1978 Monthly Notices of the Royal Astronomical Society **182** 147–156
[19] Bell A R 1978 Monthly Notices of the Royal Astronomical Society **182** 443–455
[20] Blandford R D and Ostriker J P 1978 The Astrophysical Journal Letters **221** L29–L32
[21] Krymskii G F 1977 *Akademiia Nauk SSSR Doklady* **234** 1306–1308
[22] Øieroset M, Lin R P, Phan T D, Larson D E and Bale S D 2002 Phys. Rev. Lett. **89** 195001–+
[23] Biskamp D 1993 *Nonlinear magnetohydrodynamics* (Cambridge Monographs on Plasma Physics, Cambridge [England]; New York, NY: Cambridge University Press, —c1993)
[24] Aschwanden M J 2002 Space Sci. Rev. **101** 1–227
[25] Goldman M V, Newman D L and Lapenta G 2016 Space Sci. Rev. **199** 651–688
[26] Le A, Egedal J and Daughton W 2016 Phys. Plasma **23** 102109
[27] Speiser T W 1965 J. Geophys. Res. **70** 4219–+
[28] Birn J and Hesse M 2009 *Annales Geophysicae* **27** 1067–1078 URL https://www.ann-geophys.net/27/1067/2009/
[29] Egedal J, Daughton W and Le A 2012 *Nature Physics* **8** 321–324
[30] Khotyaintsev Y V, Graham D B, Norgren C, Eriksson E, Li W, Johlander A, Vaivads A, André M, Pritchett P L, Retinò A, Phan T D, Ergun R E, Goodrich K, Lindqvist P A, Marklund G T, Le Contel O, Plaschke F, Magnes W, Strangeway R J, Russell C T, Vaith H, Argall M R, Kletzing C A, Nakamura R, Torbert R B, Paterson W R, Gershman D J, Dorelli J C, Avanov L A, Lavraud B, Saito Y, Giles B L, Pollock C J, Turner D L, Blake J D, Fennell J F, Jaynes A, Mauk B H and Burch J L 2016 Geophys. Res. Lett. **43** 5571–5580
[31] Pritchett P L 2008 *Physics of Plasmas* **15** 102105
[32] Li X, Guo F, Li H and Li G 2017 The Astrophysical Journal **843** 21
[33] Muñoz P A and Büchner J 2018 The Astrophysical Journal **864** 92 (*Preprint* 1705.01066)
[34] Che H, Drake J F, Swisdak M and Yoon P H 2010 Geophys. Res. Lett. **37** 11105–+
[35] Che H, Drake J F and Swisdak M 2011 Nature **474** 184
[36] Daughton W, Roytershteyn V, Karimabadi H, Yin L, Albright B J, Bergen B and Bowers K J 2011 Nature Physics **7** 539–542
[37] Moser A L and Bellan P M 2012 Nature **482** 379–381
[38] Eyink G, Vishniac E, Lalescu C, Aluie H, Kanov K, Bürger K, Burns R, Meneveau C and Szalay A 2013 Nature **497** 466–469
[39] Phan T D, Eastwood J P, Cassak P A, Øieroset M, Gosling J T, Gershman D J, Mozer F S, Shay M A, Fujimoto M, Daughton W, Drake J F, Burch J L, Torbert R B, Ergun R E, Chen L J, Wang S, Pollock C, Dorelli J C, Lavraud B, Giles B L, Moore T E, Saito Y, Avanov L A, Paterson W, Strangeway R J, Russell C T, Khotyaintsev Y, Lindqvist P A, Oka M and Wilder F D 2016 Geophys. Res. Lett. **43** 6060–6069
[40] Galeev A A, Kuznetsova M M and Zelenyi L M 1986 Space Sci. Rev. **44** 1–41
[41] Drake J F, Shay M A, Thongthai W and Swisdak M 2005 Phys. Rev. Lett. **94** 095001–+
[42] Che H, Goldman M V and Newman D L 2011 Phys. Plasma **18** 052109
[43] Che H 2014 Phys. Plasma **21** 062305
[44] Dahlin J T, Drake J F and Swisdak M 2014 *Physics of Plasmas* **21** 092304 (*Preprint* 1406.0831)
[45] Dahlin J T, Drake J F and Swisdak M 2016 *Physics of Plasmas* **23** 120704
[46] Dahlin J T, Drake J F and Swisdak M 2017 *Physics of Plasmas* **24** 092110
[47] Drake J F, Swisdak M, Che H and Shay M A 2006 Nature **443** 553–556

[48] Coppi B, Laval G and Pellat R 1966 Phys. Rev. Lett. **16** 1207–1210
[49] Kaw P K, Valeo E J and Rutherford P H 1979 Phys. Rev. Lett. **43**(19) 1398–1401 URL `http://link.aps.org/doi/10.1103/PhysRevLett.43.1398`
[50] Pritchett P L and Wu C C 1979 *Physics of Fluids* **22** 2140–2146
[51] Tajima T and Sakai J I 1986 *IEEE Transactions on Plasma Science* **14** 929–933
[52] Schumacher J and Kliem B 1997 *Physics of Plasmas* **4** 3533–3543
[53] Zank G P, le Roux J A, Webb G M, Dosch A and Khabarova O 2014 The Astrophysical Journal **797** 28
[54] le Roux J A, Zank G P and Khabarova O V 2018 The Astrophysical Journal **864** 158
[55] Guo F, Li H, Daughton W and Liu Y H 2014 *Physical Review Letters* **113** 155005
[56] Guo F, Li H, Daughton W, Li X and Liu Y H 2016 *Physics of Plasmas* **23** 055708 (*Preprint* `1604.02924`)
[57] Guo F, Li X, Li H, Daughton W, Zhang B, Lloyd-Ronning N, Liu Y H, Zhang H and Deng W 2016 The Astrophysical Journal Letters **818** L9 (*Preprint* `1511.01434`)
[58] Borovikov D, Tenishev V, Gombosi T I, Guidoni S E, DeVore C R, Karpen J T and Antiochos S K 2017 The Astrophysical Journal **835** 48
[59] Oka M, Phan T D, Krucker S, Fujimoto M and Shinohara I 2010 The Astrophysical Journal **714** 915–926 (*Preprint* `1004.1154`)
[60] Bian N H and Kontar E P 2013 *Physical Review Letters* **110** 151101 (*Preprint* `1302.6090`)
[61] Zank G P, Hunana P, Mostafavi P, Le Roux J A, Li G, Webb G M, Khabarova O, Cummings A, Stone E and Decker R 2015 The Astrophysical Journal **814** 137
[62] le Roux J A, Zank G P, Webb G M and Khabarova O 2015 The Astrophysical Journal **801** 112
[63] le Roux J A, Zank G P, Webb G M and Khabarova O V 2016 The Astrophysical Journal **827** 47
[64] Fermo R L, Drake J F and Swisdak M 2010 *Physics of Plasmas* **17** 010702
[65] Zhao L L, Zank G P, Khabarova O, Du S, Chen Y, Adhikari L and Hu Q 2018 The Astrophysical Journal **864** L34
[66] Drake J F, Swisdak M and Fermo R 2013 The Astrophysical Journal Letters **763** L5
[67] Sironi L and Spitkovsky A 2014 *The Astrophysical Journal Letters* **783** L21 URL `http://stacks.iop.org/2041-8205/783/i=1/a=L21`
[68] Li X, Guo F, Li H and Li G 2015 The Astrophysical Journal Letters **811** L24
[69] Werner G R, Uzdensky D A, Cerutti B, Nalewajko K and Begelman M C 2016 *The Astrophysical Journal Letters* **816** L8 URL `http://stacks.iop.org/2041-8205/816/i=1/a=L8`
[70] Holman G D, Sui L, Schwartz R A and Emslie A G 2003 The Astrophysical Journal Letters **595** L97–L101
[71] Krucker S, Kontar E P, Christe S and Lin R P 2007 The Astrophysical Journal Letters **663** L109–L112
[72] Krucker S, Oakley P H and Lin R P 2009 The Astrophysical Journal **691** 806–810
[73] James T, Subramanian P and Kontar E P 2017 *Monthly Notices of the Royal Astronomical Society* **471** 89–99 URL + `http://dx.doi.org/10.1093/mnras/stx1460`
[74] Muñoz P A, Büchner J and Kilian P 2017 *Physics of Plasmas* **24** 022104
[75] Wang L, Lin R P, Salem C, Pulupa M, Larson D E, Yoon P H and Luhmann J G 2012 The Astrophysical Journal Letters **753** L23

On the Shock Source of Sustained Gamma-Ray Emission from the Sun

N Gopalswamy[1], P. Mäkelä[2], S. Yashiro[2], A. Lara[2,3], S. Akiyama[2], and H. Xie[2]

[1]NASA Goddard Space Flight Center, Greenbelt, MD 20771, USA
[2]The Catholic University of America, Washington DC 20064, USA
[3]Instituto de Geofisica, UNAM, Mexico City, Mexico

Abstract. It has recently been shown that the spatially and temporally extended γ-ray emission in solar eruptions are caused by >300 MeV protons precipitating on the Sun from shocks driven by coronal mass ejections (CMEs). The γ-rays result from the decay of neutral pions produced in the proton-proton interaction when the >300 MeV protons collide with those in the chromosphere. The evidence comes from the close correlation between the durations of the sustained gamma-ray emission (SGRE) and the associated interplanetary (IP) type II radio bursts. In this paper, we provide further evidence that support the idea that protons accelerated in IP shocks driven by CMEs propagate toward the Sun, precipitate in the chromosphere to produce the observed SGRE. We present the statistical properties of the SGRE events and the associated CMEs, flares, and type II radio bursts. It is found that the SGRE CMEs are similar to those associated with ground level enhancement events. The CME speed is well correlated with the SGRE fluence. High CME speed is an important requirement for the occurrence of SGRE, while the flare size is not. Based on these results, we present a schematic model illustrating the spatially and temporally extended nature of SGRE related to the CME flux rope-shock structure.

1. Introduction

Morrison [1] suggested that both continuum and line emissions in the γ-ray band are expected from cosmic sources due to processes such as bremsstrahlung, radiative decay of neutral pions, de-excitation of nuclei, and electron-positron annihilation. Of these he suggested bremsstrahlung continuum and 2.223 MeV line due to neutron capture are likely from the active Sun. Lingenfelter and Ramaty [2] performed detailed calculations of γ-ray emission processes from the Sun. The 2.223 MeV line was first detected by the Gamma-ray Monitor on board the OSO-7 mission. The pion-decay continuum from both the impulsive and late phases was identified by Forrest et al. [3] during the 1982 June 3 γ-ray event using Solar Maximum Mission's Gamma Ray Spectrometer data. A typical γ-ray spectrum showing various lines below ~10 MeV, bremsstrahlung, positronium, and pion continua can be found in [4]. The pion decay continua have three components: γ-rays from the decay of neutral pions, electron and positron bremsstrahlung γ-rays from the decay of charged pions, and positron annihilation continuum from charged pions. Only about a dozen such pion-decay events were recorded by different telescopes between 1982 and 1991 [5]. A few more were added mainly from CORONAS-F/SONG [6]. The primary characteristic of the late-phase γ-ray events is that they last beyond the impulsive phase of the associated flare for minutes to many hours. Of these, only two events had durations >2 h in the pre-Fermi era. Fermi mission's Large Area Telescope (Fermi/LAT [7]) revealed that such late-phase events at energies >100 MeV are rather common [8-10], thanks to the unprecedented sensitivity of the detector in the energy range 20 MeV to 300 GeV. In some cases, the Fermi/LAT events lasted for almost a day [11-13]. These events have been known under various names: Long Duration Gamma-ray Flares (LDGRF

Content from this work may be used under the terms of the Creative Commons Attribution 3.0 licence. Any further distribution of this work must maintain attribution to the author(s) and the title of the work, journal citation and DOI.
Published under licence by IOP Publishing Ltd

[14]), Sustained Gamma-Ray Emission (SGRE [15-16,10,13]) and Late-Phase Gamma-Ray Emission (LPGRE [9]). "SGRE" truly denotes the emission lasting well past the associated flare, so we use it here. The origin of energetic protons responsible for pion production has been controversial for more than 30 years. It was already noted [3] that the protons may be accelerated by a mechanism similar to that responsible for solar energetic particles (SEPs) in space and different from the one responsible for the impulsive particles. Murphy et al. [17] showed indeed that the impulsive phase proton spectrum is too steep (inferred from γ-ray line emission) to produce pion continuum, while the flat, late-phase proton spectrum (producing the pion continuum) is consistent with the hard spectrum of SEPs in the interplanetary (IP) medium. They also suggested that the protons responsible for the late-phase emission are accelerated at coronal shocks. In fact, an intense metric type II burst, indicative of a coronal shock, was observed during the 1982 June 3 event (Solar Geophysical Data, 1982). Akimov et al. [18] agreed with the shock source for the late-phase γ-rays during the 15 June 1991 eruption observed by the GAMMA-1 telescope on board the Gamma Observatory. In a later paper, these authors abandoned the shock source suggesting that the protons from the shock may not precipitate to the Sun from a few solar radii; instead they suggested the protons may be accelerated due to reconnection in the current sheet behind the associated coronal mass ejection (CME) [19]. It has now become clear that such reconnection is long gone well before the end of Fermi/LAT SGRE events [20]. Ryan and Lee [21] suggested that both impulsive and late-phase γ-rays can be explained by energetic particles along with MHD turbulence trapped in flare loops with different turbulence characteristics in the two phases; they had to propose a third phase for SEPs in the IP space, distinct from the late-phase. Kanbach et al. [22], who reported on the longest-duration pre-Fermi-era event of 1991 June 11, also adapted the flare-loop interpretation, but required the loop to be turbulence-free to account for the >8 h duration of the SGRE event.

One of the implications of the flare-loop idea is that the acceleration region is spatially confined to the post-eruption arcade and flare ribbons where the accelerated particles precipitate. The only variability may be the ribbon separation, which is small during the impulsive phase and attains a maximum size in the late phase. The average size of a post eruption arcade is ~90,000 km, while the maximum size is ~200,000 km [23]. The maximum size corresponds to ~17° on the Sun. The spatially-confined nature of the γ-ray source predicted by the flare-loop idea was challenged when 2.223 MeV γ-ray line emission was detected during the 1989 September 29 backside event [24-25]. The active region was ~8° behind the west limb (S25W98). The 2.223 MeV line is produced deep in the chromosphere, so it cannot reach a near-Earth observer from the flare location. The eruption was associated with a fast CME (~1800 km/s) observed by the Coronagraph/Polarimeter on board SMM and a metric type II radio burst. The typical half width of a fast CMEs is ~45°, so there is no difficulty for the particles accelerated in the shock to precipitate on the frontside. Cliver et al. [25] cautioned that the mechanism may or may not apply to the SGRE emission. However, Fermi/LAT observed SGRE emission due to an eruption ~35° behind the east limb on 2014 September 1 [26-30,8,16]. The 45° CME half width is more than adequate for the shock protons to precipitate on the frontside because the shock is typically more extended than the CME flux rope. A 17°-wide post eruption arcade located ~35° behind the limb cannot be the source of γ-rays observed by the Fermi mission.

The spatially and temporally extended nature of the γ-ray source requires an accelerator that is spatially and temporally extended: the CME-driven shock. The solid proof that the SGRE is due to protons from shocks came from the finding that SGRE durations are about the same as the durations of the associated type II radio bursts [13,31], which are excellent indicators of shocks in the corona and IP medium. In this paper, we examine the close connection among SGRE, type II radio bursts, and SEP events, all related to energetic CMEs that seem to be the key solar phenomenon of interest. We also discuss some SGRE events, which seem to be in apparent contradiction to the idea that energetic protons are transported back to the Sun. We use CME observations from the Solar and Heliospheric Observatory (SOHO) and the Solar Terrestrial Relations Observatory (STEREO). The radio data are from the Wind and STEREO spacecraft. We use EUV images from the Solar Dynamics Observatory (SDO) to identify the post eruption arcades in the source regions. We also use GOES and STEREO data on SEP events. Information on the 1991 June 11 SGRE is obtained from published literature and from the Solar Geophysical Data made available by NOAA/NGDC, while the radio data are from the Ulysses mission.

Journal of Physics: Conference Series **1332** (2019) 012004 doi:10.1088/1742-6596/1332/1/012004

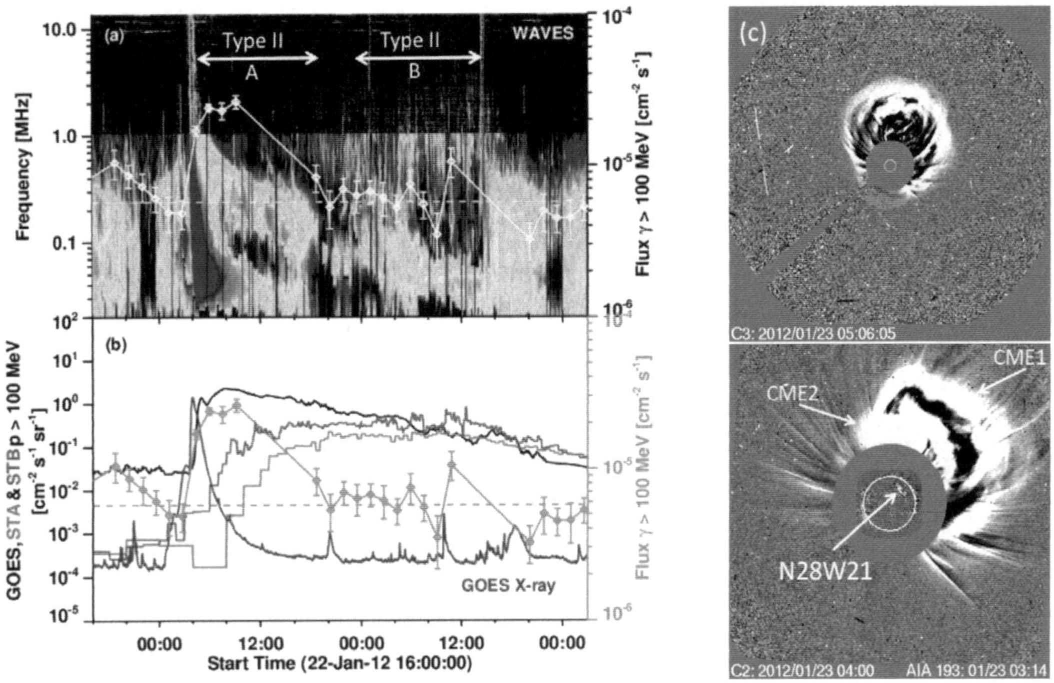

Figure 1. (a) The 2012 January 23 SGRE event and the associated type II burst, which has two phases (A, B), separated by ~6 hours. The SGRE duration coincides with the duration of phase A. The vertical feature in the beginning of the type II is a type III burst caused by energetic electrons escaping from the flare site. (b) The SGRE flux plot along GOES soft X-ray light curve and the >100 MeV SEP flux from three spacecraft: GOES, STEREO-Ahead (STA), and STEREO-Behind (STB). The e-folding drop of >100 MeV proton flux roughly occurs at the SGRE end time. (c) The CME responsible for the type II burst from N28W21 at two instances: at 04:00 UT at the onset and at 05:06 UT at the SGRE peak. The two CMEs (CME1, CME2) in the 04:00 UT image merged into a single one at 05:06 UT.

2. Fermi/LAT SGRE Events

In order to provide an overview of the SGRE events and the associated phenomena studied in this work, we have shown the 2012 January 23 event associated with a CME from NOAA active region (AR) 11402 located at N28W21. The SGRE is associated with an M8.7 GOES soft X-ray (SXR) flare that starts, peaks and ends at 03:38 UT, 03:59 UT, and 04:34 UT, respectively. High-frequency (e.g., 17 GHz) microwave bursts peak at 03:50 UT, about nine minutes before the SXR peak. This is typical of most flares, so we take the SXR peak time as the end of the impulsive phase and also as the start of the late-phase γ-rays (SGRE). The SGRE peaks around 09:12 UT, about 5 hr after the SXR peak. The eruption location is reasonably connected to Earth, so the >100 MeV proton flux (used as a proxy to the >300 MeV proton flux) sharply increases after the eruption and reaches a broad peak around the time the SGRE attains maximum. Even though STA and STB are not well connected to the eruption, they do observe the SEP event as can be seen from the plots in the lower panel. The end time of the SGRE is determined as the mid time (19:25 UT) between the last signal data point above background (18:35 UT) and the next data point (20:15 UT). Thus the SGRE duration is 15.43±0.83 hr. The CME first appears in the LASCO/C2 field of view (FOV) at 04:00 UT and merges with a preceding CME (CME1) from the same source region. The compound CME has a speed of ~2500 km/s within the LASCO FOV. Consistent with the large source latitude, the central position angle in the early images is ~ 326° but the CME soon becomes an asymmetric halo. The type II burst is observed in the decameter-hectometric (DH) domain starting at 04:00 UT, but a metric type II burst has started at 03:38 UT observed by ground-based radio telescopes. The DH type II has a large break at 19:25 UT, which is also the end time of SGRE. We have labelled this phase of the type II burst as "A" in Figure 1a. After a break of ~6 hr, the type II burst continues until the shock arrival at the Wind spacecraft the next day at 14:33 UT (marked "B" in Figure 1a). Thus, we have taken the end time of the type II between these two times, giving a duration of 24.99±9.57 hr. The SEP event is the second largest (as measured from the >10 MeV flux, 6310 pfu). The e-folding time of the >100 MeV proton flux from the peak also coincides with the SGRE end time. It must be noted that lower energy particles continue to be accelerated until the shock arrives at Earth and even beyond; the particles observed as Earth passes through the shock are the energetic storm particles. The flux of >300 MeV protons greatly diminishes beyond phase A noted above. For the same reason, weak type II bursts continue all the way to shock arrival. The SGRE, flare, CME, SEP, and type II burst characteristics derived this way are listed in Table 1 for 19 Fermi/LAT SGRE events with duration >3 hr (from [13,31]). Also included is the 1991 June 11 pre-Fermi SGRE event that satisfies the duration criterion (>3 h). There was no coronagraph operating during this event, so there is no CME data; however, a CME was observed in the IP medium [31].

Table 1. List of SGRE events with the associated CMEs, flares, and type II bursts

SGRE		**CME**			**Flare**		**SEP**	**Type II**	
Start UT	Dur. (hr)	Fl cm^{-2}	UT	V km/s	Class	Location	pfu	UT	Dur. (hr)
91/06/11 02:09	10.38 ±1.5	19.5	------	------	X12.0	N31W17	1180[c]	02:01	8.67 ±0.67
11/03/07 20:12	21.02±1.55	1.08	20:00	2223H	M3.7	N31W53	50	20:00	13.17±0.67
11/06/2 07:46	6.98 ±1.47	0.12	08:12	1422H	C3.7	S19E25	0.10	08:00	4.47 ±0.02
11/06/7 06:41	3.08 ±1.67	0.26	06:49	1321H	M2.5	S21W54	72	06:45	10.93 ±0.32
12/01/23 03:59	15.43±0.84	0.69	04:00	2511H	M8.7	N28W21	6310	03:38	24.98±9.57
12/01/27 18:37	3.59 ±0.82	0.18	18:27	2541H	X1.7	N27W71	795	18:30	10.23 ±0.33
12/03/05 04:09	4.25 ±0.80	0.06	04:00	1627H	X1.1	N17E52	100B[b]	04:00	6.99 ±1.08
12/03/07 00:24	20.47 ±0.84	24.1	00:24	3146H	X5.4	N17E27	6350B	00:30	27.78±6.61
12/03/09 03:53	8.85 ±0.80	0.16	04:26	1229H	M6.3	N15W03	600B	04:00	3.79±1.71
12/03/10 17:44	11.62 ±1.59	0.05	18:00	1638H	M8.4	N17W24	120B	17:55	7.42±0.42
12/05/17 01:47	3.08 ±0.80	0.10	01:48	1596H	M5.1	N11W76	255[c]	01:40	4.73 ±0.20
13/05/13 02:17	7.47 ±1.62	0.13	02:00	1270H	X1.7	N11E90	25[b]	02:20	6.13±0.28
13/05/13 16:05	8.69 ±0.95	0.40	16:07	1852H	X2.8	N11E85	400B[b]	16:15	8.58±0.17
13/05/14 01:11	5.99 ±0.82	0.35	01:25	2645H	X3.2	N08E77	608B[b]	01:16	6.23±0.83
13/05/15 01:48	3.61 ±0.81	0.03	01:48	1408H	X1.2	N12E64	42[b]	01:49	6.41±0.28
14/02/25 00:49	8.46 ±1.59	14.9	01:25	2153H	X4.9	S12E82	400[b]	00:56	10.32±0.75
14/09/01 11:11	3.92 ±0.76	7.94	11:12	2017H	X2.4[a]	N14E127	3083[b]	11:12	7.5±2.25
15/06/21 02:36	14.06 ±1.61	27.1	02:36	1740H	M2.6	N12E16	1066	02:33	20.95±1.50
17/09/06 12:02	18.43 ±0.80	1.01	12:24	1884H	X9.3	S08W33	844B	12:05	20.92±1.00
17/09/10 16:06	15.18 ±0.81	13.6	16:00	3165H	X8.2	S09W92	1490B[c]	16:02	15.72±0.92

[a]estimated; [b]from STEREO-B; B – the SEP event had high background; [c]GLE;1 pfu = 1 particle/cm^2.s.sr

For each SGRE event, the duration and fluence (Fl) are listed in columns 2 and 3. The CME speeds are three-dimensional speeds obtained from the graduated cylindrical cell (GCS [33]) model that fits a flux rope and a shock to the CMEs. The CMEs are observed by SOHO's Large Angle and Spectrometric Coronagraph (LASCO [32]) and STEREO's Sun Earth Connection Coronal and Heliospheric Investigation (SECCHI [34]) observations. The suffix 'H' to the speed values indicates that the CME is a full halo. The listed soft X-ray flare sizes and the heliographic coordinates of the eruption location are from NOAA's Space Weather Prediction Center. The SEP intensity listed is the maximum value in pfu attained by >10 MeV protons. These values are obtained from GOES data compiled at the CDAW Data Center (https://cdaw.gsfc.nasa.gov/CME_list/sepe/). For eastern events, the SEP intensity is computed from the STEREO high-energy telescope (HET) and low-energy telescope (LET) data. The equivalent >10 MeV intensity is obtained by first computing SEP spectra at energies below 100 MeV and extrapolating to higher energies (see [35] for details). The type II burst start time and duration are from the Radio and Plasma Wave (WAVES [36]) experiment on Wind compiled at the CDAW Data Center (https://cdaw.gsfc.nasa.gov/CME_list/radio/waves_type2.html), except that the 2014 September 1 event data are from STEREO/WAVES. Table 1 is the basis for many plots and results presented in this paper.

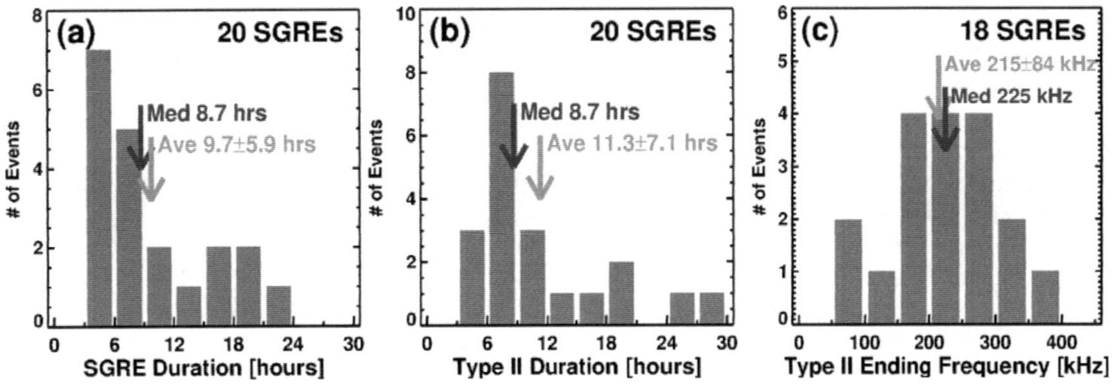

Figure 2. Distributions of SGRE durations (left), type II burst durations (middle) and ending frequencies of type II bursts (right). The median (Med) and average (Ave) values are noted on the plots.

3. SGRE and type II burst durations

The close relation between SGRE and type II bursts is evidenced by their durations. The type II burst duration is closely related to the burst's ending frequency. Since the type II burst is emitted at the local plasma frequency or its harmonic in the vicinity of the shock, the ending frequency corresponds the distance from the Sun where such plasma frequencies (and hence electron densities) prevail. Figure 2 shows distributions of the SGRE durations, type II durations, and ending frequencies for the events in Table 1. The range and average durations of the two electromagnetic emissions are similar. The average duration of type II bursts is about 1.5 hr longer mainly because of the two intense events that had a break in the dynamic spectrum but the type II burst continued intermittently until the shock arrived at the observing spacecraft (one of them is in Fig. 1 and the other is the 2012 March 7 event). In these two cases, the error bars were very generous resulting in the higher average duration.

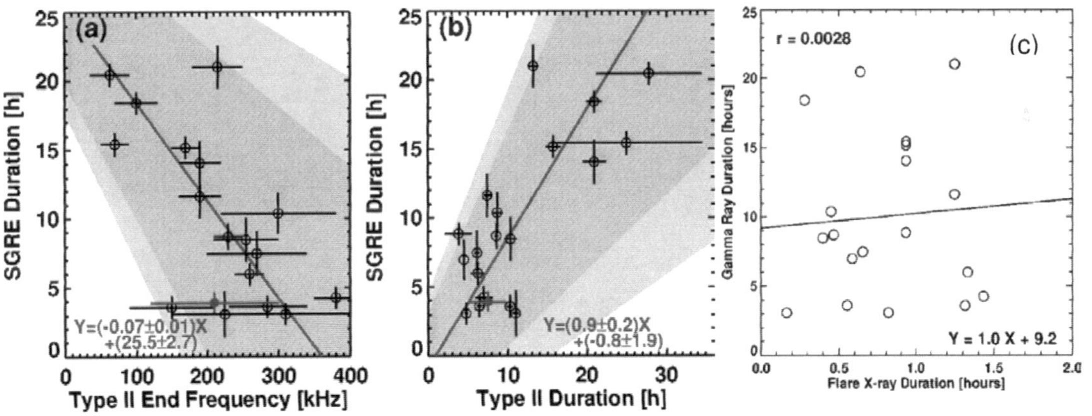

Figure 3. The scatter plot of SGRE durations with the ending frequencies (a) and durations (b) of type II bursts. The blue data points correspond to the 2014 September 1 backside event (not included in the correlation). The blue and yellow shaded areas denote the 95% and 99% confidence levels. (c) The scatter plot between SGRE duration and the SXR flare duration showing no correlation.

The close relationship of SGRE durations with type II ending frequencies and durations found for SGRE events with >5 h duration [13] also holds when we included events with >3 durations. Figure 3 shows the corresponding scatter plots. The correlations indicate a clear linear relationship: for longer duration SGRE events, the underlying shock travels a larger distance from the Sun, and the SGRE ends when type II burst ends. This result is remarkable because the energetic particles responsible for the two types of emissions originate from the same shock, but the electromagnetic emissions are produced at different locations: type II bursts in the vicinity of the shock and SGRE at the Sun where the >300 MeV protons precipitate after traversing the Sun-shock distance. On the other hand, the SXR flare duration has no correlation with the SGRE duration.

The SGRE scenario inferred from the SGRE – Type II burst correlations is illustrated in Fig. 4. Protons diffusing back to the Sun along open magnetic field lines threading the shock can produce both neutral and charged pions. At photon energies >100 MeV, the main contribution comes from the decay of neutral pions into γ-rays. Charged pion decay results in electron and positron bremsstrahlung as well as positron annihilation γ-rays, but the flux of these photons is more than an order of magnitude weaker [17]. There may also be a small contribution to the electron bremsstrahlung from high-energy electrons diffusing back to the Sun from the shock. One of the implication of the model in Fig. 4 is the spatially extended nature of the γ-ray source. The spatial extent is determined by the extent of the nose region of the shock where the highest energy particles are accelerated [37] and the field lines that are accessed by these particles. The flux rope – shock structure requires that particles precipitate along open field lines that lie on the outskirts of the flux rope because only these field lines connect to the shock nose. This suggests that the largest spatial extent of the source should be in the direction connecting the foot-points of the flux rope. For example, for a high-inclination flux rope, the γ-ray source is extended in the north-south direction; for a low-inclination flux rope, it is extended east-west [38]. Figure 4b shows the post eruption arcade and the core dimmings during the 2015 June 21 SGRE indicating a high-inclination flux rope (see [39]). Spatial localization of SGRE from this event needs to be done to verify the asymmetry in spatial extension. Another important point is the field strength in the core dimming region is close to the quiet Sun values, not the active region values [39].

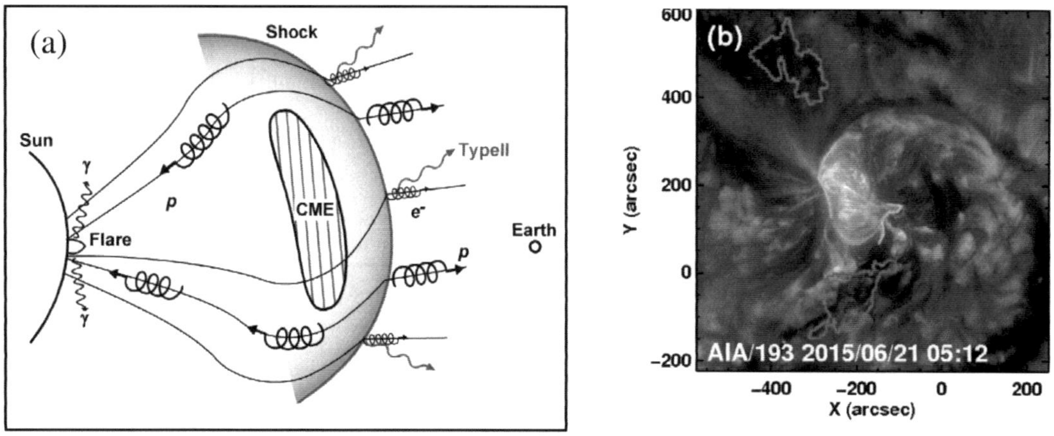

Figure 4. (a) Schematic model showing the propagation of protons (p) and electrons (e^-) from the CME-driven shock toward and away from the Sun. Low-energy electrons (~10 keV) produce type II bursts in the vicinity of the shock. Energetic protons and electrons can be observed as SEP events when the observer is connected to the field line on which these particles propagate. A representative flare loop is shown at the Sun. (b) Two types of magnetic structures during the eruption associated with the 2015 June 21 SGRE event: the post eruption arcade (feet marked in green) and the CME flux rope (feet marked by the red contours). The feet of the flux rope correspond to the core dimmings in the 193 Å EUV image obtained by SDO's Atmospheric Imaging Assembly (AIA).

Figure 5 shows the CME associated with the 2014 February 24 SGRE event as observed by STEREO coronagraph COR2 and Heliospheric Imager (HI). When the SGRE ended, the shock is at a distance of ~70 Rs; the CME also slows down to ~1200 km/s. Thus, the source of protons is also located up to tens of solar radii by the time SGRE ends. In another well-observed case (2015 June 21), the shock distance was obtained as ~90 Rs from the drift rate of the associated type II burst [13,39]. Such distances are consistent with the ending frequency of type II bursts, which is the local plasma frequency at such distances (see Fig. 3).

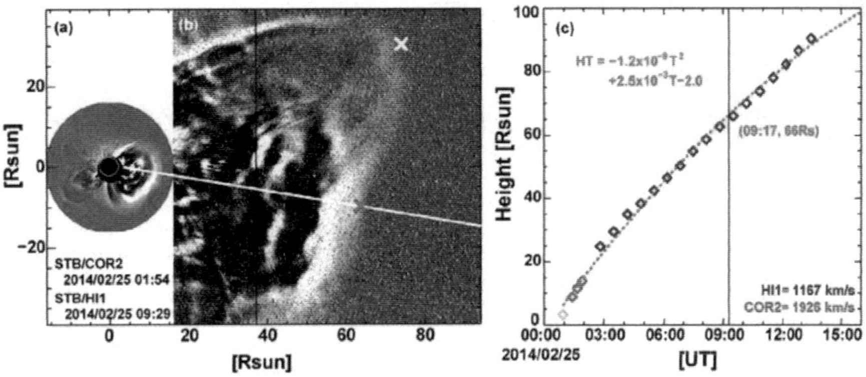

Figure 5. The 2014 February 25 CME in the COR2 FOV (a), in the HI-1 FOV (b) and the height (HT) -time (T) plot (c) measured along the white line in (b). Images (a) and (b) correspond to SGRE peak and end. When SGRE ended, the CME leading edge was at ~66 Rs from the Sun's center. The yellow cross marks probably the shock location around 70 Rs.

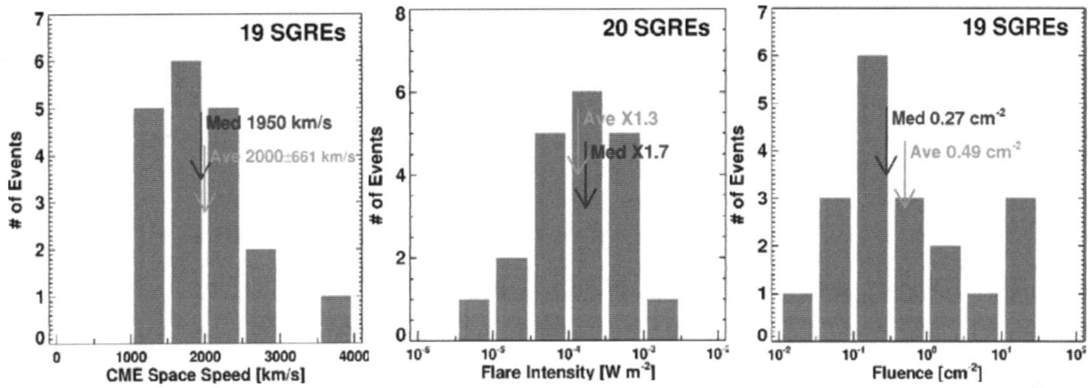

Figure 6. Distribution of deprojected (space) CME speeds (left), SXR flare sizes (middle) and fluence (right) of SGRE events in Table 1. The median (Med) and average (Ave) values are noted.

4. CMEs and flares associated with SGRE events

The CME speeds listed in Table 1 are averaged over the coronagraph FOV. The sky-plane speed measured from LASCO alone can also be deprojected using a cone model and the results are not too different (see https://cdaw.gsfc.nasa.gov/CME_list/halo/halo.html). We see that SGRE CMEs are all very fast (column 5 in Table1and Fig. 6), with an average speed of 2000±661 km/s. Such high speeds (10.3 Rs/hr), when combined with the average SGRE duration (9.7 hr), gives a shock distance of ~100 Rs (see Fig. 5) by the time the SGRE ends. The SGRE CME speed is similar to the average speed of CMEs responsible for ground level enhancement (GLE) in SEP events [40]. In addition, all these CMEs are full halo CMEs as indicated by the suffix "H" to the speeds in Table 1. Halo CMEs are known to be inherently more energetic compared to the general population [41-43]. The larger the halo fraction the higher is the average kinetic energy of a CME population. All SGRE CMEs being halos represents the highest halo fraction even higher than that in GLE CMEs. Shocks of such CMEs are known to accelerate particles to energies >1 GeV, so the presence of >300 MeV protons required for producing the pion-continuum is guaranteed. It must be noted that both the GLEs of solar cycle 24 (2012 May 17 and 2017 September 10) are SGRE events listed in Table 1.

The SXR flare sizes of the SGRE events (see Table 1 and Fig. 6, middle) are generally large: 12 X, 7 M, and 1 C. While it is known that faster CMEs are generally associated with larger flares, it is remarkable that ~42% of the flares are not X-class flares. However, as noted above, all the events were associated with fast halo CMEs. Such a distribution of flare sizes was also noted in GLE events [40]. We computed the γ-ray fluences from the light curves considering 5-min bins by extrapolating between Fermi/LAT data points. The γ-ray fluence varies over three orders of magnitude with an average value of ~0.5 cm^{-2} (see Fig. 6, right). The scatter plots of SGRE fluence with CME speed and SXR flare fluence are reasonably correlated with similar correlation coefficients (Fig. 7). These correlations are statistically significant as the correlation coefficients exceeded the Pearson's critical correlation coefficient. The probability (p) that such high correlations are obtained by chance is 0.005.

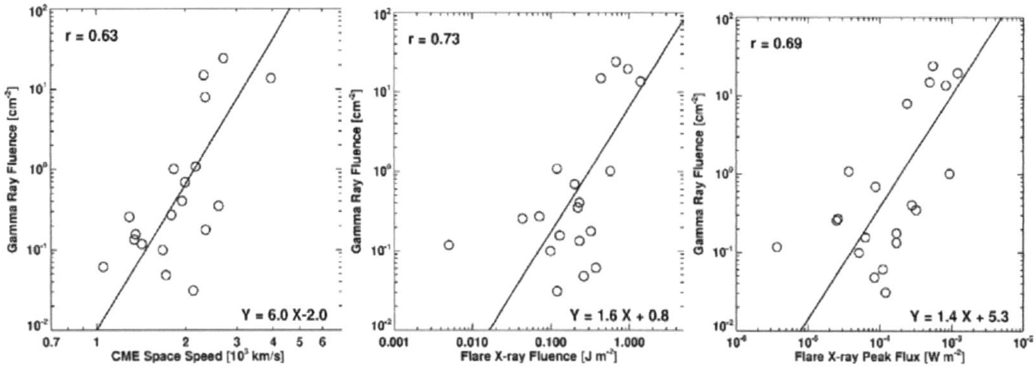

Figure 7. The scatter plot of SGRE fluence with CME speed (left), SXR flare fluence (middle), and SXR peak flux (right). The fluence is calculated by interpolating linearly between >100 MeV γ-ray flux data points. In the regression equations, X denotes (in log units) CME speed in 10^3 km/s (left), SXR fluence in J m^{-2} (middle) and SXR peak flux in W m^{-2} (right); Y is the gamma ray fluence in cm^{-2} in log units. All correlations are statistically significant: The Pearson critical correlation coefficient is 0.575 (p=0.005 for 19 CMEs and peak SXR fluxes) and 0.590 (p =0.005 for 18 flare fluences). The correlations are slightly different since the 1991 June 11 event is not included in the speed plot (no CME data), while the SXR fluence and peak flux are available from Solar Geophysical Data.

5. SGRE CME source locations

The source locations of SGRE eruptions are all in the active region belt (see Table 1, column 7), where the magnetic field strengths are known to be high and required to power the energetic CMEs. The latitudes are higher early in the solar cycle (e.g., events in 2011) and become lower as the cycle progresses (e.g., events in 2013 onward) reflecting the sunspot butterfly diagram. The sources are at all longitudes, with no preference for the eastern or western hemisphere. This is contradictory to the report by Winter et al. [44], who claimed that SGRE sources occur preferentially from the north-eastern quadrant of the Sun. In fact, the SGRE events during March 5-10, 2012 have source locations ranging from E52 to W24. Similarly, the source regions of the 1991 June 4-15 SGRE events in the pre-Fermi era started in the eastern hemisphere (E70 on June 4) and ended in the western hemisphere (W69 on June 15, see [45] for the evolution of the source active region 6659). The June 4-9 SGRE events occurred in the eastern hemisphere, while the intense ones on June 11 and 15 occurred in the western hemisphere. Among the 38 Fermi/LAT SGRE events listed in [44], 34 have known source locations, which are evenly split between the eastern and western hemispheres, just like our events in Table 1. Even the X-class flares without SGREs in [44] are evenly distributed between the eastern and western hemispheres. We do see more source locations in the northern hemisphere (15 out of 20 events in Table 1). This is due to the double-peak nature of the sunspot cycle 24 in which the northern (southern) hemisphere had more active regions during the first (second) peak [46].

6. SEP events and SGRE

Table 1 shows that all SGRE events are associated with large SEP events either at Earth or at STEREO-Behind (STB) as indicated by the >10 MeV proton intensity in particle flux units (pfu). STB is well connected to SGRE events originating from the eastern hemisphere in Earth view. There is one

Journal of Physics: Conference Series **1332** (2019) 012004 doi:10.1088/1742-6596/1332/1/012004

exception, the 2011 June 2 event that has no significant SEPs, although there is a type II burst indicative of particle acceleration. This event will be discussed later. Many SEP events occur when the background is high due to preceding events. If a shock has access to the background particles, one expects an efficient acceleration of these particles. It has been noted that only about half of the SEP events are associated with SGRE [47]. Not all large SEP events are expected to have SGRE because some have soft spectra. The soft spectrum can result when the underlying CME forms shocks at large distances from the Sun [35,48], or when the shock nose is not connected to the observer [37]. SEPs from CMEs associated with filament eruptions outside active regions have a soft spectrum because of the low acceleration efficiency at larger distance from the Sun due to the decreasing magnetic field. The shock is the strongest at the nose, so the highest-energy particles are accelerated there. On the other hand, lower energy particles can be accelerated over the whole shock surface. The SEP event associated with the 2011 March 7 SGRE has a soft spectrum (fluence spectral index is 4.70 compared to 2.68 for GLE events, see [35]). This event has been considered as a counter example for the SEP cause of SGRE because it lacks high-energy SEPs. Figure 8 shows the GOES proton intensity plots at various energy channels and a snapshot of the associated CME heading in the north-west direction at a central position angle of 313°. The direction is consistent with the far-north source location (N31W53) and the solar B0 angle (-7°.25). This means the shock nose is not connected to GOES 13 in the ecliptic and hence the high-energy particles are not expected to be detected, although they precipitate at the Sun and produce SGRE. The shock flank, where only low-energy particles are accelerated, is connected to GOES resulting in the observed soft spectrum. The situation is similar in the case of the 2012 January 23 event, but the spectrum is slightly harder. This is likely because the nose extent of this CME is larger than that of the 2011 March 7 CME.

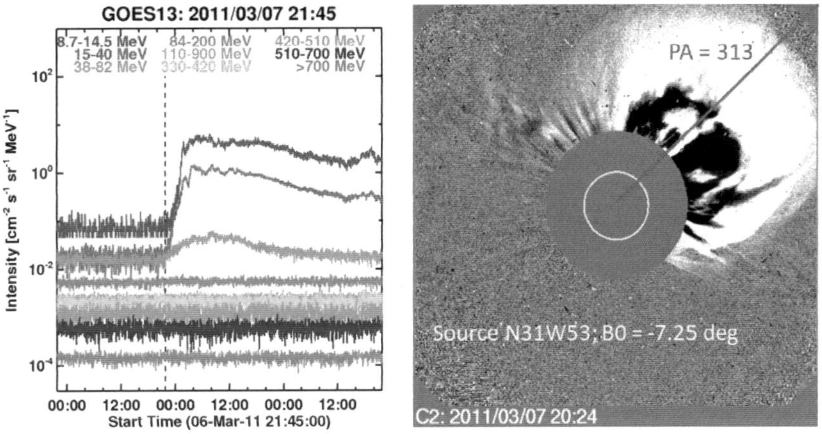

Figure 8. (left) GOES 13 SEP data showing no particles in energy channels higher than 38-82 MeV. (right) The underlying CME heading along position angle (PA) 313°, which is 43° above the ecliptic. Given the source location at N31W53 and a solar B0 angle of -7°.25, it is clear that the shock nose is ~38° from the ecliptic, which too far for the high-energy particles to arrive at Earth.

7. SGRE clustering and preconditioning of the ambient medium

In column 1 of Table 1 we have listed several SGRE events in bold font. The events occurred in clusters (originating from the same active region). There are 4 clusters involving 12 SGRE events (or 63%) among the 19 Fermi/LAT events. The 2012 March events and 2013 May events form the two largest clusters (4 SGREs in each). The longitudinal range over which the CMEs erupted is 38° and 36° for these two clusters. Note that these values are typically less than the half width of energetic CMEs. The 2012 January 23 and 27 events from the same active region are separated by a longitudinal extent of ~50°, while the September 2017 events are separated by ~59°. Even the 1991 June 4-15 events [49] occurred in a cluster of 5, two of them lasting for >2 h (June 11 and 15). The clustering implies that the ambient medium is highly disturbed and filled with energetic particle populations.

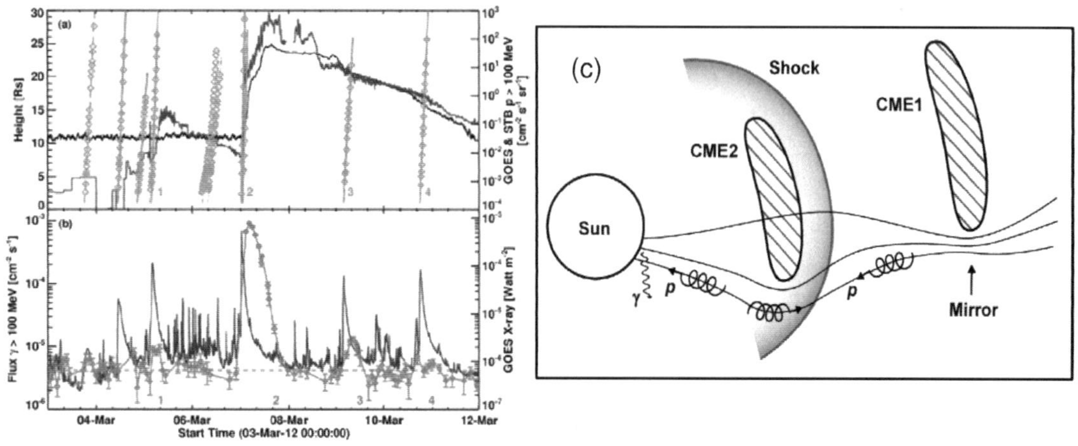

Figure 9. One of the SGRE clusters with 4 events (numbered 1-4): (a) CME height-time plots (orange) over-plotted on >100 MeV proton flux curves (blue –STB, black – GOES) and (b) >100 MeV SGRE flux and SXR flux showing the flares. (c) A scenario in which a magnetic mirror is formed near CME1 that reflects most of the particles from CME2's shock back to the Sun.

Figure 9a,b show the SGREs, CMEs, SEPs, and flares during the 2012 March cluster. In addition to the CMEs that caused the SGRE events, other CMEs from the same active region are also shown in the height-time plots. The >100 MeV proton flux is also shown over the period of the SGRE cluster. It is clear that the >100 MeV proton flux is the lowest for SGRE1 and the highest for SGRE2. The proton flux is declining after SGRE2, but boosted slightly higher due to CME3 and CME4 responsible for SGRE3 and SGRE4, respectively. The CME of SGRE1 is preceded 3 CMEs, while that of SGRE2 is preceded by 2 CMEs from the same source region. SGRE2 is actually associated with two CMEs in quick succession, the first one being faster (3146 km/s) than the second (2160 km/s) as can be seen from the slightly different slopes of the closely-spaced height-time plots. Both CMEs are shock driving, but there is no chance of the second CME overtaking the first one. Such a combination seems to have resulted in a copious supply of >300 MeV protons precipitating at the Sun: the first CME helps reflect particles accelerated by the second CME. The 2012 March cluster shows later CMEs in the cluster are launched into a medium highly disturbed by preceding CMEs, energetic particles, and possibly enhanced turbulence. Such a situation is termed as preconditioning and thought to influence the intensity of SEP events [50-51]. The ambient medium may take about 2-5 days to relax back after the passage of a large CME [52], so the preconditioning is likely to be a source of variability in SGRE events. The type of effect one expects depends on the relative timing, kinematics, and distance between the primary and preceding CMEs.

Now we consider the 2011 June 02 SGRE event, which has a duration >6 hr, but there is no associated SEP event. The CME is associated with a DH type II radio burst, indicative of particle acceleration by the shock. The speed (~1100 km/s) of the underlying CME at 08:12 UT is well below the average (~2000 km/s) but similar to the average speed of CMEs associated with SEP events and DH type II bursts. The source, located at S19E25, is not well connected to Earth. For STB (located at E93), the source is at W68 and hence well connected, but the >10 MeV SEP flux is negligible at STB also. This can happen if particles accelerated at the shock of the primary CME are reflected back to the Sun by the mirror formed near the preceding CME at 18:26 UT on June 01. The expected scenario is shown in Fig. 9c, which can be thought of as an extreme case in which the preceding CME does not allow any SEPs to go past the preceding CME to be detected by the observer.

While the mirror scenario in Fig. 9c explains the lack of an SEP event, there is another mirroring issue often brought in against the shock scenario [28]: the strong mirror force may prevent protons from the shock source from precipitation. However, this is a common problem for closed loops in post eruption arcades and open filed lines in the periphery of the CME flux ropes. The mirror ratio is likely to be the same in both cases: flare loops have high field strengths at the reconnection region and foot points, while the open field lines have weaker field both at the shock and at the quiet chromosphere in the flux rope foot pints located in the outskirts of the active region (see Fig. 4b).

8. Summary and conclusions

We investigated how the >100 MeV SGRE events with duration >3 hr are related to IP type II radio bursts, large SEP events, and white-light CMEs in order to understand the source of >300 MeV protons responsible for the observed γ-ray continuum. We developed a schematic model based on the observational result that the SGRE ends around the same time the type II burst ends. The ultimate energy source is the magnetic energy release at the Sun in the form of fast and wide CMEs that drive a shock in the corona and IP medium. The same shock accelerates ~10 keV electrons that produce type II radio bursts and >300 MeV protons responsible for the observed γ-rays. Although the precipitating protons produce charged and neutral pions, at photon energy >100 MeV, the dominant contribution is from the decay of neutral pions. Comparison of the radio dynamic spectrum with the CME imagery indicates that the CME-driven shock may be located roughly half way between the Sun and Earth when SGRE ends in many cases. The ending frequencies of the type II bursts are consistent with the local plasma frequency prevailing at such distances. CMEs with such low ending frequencies for type II bursts are known to be very energetic, consistent with 100% halo fraction and ~2000 km/s average speed of the SGRE CMEs. Another strong support for the proposed shock scenario is that the SGRE CMEs have the same properties as GLE CMEs confirming the acceleration of GeV particles and hence >300 MeV particles. The SGRE CME speed and SXR fluence are well correlated with the SGRE fluence, suggesting major energy release in SGRE source regions. The majority of SGRE events occurred in clusters suggesting the availability of seed particles for later shocks in a given cluster.

The association between SEP and SGRE events is less than perfect because of the magnetic connectivity and the spectral hardness requirements. Inherently soft-spectrum events are not expected to produce SGREs because they lack the required high-energy particles (e.g., the SEP events associated with CMEs from filament eruption regions outside active regions). Poorly connected SEP events may still produce SGREs, but the particles are not detected by the observer. The poor cadence of Fermi/LAT may miss some SGRE events even though there are large SEP events with >300 MeV protons [9]. Finally, the exceptional case of an SGRE event with no SEP event can be explained by a propagation issue in that a magnetic mirror formed near a preceding CME reflects particles back to the Sun to produce γ-rays, but no particles escape to be detected as SEP events.

Acknowledgments

We benefited from the open data policy of Fermi, SOHO, Wind, STEREO, SDO, and Wind missions. This work was supported by NASA's Living With a Star program. PM was partially supported by NASA grant NNX15AB77G. HX was partially supported by NASA grant NNX15AB70G.

9. References

[1] Morrison P 1958 *Ncim.* **7** 858
[2] Lingenfelter R E, Ramaty R 1967, In: High Energy Nuclear Reactions in Astrophysics, B S P Shen (ed), New York, W. A. Benjamin, p. 991
[3] Forrest D J, Vestrand W T, Chupp E L et al. 1985, *19th ICRC* **4** 146
[4] Ramaty R, Mandzhavidze N. 1998 ArXiv Astrophysics e-prints, arXiv:astro-ph/9810089
[5] Chupp E L, Ryan J M 2009, *RAA* **9** 11
[6] Kuznetsov S N, Kurt V G, Yushkov B, Myagkova I, Galkin V I, Kudela K. 2014 *ASSL* **400** 301
[7] Atwood W B, Abdo A A, Ackermann M, et al. 2009 *ApJ* **697** 1071
[8] Ackermann M, Allafort A, Bardini L, et al. 2017 *ApJ* **835** 219
[9] Share G H, Murphy R J, White S M, et al. 2018 *ApJ* **869** 182
[10] Klein K-L, Tziotziou K, Zucca P, et al. 2018 *ASSL* **444** 133

[11] Ajello M, Albert A, Allafort A, et al. 2014 *ApJ* **789** 20
[12] Omodei N, Pesce-Rollins M, Longo F, Allafort A, Krucker S 2018 *ApJL* **865** L7
[13] Gopalswamy N, Mäkelä P, Yashiro S, et al. 2018 *ApJL* **868** L19
[14] Ryan J M 2000 *SSRv* **93** 581
[15] Share G H, Murphy R J, Tolbert A K, et al. 2017 *ArXiv* 1711.01511
[16] Plotnikov I, Rouillard A P, Share G H 2017 *A&A* **608** A43
[17] Murphy R J, Dermer C D, Ramaty R 1987 *ApJS* **63** 721
[18] Akimov V, Afanassyev V G, Belaousov A S et al. 1991 *22nd ICRC* **3** 73
[19] Akimov V V, Ambrož P, Belov A V et al. 1996 *SoPh* **166** 107
[20] Kahler S W, Cliver E, W, Kazachenko M 2018 *ApJ* **868** 81
[21] Ryan J M, Lee M A 1991 *ApJ* **368** 316
[22] Kanbach G, Bertsch D L, Fichtel C E, et al. 1993 *A&AS* **97** 349
[23] Yashiro S, Gopalswamy N, Mäkelä P, Akiyama S 2013 SoPh **284** 5
[24] Vestrand W T, Forrest D J 1993 *ApJL* **409** L69
[25] Cliver E W, Kahler S W, Vestrand W T 1993 *23rd ICRC* **3** 91
[26] Pesce-Rollins M, Omodei N, Petrosian V, et al. 2015 *ApJL* **805** L15
[27] Pesce-Rollins M, Omodei N, Petrosian V et al 2015 *34th ICRC* 128
[28] Hudson H S 2018 *IAUS* **335** 49
[29] Grechnev V, Kiselev V I, Kashapova L K et al. 2018 *SoPh* **293**, 133
[30] Gopalswamy N, Mäkelä P, Akiyama S, Yashiro S 2019 under preparation
[31] Gopalswamy N, Mäkelä P, Yashiro S et al 2018, *ArXiv* 1810.08958
[32] Brueckner G E Howard R A, Koomen M J et al *SoPh* **162** 357
[33] Thernisien A 2011 *ApJS* **194** 33
[34] Howard R A Moses J D Vourlidas A 2008 *SSRv* **136** 67
[35] Gopalswamy N, Yashiro S, Thakur N, et al 2016 *ApJ* **833** 216
[36] Bougeret J-L, Kaiser M L, Kellogg P J 1995 *SSRv* **71** 231
[37] Gopalswamy N, Xie H, Akiyama S, Mäkelä P, Yashiro S *EPS* **66** 104
[38] Gopalswamy N, Mäkelä P, Yashiro S, Xie H, Akiyama S, Thakur N 2015 *JPhCS* **642** 012012
[39] Gopalswamy N, Mäkelä P, Yashiro S, Xie H, Akiyama S, Thakur N 2018 *JASTP* **179** 225
[40] Gopalswamy N, Xie H Yashiro S, Akiyama S, Mäkelä P, Usoskin I G 2012 *SSRv* **171** 23
[41] Lara A, Gopalswamy N, Xie H et al. 2006 *JGR* **111** A06107
[42] Gopalswamy N Yashiro S Michalek G et al 2010 *SunGe* **5** 7
[43] Gopalswamy N, Akiyama S, Yashiro S, Mäkelä P 2010 *ASSP* **19** 289
[44] Winter L M, Bernstein V, Omodei V, Pesce-Rollins M 2018 *ApJ* **864** 39
[45] Schmieder B, Hagyard M J, Guoxiang A 1994 *SoPh* **150** 199
[46] Gopalswamy N, Yashiro S, Akiyama S, et al 2016 *ApJ* **823** L15
[47] Bazilevskaya G A 2017 *JPhCS* **798** 012034
[48] Gopalswamy N, Mäkelä P, Akiyama S, et al 2015 *ApJ* **806** 8
[49] Schneid E J, Bertsch D L, Dingus B L et al 1996 *A&AS* **120** 299
[50] Gopalswamy N Yashiro S, Lara A et al 2003 *GRL* **30** 8015
[51] Gopalswamy N, Yashiro S, Krucker S, Stenborg G, Howard R A 2004, *JGR* **109**, A12105
[52] Temmer M, Reiss M A, Nikolic L, Hofmeister S J, Veronig A M 2017 *ApJ* **835** 141

Radial evolution of the properties of small-scale magnetic flux ropes in the solar wind

Qiang Hu, Yu Chen, and Jakobus le Roux

Department of Space Science and CSPAR, The University of Alabama in Huntsville, Huntsville, AL 35805, USA

E-mail: `qiang.hu@uah.edu`

Abstract. Small-scale magnetic flux ropes, most with duration $\leq$ 1 hour at 1 AU, are found to be ubiquitous in the solar wind from in-situ spacecraft measurements. We have built an event database (`fluxrope.info`) for these structures identified and summarized their main physical properties. Quantitative analysis provided strong evidence in support of the view of their generation through turbulence cascade processes in space plasmas. We extend such analysis to spacecraft measurements at larger radial distances, mainly those from the Ulysses spacecraft mission. We perform comparison of relevant properties of these structures and reveal their radial evolution near the ecliptic plane. Meanwhile we also try to derive, quantitatively, the parameters important for the underlying processes including magnetic reconnection, intrinsic to turbulence cascade, in order to aid in relevant theoretical investigations. We also supplement the largely statistical analysis results with individual case studies to illustrate the process of flux rope merging and associated particle energization signatures through an observational approach. We speculate on the implications of our results and foresee future investigations that can help improve our current understanding of the origin and evolution of these small-scale magnetic flux ropes throughout the heliosphere.

1. Introduction

Small-scale magnetic flux ropes, usually identified from in-situ spacecraft measurements with duration ranging from tens of minutes to a few hours, constitute an important type of space plasma structures. They are found to be present in interplanetary medium, embedded in solar wind streams, especially near the heliospheric current sheet (HCS) in the ecliptic plane [1, 2, 3, 4, 5, 6]. They arguably can go further down to even smaller scales, e.g., to duration of a few seconds, provided that the cadences of in-situ measurements for both magnetic field and plasma parameters are sufficiently high. For more sophisticated analysis and modeling approaches, employing the in-situ spacecraft data, both data sets are needed to achieve a more general characterization of the flux rope structures in both their configurations and physical properties, especially in a highly quantitative manner. In addition, auxiliary but increasingly available measurements on suprathermal electrons and elemental composition/charge states are also desirable for discerning the characteristics of the underlying structures and the associated interpretation of their origin [7, 8].

One method that distinguishes itself from the others is the so-called Grad-Shafranov (GS) reconstruction technique, based on the Grad-Shafranov equation describing a two-dimensional (2D) plasma configuration in magnetohydrostatic equilibrium. The main advantage of the GS

Content from this work may be used under the terms of the Creative Commons Attribution 3.0 licence. Any further distribution of this work must maintain attribution to the author(s) and the title of the work, journal citation and DOI.
Published under licence by IOP Publishing Ltd

reconstruction is the full characterization of the 2D configuration in cylindrical geometry by a single scalar function, A, which represents the nested magnetic flux surfaces enclosing a central cylindrical axis for a flux rope configuration. This enables the reconstruction of the distribution of the flux function over the cross section plane, given the measured values along the spacecraft path across the nested flux surfaces. The standard output from the GS reconstruction includes the geometric parameters of the structure, namely, the cylindrical axis orientation as well as the spatial extent of the cross section, and the set of physical quantities characterizing the magnetic field configuration by yielding the spatial distributions of magnetic field components and other derived quantities such as the current density.

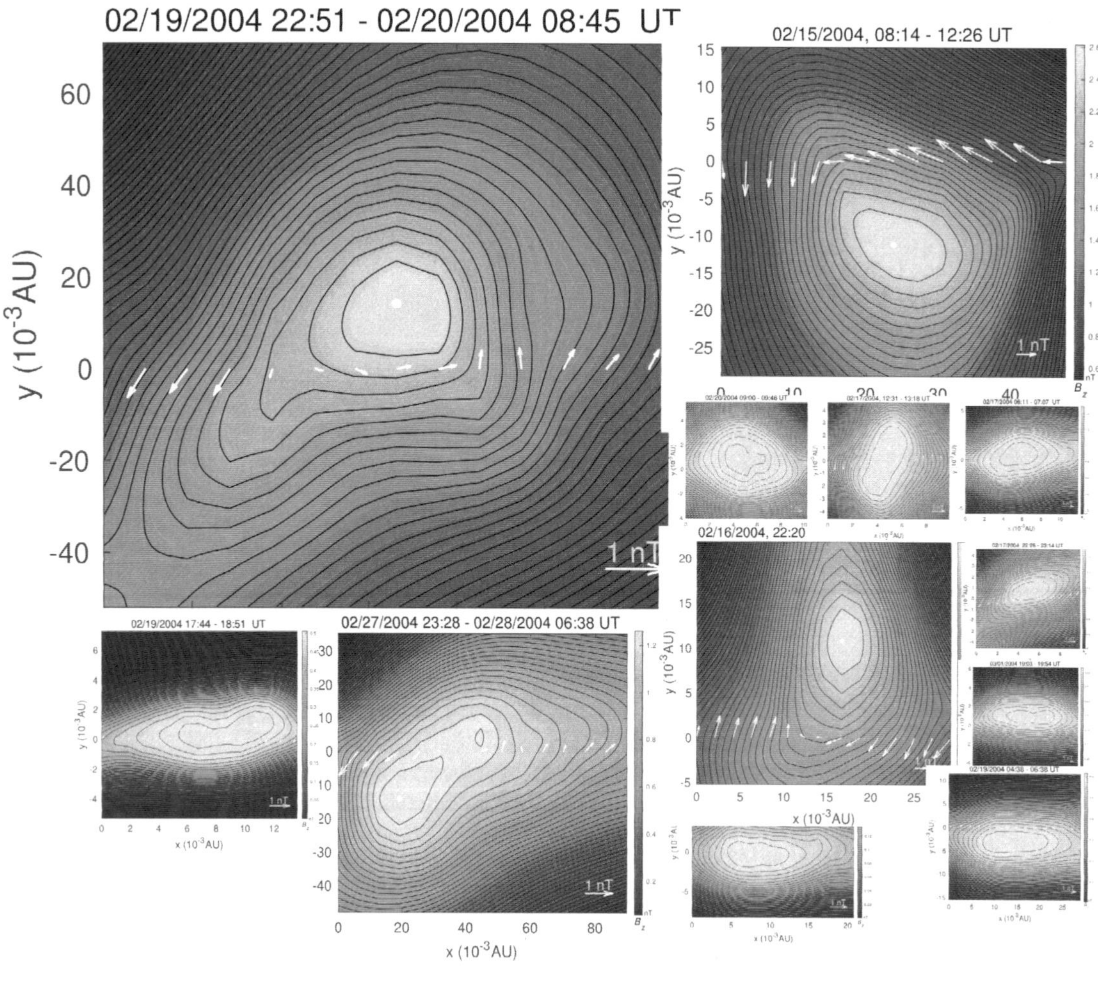

Figure 1. A collage of cross sections of cylindrical flux ropes obtained from the quantitative GS reconstruction results (approximately to scale; adapted from [9]). Each rectangle represents a spatial domain over the cross section of the flux rope, on a plane perpendicular to its cylindrical axis with contours of the magnetic flux function A (equivalent to the transverse field lines) and the axial field component in color. See Figure 8 for detailed descriptions of flux rope configuration from GS reconstruction result. The center usually possesses the strongest axial field. Note the axis orientation varies from one flux rope to the other.

The procedures of the GS reconstruction have been well established and applied to various space plasma environment. We refer the readers to a recent technical review [10] for details. Recent development most relevant to this report is the implementation of the GS-based automated detection algorithm of small-scale magnetic flux ropes with demonstrated success [4, 11]. An unprecedented number of events with significant amount of variability is identified and archived, which has enabled meaningful statistical analysis and pursuit of inter-connections to other relevant studies. The technical details of the implementation are provided in a recent publication [4], documenting the database built based on this approach. As an ongoing effort, the identification and characterization of small-scale magnetic flux ropes are carried out for various spacecraft missions. The results, in the form of event lists with a set of concise but essential information, are summarized and archived on the event database website, `https://fluxrope.info`.

To get a glimpse into the capability of the GS reconstruction technique and the wealth of information it generates, we show, in Figure 1, an example of a collection of selected flux rope configurations obtained from the Ulysses spacecraft in-situ measurements at a heliocentric distance $\sim$ 5 AU over a time period of about two weeks. These are the actual GS reconstruction output for a dozen data intervals (adapted from [9]), showing the cross section configurations typical of magnetic flux ropes with nested flux surfaces on which the field lines are winding around the central axis (owing to the non-vanishing axial field component). Despite such a common flux rope configuration, they show great variability in overall shape, scale size and other properties. Here the scale sizes, approximately to scale as illustrated, vary from about 0.01 AU to 0.1 AU. Generally the distributions of their duration and scale sizes exhibit power laws, based on the statistical analysis of our database, containing tens of thousands of events [4, 6], with more numbers of events with smaller values. Whether there exists a smooth transition to or an indication for a distinct population of the large-scale flux ropes (with duration of tens of hours to days, i.e., the magnetic clouds) remains unclear, partially due to the relatively small event counts for that range. It is interesting to note that although these flux ropes differ in their axial orientations, there does exist a general trend for the flux rope axes to align with the local Parker spiral direction, especially for events occurring near the ecliptic plane [4, 6].

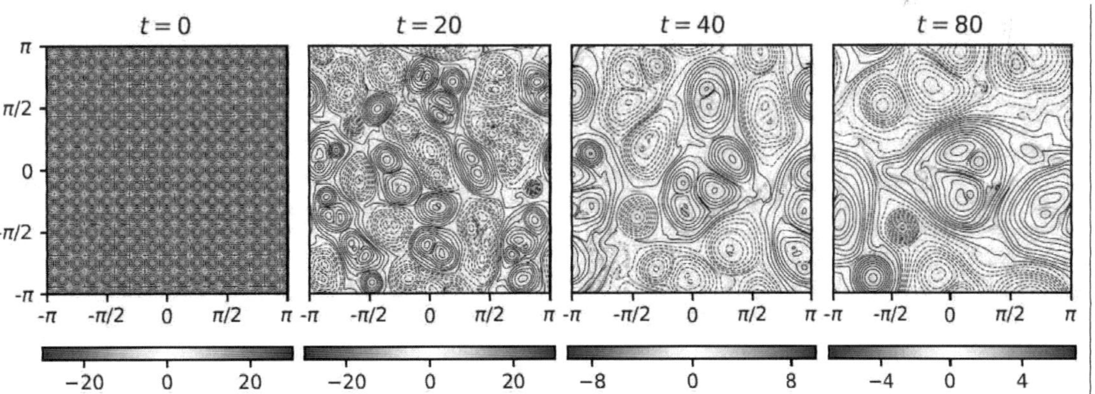

Figure 2. The numerical simulation for the evolution of 2D flux rope merging process from $t = 0$ with idealized distribution of flux ropes with equal sizes and alternating signs of helicity. The contours represent the magnetic flux function ψ and the colors the (axial) current density (adapted from [12]).

One benefit of the GS reconstruction technique is to enable a direct comparison of the physical quantities with the corresponding numerical simulation results. For example, the GS output of the 2D magnetic flux function A and the axial field and current density (j_z, also a function of A) corresponds to the dependent variables employed in many 2D magnetohydrodynamic (MHD) simulations of turbulence [13, 14, 15, 16]. As an example, Figure 2 [12] illustrates exactly the same physical quantities as in Figure 1 for a 2D MHD simulation of inverse cascade process. These physical quantities have direct one-to-one correspondence to the GS reconstruction output, i.e., $A \equiv \psi$. Therefore a direct and physical connection can be made between the two realms. One testament to this approach is a recent study by Zheng and Hu [11] in which a direct comparison between the distributions of the axial current density in flux rope/island structures from the GS-based observational analysis and 2D MHD simulations was made to demonstrate the non-Gaussian features present in both results. Such a study provided observational evidence for the view of the correspondence of these structures to 2D MHD turbulence in the inertia range [17, 18]. We will further pursue this type of investigation by comparing our observational result with the numerical simulation study of [12] in Section 2.2.

Motivated by these recent development, we have expanded the ongoing analysis to additional spacecraft missions. In Section 2, we present a comparison of a set of selected properties for flux ropes identified at two separate ranges of heliocentric radial distances near the ecliptic plane. We explore the possible underlying mechanism responsible for the radial evolution of these properties, in the context of flux rope merging (inverse cascade), a scenario depicted in Figure 2. We then attempt to zoom in and focus on one particular case study, hinting at flux rope merging between two adjacent flux ropes. We will present a few key characteristics including associated energetic particle signatures. Lastly, we conclude and offer some additional thoughts in the last section.

2. Radial Dependence of Flux Rope Properties

Table 1. Radially distributed flux rope events from ACE and Ulysses spacecraft.

Database	Latitudes	Radial Distances	Walén test slope	Event Counts
ACE	$0°$	1 AU	≤ 0.3	17,602
Ulysses	$< 30°$	>3.5 AU	≤ 0.5	10,124

We have applied the automated detection algorithm to a number of spacecraft missions and established small-scale flux rope databases with significant number of events. A unique database is from the Ulysses mission which had a unique polar orbit over the poles of the Sun that covered a wide range of heliographic latitudes and heliocentric radial distances (from ~1 to 5 AU). We describe here briefly a subset of events from the Ulysses database corresponding to the events detected in low latitudes, but at larger radial distances beyond 3.5 AU, and compare with the corresponding results detected at 1 AU. For a comprehensive report on all the events detected throughout the entire Ulysses mission, see Chen et al. [6].

2.1. Comparison Between ACE and Ulysses Results

We specifically focus on the radial evolution of flux rope properties by utilizing the simultaneous flux rope detection results at 1 AU from the ACE spacecraft for the time periods, 1998-02-05~1999-07-14, and 2002-10-12~2005-09-14, when Ulysses was spending extensive periods of time near the ecliptic plane. To better facilitate the inter-comparison between ACE and Ulysses, we use exactly the same duration range, $9 \sim 2165$ minutes, for the flux rope intervals contained in the two event sets. The different conditions and search criteria are summarized in Table 1. The different Walén test slopes have little impact on the search result, since reducing the slope to 0.3 for Ulysses results in very little change in the detection rate [6].

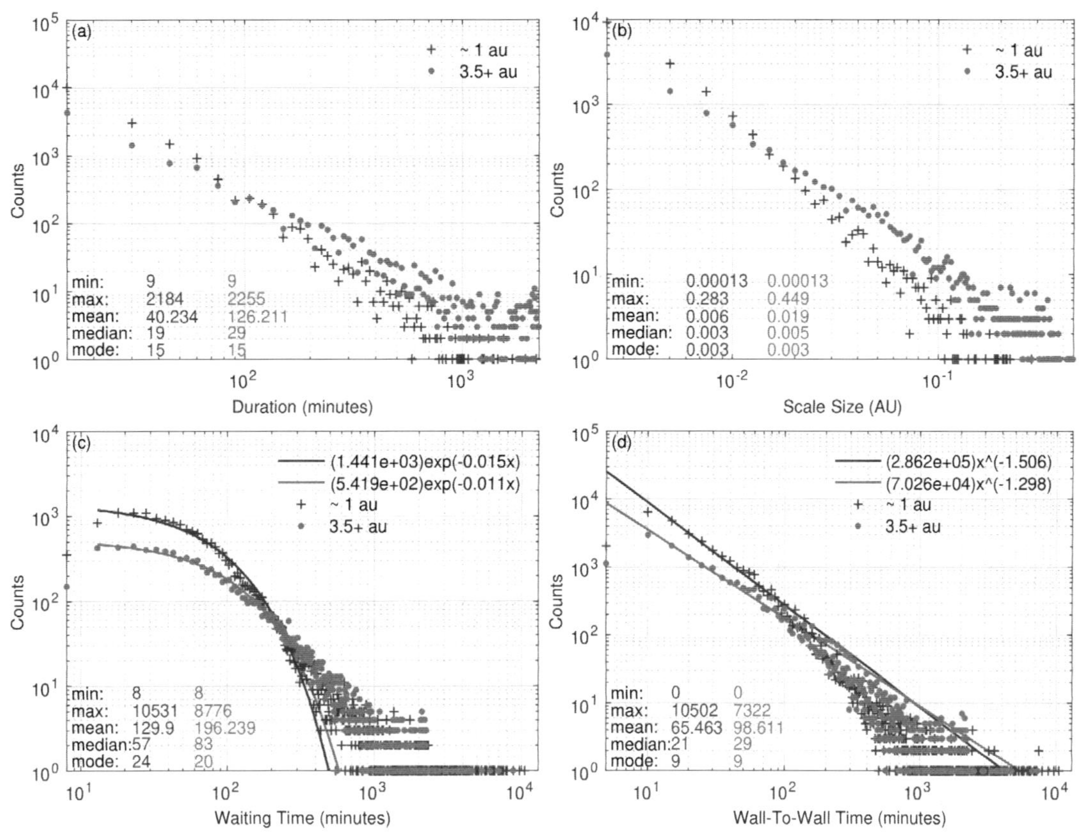

Figure 3. The distributions of flux rope properties for the radially aligned events from ACE and Ulysses. These properties are (a) duration, (b) scale size, (c) waiting time, and (d) wall-to-wall time. In (c) and (d), the exponential and power-law fitting curves are drawn for the lower-range property values with the fitting parameters denoted.

There is a clear decrease in the number of event counts from a radial distance at 1 AU to increased distances, for the same time periods and the same range of duration. When comparing the two event sets, there are clear discrepancies in plasma and magnetic field parameters, simply due to the radial evolution of solar wind plasmas, in terms of radially decaying plasma density, temperature, and magnetic field magnitude (not shown). For the other selected flux rope properties, shown in Figure 3, the difference is seen in the overall distributions from the two sets. For the distributions of duration and scale size, they exhibit clear power laws, but possess different power-law indices. For the distribution of waiting time, defined as the elapsed time between the beginning times of two adjacent flux rope intervals, an exponential functional seems to fit each distribution well for the smaller value portion, but with different fitting parameters. Figure 3d shows the wall-to-wall time distribution, which is considered to be equivalent to the waiting time distribution of current sheets near the flux rope boundaries [11]. They were found to be power-law like in the lower-value range below certain break point, corresponding to the correlation length scale of the MHD turbulence in the inertial range [14]. The power-law fittings are given, respectively, for the two distributions at different radial distances. The power-law indices are different and the break points are different as well with one around ~ 80 minutes for the distribution at 1 AU and the other ≥ 200 minutes, respectively.

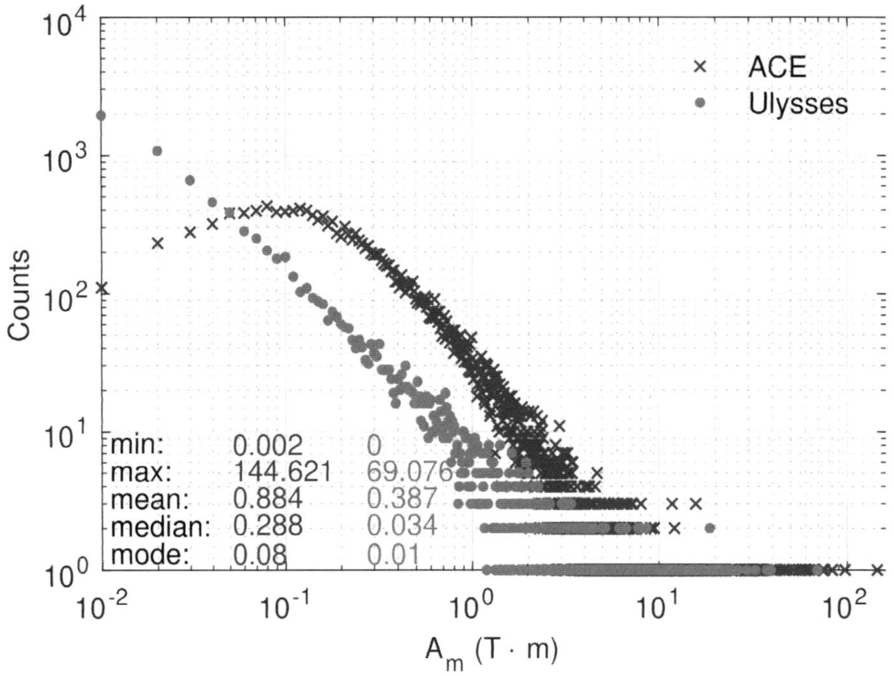

Figure 4. The distributions of the poloidal magnetic flux A_m for both ACE and Ulysses event sets.

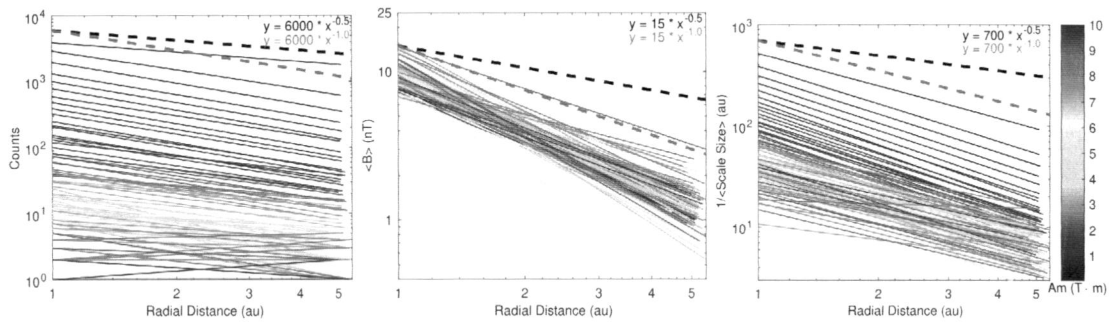

Figure 5. The radial variations of the number of event counts N, the average field magnitude $\langle B \rangle$, and the inverse scale size k (from left to right panels) in log-log scales, respectively. Each colored line connects the values at ACE and Ulysses, corresponding to the events with approximately the same A_m as indicated by the colorbar. The dashed lines provide the guide lines with power-law slopes -1 and -0.5, respectively.

2.2. Radial Evolution Owing to Flux Rope Merging

We attempt to examine the possible radial evolution of flux rope properties, making use of the databases obtained from ACE and Ulysses spacecraft. Although the flux ropes could be primarily generated through local processes independently at different radial distances, one possible scenario is the evolution of these structures in time due to interactions while convecting with the solar wind. Owing to the process of merging between adjacent flux ropes, this will lead to the corresponding radial evolution that may be present in our databases. One such process was proposed and studied by numerical simulations, considering the successive merging between adjacent flux ropes, as illustrated in Figure 2 [12]. The temporal evolution depicted corresponds to an inverse cascade process intrinsic to 2D MHD turbulence, resulting in larger scale size flux ropes with decaying magnetic field strength and fewer counts when time progresses. Based on simple "hierarchical merging" argument and the principle of (poloidal) magnetic flux conservation during flux rope merging (see also [19]), a number of scaling laws with respect to time t is deduced for the event counts N, the average field magnitude $\langle B\rangle$, and the inverse scale size k [12]:

$$N \sim t^{-1}, \quad \langle B\rangle \sim t^{-0.5}, \quad k \sim t^{-0.5},$$

under the condition that the poloidal magnetic flux $\psi(t) = \psi(0)$ is conserved for each flux rope.

From our observational point of view, we can translate t into the radial distance r, and obtain the aforementioned quantities from our databases at discrete radial distances (only two at the present time). Then we bin them according to the corresponding poloidal flux content in each flux rope, complying with the condition of flux conservation. Figure 4 shows the distributions of the amount of the poloidal magnetic flux, $A_m \equiv \psi$, representing the absolute extreme value in A with the flux rope boundary defined with $A_b \approx 0$. The range of A_m for both event sets is $[0.05, 10]$ T·m, excluding the greater values corresponding to the large-scale flux ropes which may have clear solar origin. Then we bin the three quantities for all the events based on their corresponding A_m value, and present each bin containing two points at two separate radial distances, in Figure 5. Therefore each bin (colored line) represents the radial change in the corresponding quantity subject to the possible merging process discussed above. It is seen that in general these quantities all decay with r and with power-law indices close to -1, or somewhere in-between -0.5 and -1 at best, not strictly following the prediction of [12]. Of course one caveat is that we only have two distinct points in r. It would be desirable to extend the analysis to more datasets at additional radial distances.

3. Flux Rope Merging: A Case Study

It is also informative to pursue in the direction of individual case study, from the perspective of detailed and quantitative characterization of flux rope merging between a pair of adjacent flux ropes, via the approach of GS reconstruction. We present one such case observed by both the ACE and Wind spacecraft in March 1999 when they were separated by about 250 Earth radii mostly along the Sun-Earth line. Therefore the same solar wind streams passed both spacecraft with a time delay of about an hour with ACE being further upstream.

Figure 6 shows the time series of the solar wind magnetic field and plasma parameters, together with the energetic ions flux (the bottom panel) from ACE. A number of flux ropes was identified downstream of the interplanetary shock, at 00:40 UT as marked by the vertical line. The flux rope intervals are shaded and numbered in numeric order. The intervals denoted "7" and "8" are of interest and are taken for a GS reconstruction study, partially for the reason of simultaneous ions flux enhancement, especially in the energy range $\sim$ 50 - 200 keV. The same set of measurements for the corresponding time period from Wind spacecraft is shown in Figure 7, where the flux rope interval "7" is believed to correspond to the adjacent intervals "7" and "8" in Figure 6. Such a coincidence of radial correspondence can be justified by the timing analysis of similar structures between ACE and Wind. For example, the preceding shock, the similar high-speed stream downstream, the similar magnetic field profiles during those flux rope intervals, and particularly the distinct features of the discontinuities bounding the flux rope intervals are all present in both measurements. They all seem to be consistent with a rapid transit ($\sim$ 1 hour) of these structures between the two spacecraft.

The corresponding GS reconstruction results of the cross section maps for both ACE and Wind intervals are shown in Figure 8, approximately to scale. From the ACE 2D cross

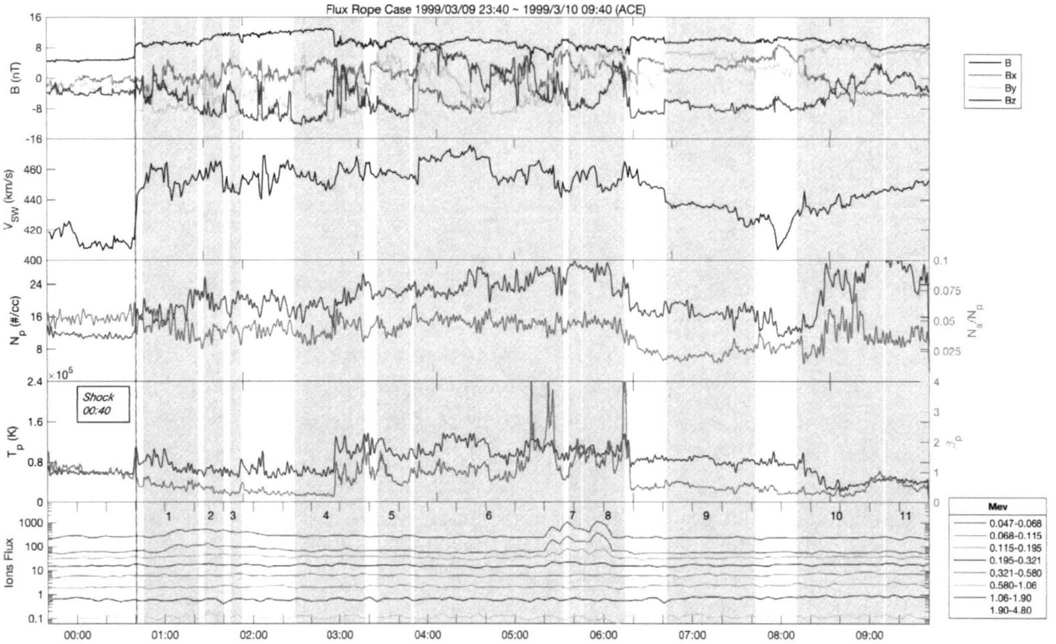

Figure 6. The time-series data from ACE spacecraft for the time period (in UT) noted on top. The panels, from top to bottom, are the magnetic field magnitude and components, the solar wind speed, the proton number density (left axis) and the alpha particle to proton number density ratio (right axis), the proton temperature (left axis) and proton β (right axis), and the ion fluxes with the corresponding energy channels shown to the right. The vertical line marks the passage of an interplanetary shock. Shaded areas represent identified flux rope intervals and are numbered.

section map, there exist two sets of cylindrical flux surfaces with parallel axes, undergoing reconnection and forming an X-point (X-line extending in the z dimension). From the Wind result, corresponding to about 1 hour later in time, a single set of nested flux surfaces emerges, centering around a central axis. Although each map is considered a snapshot of the quasi-static structure not varying in time, the morphology of the field configuration represented by these two maps obtained from real data at two closely separated times hints at the possible merging process due to the magnetic field topology favorable for magnetic reconnection. This result provides a unique and quantitative proof of such a process, provided that the dynamic evolution time scale is much longer than the transit time of the structure across the observing platforms.

For this to be satisfied in the current case, the dynamic evolution time may need to be in the order of 10^4 seconds, which can be considered to be the time taken for the two flux ropes to fully merge via magnetic reconnection. Thus the reconnected magnetic flux during the process would be equal to the amount of the poloidal flux involved. If we take the typical value 1 T·m, then the time τ it takes for the complete merging process is related to the reconnection rate or the reconnection electric field E_r via $\tau \approx 1/E_r$. For $\tau = 10^4$ seconds, the reconnection rate is approximately 10^{-4} V/m or 0.1 mV/m, which is not unreasonable for the solar wind, given that a value on the order of 1 mV/m was typically found for the reconnection events at the Earth's magnetopause [21].

Table 2 lists the physical parameters deemed important for the merging process, which has yet

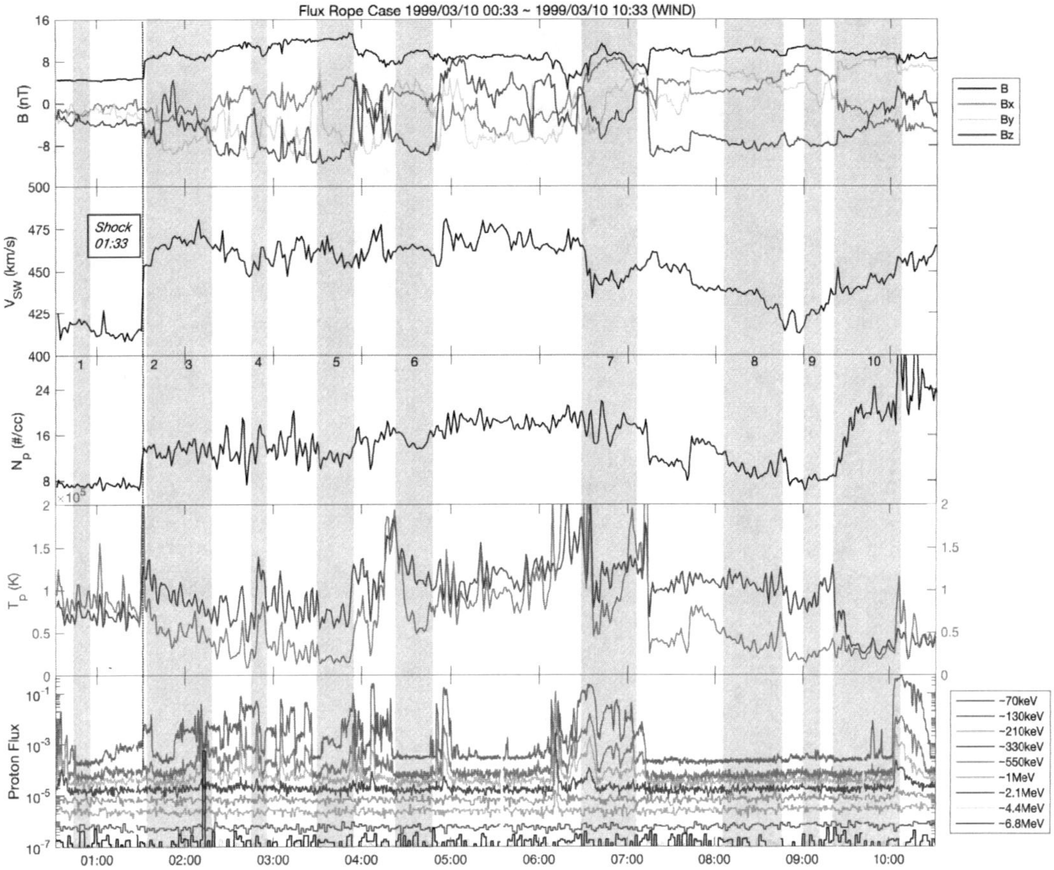

Figure 7. The time-series data from Wind spacecraft for the time period (in UT) noted on top. The format is the same as Figure 6.

Table 2. Main flux rope parameters of the GS reconstruction results from ACE (second row) and Wind (third row) spacecraft for the March 10, 1999 case study.

Φ_p (T·m)	$\Phi_t(10^9$ Wb)	B_{z0} (nT)	$j_{z0}(10^{-11}$ A/m^2)	$I_z(10^9$ A)
1.68	2.68	8.4	-4.7	-0.014
1.78	2.48	9.9	-4.5	-0.010

to be better related to theoretical investigations. The amount of poloidal flux (per unit length) Φ_p is nearly identical for the two spacecraft, corresponding to the values at the beginning or in the middle, and at the end of the merging process, respectively. The other values, including the toroidal flux Φ_t, the maximum axial field B_{z0}, the maximum axial current density j_{z0}, and the total current I_z, are typical for the flux rope of such a scale size.

4. Conclusions and Discussion

In conclusions, we have applied the automated flux rope detection algorithm based on the GS reconstruction to a number of in-situ spacecraft missions in the solar wind, and generated event databases with the event occurrence rates of a few hundreds per month on average [4, 6]. In

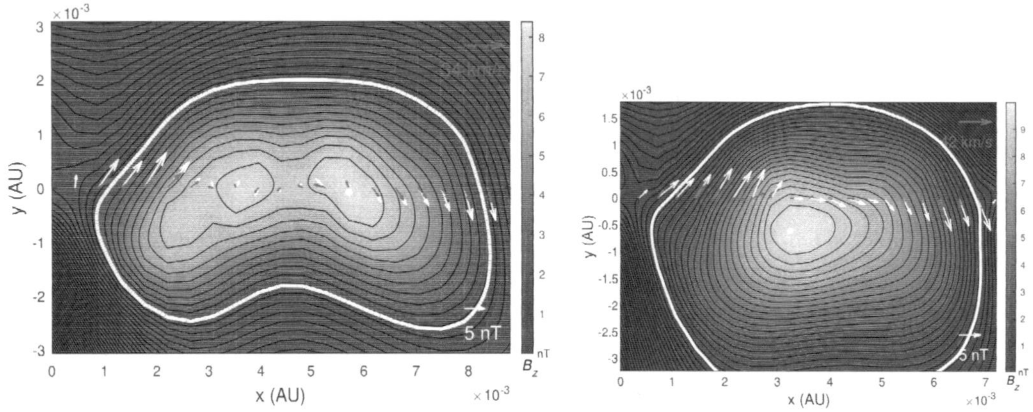

Figure 8. The GS reconstruction result of the cross sections of the flux rope from ACE (left panel) and Wind (right panel) spacecraft measurements, respectively. On each cross section map, the black contours represent the iso-surfaces of the magnetic flux function and the colors the axial magnetic field component with scales given by the colorbar in nT. The white arrows are the measured transverse field components along the spacecraft path ($y = 0$), while the green arrows the corresponding remaining transverse plasma flows in the frame moving with the structure. Two reference arrows of the magnitude "5 nT" and the local average Alfvén speed are drawn to scale for the transverse magnetic field magnitude and remaining flow speed, respectively. The thick white contour line marks the flux rope boundary [20].

particular, we have extended our analysis to the in-situ measurements by the Ulysses mission at the heliocentric radial distances greater than 1 AU, near the ecliptic plane, and compared with the corresponding analysis result at 1 AU.

We find that the main flux rope properties such as the duration, scale size and wall-to-wall time all follow power-law distributions, but possess different power-law indices for the event sets at different radial distances. The waiting time distributions, on the other hand, exhibit the exponential function behavior with different exponents. The wall-to-wall time distribution for the event set from Ulysses at greater radial distances exhibits a break point at $\sim$ 200 minutes. A preliminary analysis on the radial evolution of selected flux rope properties including the event count, the average magnetic field, and the inverse scale size does not yield the scaling laws with respect to the radial distance, predicted by a simple "hierarchical merging" argument, albeit it should be cautioned that only two distinct radial distances are considered. A detailed case study is performed for the flux rope intervals traversed by the radially aligned ACE and Wind spacecraft. The reconstructed cross-section maps of the flux function at the two spacecraft locations / instances (about 1 hour apart) reveal the presumed merging process, i.e., a pair of adjacent single flux ropes at ACE becoming one single flux rope at Wind. During the process, it is also found that the poloidal magnetic flux is conserved, a quantitative result envisaged by some theoretical works. There are also simultaneous energetic ions flux enhancement coinciding with the flux rope intervals. The implications of these coincident or co-spatial signatures have yet to be explored, especially in a more quantitative manner.

Motivated by the recent development in theoretical investigations of particle energization by flux rope dynamics including merging (e.g., [22]), it is also attempting to try to correlate the topological change revealed for the flux rope merging, in this study, to the associated particle signatures. However it is still challenging to make direct comparisons, considering the disparity in scale sizes in respective investigations. It also remains unclear how energetic electrons behave, given that the relevant observational analysis is rare (e.g., [23]). These are topics to be further pursued.

Acknowledgments
We are grateful to our colleagues at SPA/CSPAR, UAH, Drs. Laxman Adhikari, Gang Li, Gary Webb, Gary Zank, and Lingling Zhao for on-going collaborations. The in-situ spacecraft data are accessed via the Coordinate Data Analysis Web (CDAWeb) of NASA GSFC and the ACE Science Center. We acknowledge NASA grants NNX15AI65G, NNX17AB85G, 80NSSC18K0623, 80NSSC19K0276, and subaward SAO SV4-84017, and NSF grant AGS-1650854 for support. Special thanks also go to the SCOSTEP/VarSITI program for support of the development and maintenance of the on-line small-scale flux rope database at `http://fluxrope.info`.

References
[1] Moldwin M B, Phillips J L, Gosling J T, Scime E E, McComas D J, Bame S J, Balogh A and Forsyth R J 1995 *Journal of Geophysical Research: Space Physics* **100** 19903–19910 ISSN 2156-2202 URL `http://dx.doi.org/10.1029/95JA01123`
[2] Moldwin M B, Ford S, Lepping R, Slavin J and Szabo A 2000 *Journal of Geophysical Research: Space Physics* **27** 57–60
[3] Cartwright M L and Moldwin M B 2010 *Journal of Geophysical Research (Space Physics)* **115** A08102 ISSN 2156-2202 a08102 URL `http://dx.doi.org/10.1029/2009JA014271`
[4] Hu Q, Zheng J, Chen Y, le Roux J and Zhao L 2018 *Astrophys. J. Supp.* **239** 12 (*Preprint* `1810.01308`)
[5] Chen Y, Hu Q, le Roux J and Zheng J 2018 *Journal of Physics: Conference Series* **1100** 012006 URL `https://doi.org/10.1088/012006`
[6] Chen Y, Hu Q and le Roux J A 2019 *The Astrophysical Journal* **submitted** (*Preprint* `1905.00986`)
[7] Feng H Q, Zhao G Q and Wang J M 2015 *Journal of Geophysical Research: Space Physics* **120** 10,175–10,184 ISSN 2169-9402 2015JA021643 URL `http://dx.doi.org/10.1002/2015JA021643`
[8] Huang J, Liu Y C M, Peng J, Qi Z, Li H, Klecker B, Song H, Zheng J and Hu Q 2018 *Journal of Geophysical Research (Space Physics)* **123** 7167–7180
[9] Zhao L L, Zank G P, Chen Y, Hu Q, le Roux J A, Du S and Adhikari L 2019 *The Astrophysical Journal* **872** 4 URL `https://doi.org/10.3847/1538-4357/aafcb2`
[10] Hu Q 2017 *Sci. China Earth Sciences* **60** 1466–1494
[11] Zheng J and Hu Q 2018 *Astrophys. J. Lett.* **852** L23 (*Preprint* `1801.01771`)
[12] Zhou M, Bhat P, Loureiro N F and Uzdensky D A 2019 *arXiv e-prints* (*Preprint* `1901.02448`)
[13] Servidio S, Matthaeus W H and Dmitruk P 2008 *Physical Review Letters* **100** 095005
[14] Greco A, Matthaeus W H, Servidio S, Chuychai P and Dmitruk P 2009 *The Astrophysical Journal Letters* **691** L111 URL `http://stacks.iop.org/1538-4357/691/i=2/a=L111`
[15] Wan M, Matthaeus W H, Servidio S and Oughton S 2013 *Physics of Plasmas* **20** 042307
[16] Rappazzo A F and Parker E N 2013 *The Astrophysical Journal* **773** L2 URL `https://doi.org/10.1088/2041-8205/773/1/12`
[17] Matthaeus W H, Bieber J W, Ruffolo D, Chuychai P and Minnie J 2007 *The Astrophysical Journal* **667** 956–962
[18] Zank G P, Adhikari L, Hunana P, Shiota D, Bruno R and Telloni D 2017 *The Astrophysical Journal* **835** 147 URL `http://stacks.iop.org/0004-637X/835/i=2/a=147`
[19] Drake J F, Swisdak M, Che H and Shay M A 2006 *Nature* **443** 553–556
[20] Hu Q, Smith C W, Ness N F and Skoug R M 2004 *Journal of Geophysical Research: Space Physics* **109** ISSN 2156-2202 a03102 URL `http://dx.doi.org/10.1029/2003JA010101`
[21] Hasegawa H, Sonnerup B U and Nakamura T K M 2010 *Journal of Geophysical Research: Space Physics* **115** (*Preprint* `https://agupubs.onlinelibrary.wiley.com/doi/pdf/10.1029/2010JA015679`) URL `https://agupubs.onlinelibrary.wiley.com/doi/abs/10.1029/2010JA015679`
[22] Du S, Guo F, Zank G P, Li X and Stanier A 2018 *The Astrophysical Journal* **867** 16 URL `https://doi.org/10.3847/1538-4357/aae30e`
[23] Chen L J, Bhattacharjee A, Puhl-Quinn P A, Yang H, Bessho N, Imada S, Mühlbachler S, Daly P W, Lefebvre B, Khotyaintsev Y, Vaivads A, Fazakerley A and Georgescu E 2008 *Nature Physics* **4** 19–23

Implications of the Voyager 1 and 2 Particle and Field Observations around their respective Heliopause Crossings

D S Intriligator[1], W D Miller[1], J Intriligator[1,2], and W Webber[3,4]
[1]Carmel Research Center, Inc., Santa Monica, CA
[2]Tufts University, Medford, MA
[3]New Mexico State University, Las Cruces, NM
[4]Deceased

E-mail: devriei@aol.com

Abstract. The numerous contrasts between the Voyager 1 particle and field observations around its heliopause crossing in 2012 and the corresponding observations by Voyager 2 in late 2018 and the beginning of 2019 are consistent with understanding the heliopause itself as the separator between the solar system magnetic field and the interstellar medium magnetic field, at a location that varies in response both to short-term effects of plasma instabilities and longer-term bulk motions of the heliosheath. Plasma flows in the respective regions provide a framework for explaining the observed decreases with heliopause distance of the count rates of both solar system solar and anomalous cosmic rays diffusing outward from the heliopause and galactic cosmic rays diffusing inward. Formation of excited hydrogen atoms by charge-exchange collisions in interstellar plasma displaced and accelerated (as described in the discussion of solar system cosmic rays) by the motion of the heliosphere provides a mechanism for producing recently recognized anomalous Lyman alpha emissions around the front of the heliosphere. We discuss the consequent possibility of previously unobserved Lyman alpha emissions associated with shock propagation in the local interstellar medium.

1. Introduction

It is a pleasure to recognize and honor Randy Jokipii (Figure 1) and to acknowledge his many significant predictions and key analyses in space physics and astrophysics. Jokipii has been a friend and inspiration worldwide to many researchers on plasmas and energetic particles, including at Carmel Research Center, Inc.

Figure 1. We are indebted to Randy Jokipii for his many important contributions, including predictions of phenomena that might be found within the heliosphere, beyond the heliopause, and analyses of Voyager and other spacecraft observations throughout these regions.

The Voyager 2 (V2) data publicly available on the web show that evidence for the V2 approach to the heliopause (HP) started in the plasma system (PLS) in May and in the cosmic ray system (CRS) and magnetometer in August 2018. The count rate of particles with energies >70 MeV/nucleon (the most convenient aggregate measure of the local density of galactic cosmic rays (GCRs)) began to creep up similarly to the observations when Voyager 1 (V1) approached the HP. The GCR inflow may have enhanced V2's observed plasma slowing, like the GCR effect at the V2 TS crossing (Florinski, et al. [17]). Various plasma, energetic particle, and magnetic field data show that the V2 crossing of the HP occurred in early November, 2018, after a much more prolonged approach than was observed at V1.

The decline in the count rate at V2 of solar system particles with energies >0.5 MeV/nucleon (solar and anomalous cosmic rays (ACRs) well below 70 MeV/nucleon in energy) after the HP crossing reached the GCR rate, as after V1's HP crossing, so then only GCRs were present. The decline at V2 was slower and less regular than the decline at V1, but the energetic particle and magnetic field data show that V1 saw three forward HP crossings and two reverse ones, implying two brief emergences into the local interstellar medium (LISM) before the final permanent one, while the V2 emergence was unique. V1 and V2 both crossed near 120 AU but respectively near (174, 35N) and (218, 32S).

Content from this work may be used under the terms of the Creative Commons Attribution 3.0 licence. Any further distribution of this work must maintain attribution to the author(s) and the title of the work, journal citation and DOI.
Published under licence by IOP Publishing Ltd

Journal of Physics: Conference Series **1332** (2019) 012006 doi:10.1088/1742-6596/1332/1/012006

Comparing the V1 and V2 data show that at V1 the magnetic field magnitude changed significantly at each forward and reverse HP crossing but unexpectedly the field direction changed very little, prompting doubts that these events were genuine HP crossings, versus crossings of a boundary between differing plasma regions within the heliosheath. However, the available preliminary V2 "Quick Look" (QL) magnetic field parameters indicate a brief period of large perturbations in field magnitude around the HP crossing, after which the field magnitude was little different from before. The QL data also show a sharp direction change at the HP crossing but are not adjusted by calibration, hence may not show a direction change when corrected. The V2 QL data agree with the V1 mag evidence that the HP itself is a current sheet. The LISM magnitudes at both spacecraft are similar.

The following sections describe how these observations are consistent with a conceptual framework derived from a combination of the V2 plasma observations, energetic neutral atom observations from other spacecraft, and recent theoretical computations. Furthermore, the inferred plasma dynamics of the interaction of the front of the heliosphere with the LISM serendipitously provides a possible explanation for anomalous Lyman alpha observations recently reported from the V1 ultraviolet spectrometer and the ALICE instrument on New Horizons, and combining this explanation with our previous analyses of shock propagation in the local interstellar medium in turn raises the possibility of previously unobserved Lyman alpha emissions associated with such shocks.

2. Comparing the V1 and V2 Crossing Observations

Stone, et al. [1] and at the Voyager CRS website [2], describe how the CRS telescopes are stacks of solid-state detectors, mounted so that particles can enter either end of each stack, with a guard detector surrounding each stack and coincidence logic incorporated into the data system. This logic identifies simultaneous energy deposition events in various combinations of detectors and thereby distinguishes counts caused by electrons from counts caused by positive ions. Among positive ions, it discriminates between protons, alpha particles, and heavier nuclei, as long as the guard detector is not triggered. Triggering this detector indicates that a particle came through the side of the telescope instead of from the desired field of view; thus readings from other detectors would be an incomplete or mistaken characterization of the particle, and are therefore rejected. See website [2] for more details.

Figure 2(a) shows the CRS count rates for particles with energies >0.5 MeV/nucleon (solar system cosmic rays (SSCR), abbreviated to ">0.5 MeV" for the plot labeling); particles with energies >70 MeV/nucleon (GCR, likewise abbreviated) and B magnitude from the magnetometer system for the period when the spacecraft was crossing the HP. The base-2 logarithmic scale on the left axis allows displaying both the SSCR count rate and the B magnitude in nT to show the simultaneity of the changes in the SSCR and B magnitude with each other and with the much smaller percentage changes in the GCR count rate, given by the linear scale on the right axis, as the first forward crossing occurred at about day 211 (July 29), followed by the first reverse crossing at about day 214 (August 1), the second forward on day 226 (August 13), the second reverse on day 233 (August 20) and the third and final forward on day 238 (August 25) of 2012. Green stars show the HP crossings in Figures 2-4.

Figure 2(b) shows the corresponding data for V2 near its crossing late on November 4 or early on November 5 of 2018. Due to the difficulty in processing the damaged V2 magnetometer (Burlaga, et al. [3]) readings only preliminary "Quick Look" (QL) data are now publicly available. Good-quality QL data were only posted after October 22, 2018, so the B data series starts then, while the CRS count rates from early September show how the rise in the GCR count rate signaled the impending HP crossing, much more slowly than the corresponding rises preceding the crossings seen on V1. The QL data are the best mag data available now but calibration and verification adjustments are likely.

Figure 2(a) also shows that after the final crossing on August 25 the SSCR count rate continued to decline for approximately another two weeks, stabilizing at around 2.1 after day 250. As this count rate for particles >0.5 MeV is less than the reading of around 2.6 for particles >70 MeV, evidently no particles in the energy range 0.5 - 70 MeV were being detected, so that the difference in the rates for what are actually GCRs is attributed to geometrical factors for which these raw rates are not adjusted.

Correspondingly, Figure 2(b) shows that after a decline that was far slower and more irregular than what was seen at V1, the >0.5 MeV count rate at V2 has stabilized at almost exactly 2.0, while the >70 MeV count rate is a little under 2.5, implying that V2 has reached a region with no detectable particles in the range 0.5 - 70 MeV, with nearly the same geometric factor effect on the count rates as at V1.

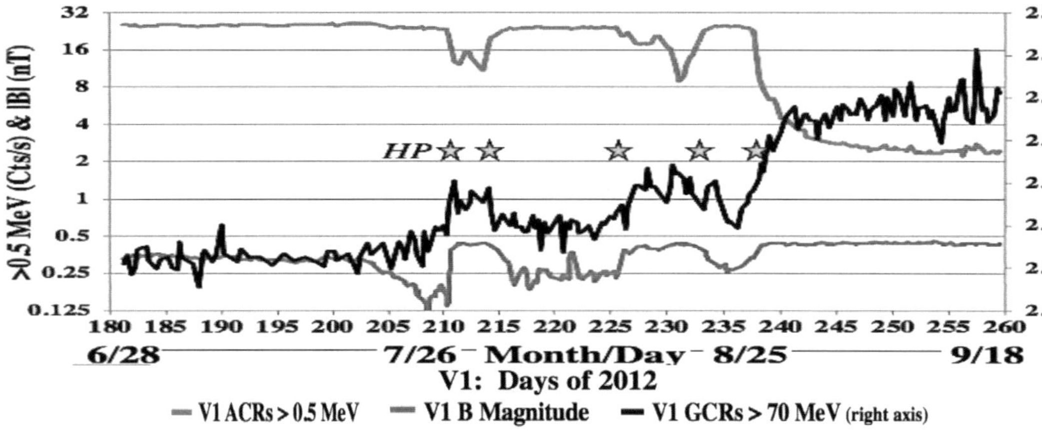

Figure 2(a). V1 HP-epoch SSCR (>0.5 MeV), GCR (>70 MeV), and B magnitude.

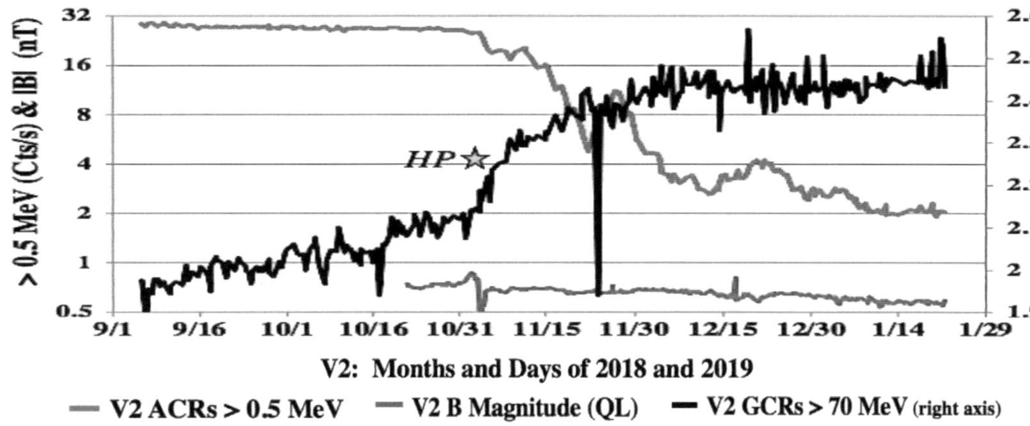

Figure 2(b). V2 HP-epoch solar CR (>0.5 MeV), GCR (>70 MeV), and B magnitude (QL).

Figure 3 shows how, as discussed previously by many authors, the direction of the B field observed by V1 did not change significantly during the HP crossings, although a sector boundary was crossed only a few days before the first HP crossing. By contrast, in Figure 4 the V2 QL, which has not been calibrated and verified yet, shows at the HP crossing a quick and long change of ~20 degrees in the azimuthal angle and an initial change of ~20 degrees in the polar angle later declining to ~10 degrees.

Although many other researchers will discuss these magnetometer observations in far more detail, the results in Figures 3 and 4, along with the correspondences between the particle observations in Figure 2, indicate that although many physical phenomena are associated with the HP, the HP itself is the current sheet separating the heliospheric and interstellar magnetic fields, and that the spacecraft successfully observed it, even though the change was manifested primarily as a change in field magnitude at V1 but as a change in field direction at V2. In the next section combining several recent simulation projects allows interpreting many other aspects of the observations in Figures 2-4.

3. Comparison with Simulations: HP Location Variations

Figure 5 is reproduced from Figure 5 of Washimi, et al. [4], in which they used OMNI data from 1 AU to estimate how long-term changes in the solar wind pressure may have affected the location of the HP along the trajectories of V1 and V2 since 2006, continuing the calculation to predict what might happen through the end of 2019. The figure shows their conclusion that in 2012 along the V1 trajectory the HP was nearly stationary, so this simulation result matches the relatively quick crossing observed by V1.

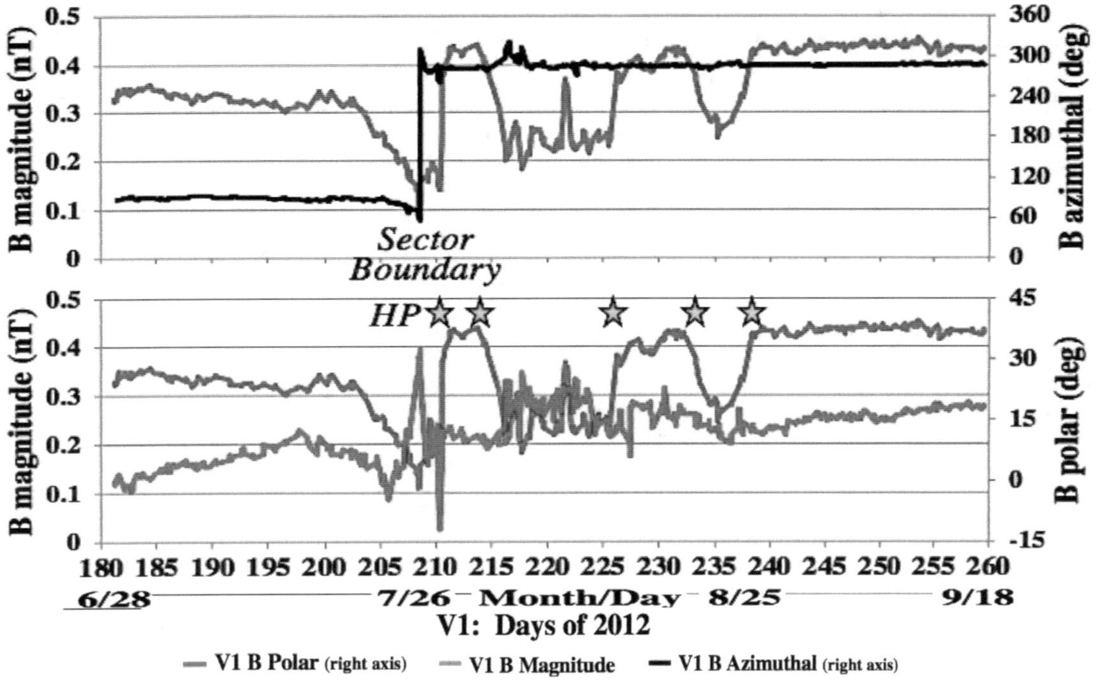

Figure 3. V1 HP-epoch magnetic observations.

However, the figure predicted that from the second half of 2017 (after completion of the paper) through the end of 2019 along the V2 trajectory the HP would be moving outward at about 2 AU/yr, so that the spacecraft, moving outward at about 3 AU/yr, would have a much slower speed relative to the HP than V1 did. This is consistent with the way that the GCR count rate increased as V2 approached the HP and the SSCR count rate decreased after V2 crossed, thus qualitatively agreeing with the observations when V1 crossed, but the count rate changes at V2 both before and after the crossing were much slower than the corresponding ones at V1. The crossing in early November, 2018, suggests that the actual HP locations in 2017-18 were about 1.5 AU farther from the Sun than the figure shows, which in view of the approximations in the model was actually a small discrepancy.

On the other hand, the observations around the crossings at both V1 and V2 showed short-term variations that are not addressed by the dynamic pressure simulation of Washimi, et al, but the plasma instability simulations of Pogorelov, et al. [5] and Borovikov, et al. [6] provide plausible hypotheses.

Figure 6 is reproduced from panel (d) of Figure 7 of [5] and also corresponds to the bottom left panel of Figure 2 of [6]. The figure shows these researchers' estimate of the field magnitude in a 2-D slice of the heliosphere plasma environment containing the Voyager trajectories (the upper diagonal line is V1 and the lower is V2) around the time of V1's crossing in 2012, with colors from blue to red representing field magnitudes from 0 to 0.7 nT. The figure indicates that the instabilities used by these authors in their simulations predict a more or less fractal structure for the HP, with wrinkles and lumps and dips on a wide range of scales.

However, since the tick marks on the horizontal and vertical axes of this figure represent increments of 10 AU, the depicted HP fluctuations are too large to be direct predictions of variations on scales that would explain the Voyager observations. They also would move too slowly, since, for example, in other panels in the figures in [5] and [6] the lobe extending in front of the V1 trajectory is little different two years later in the simulation, and in text around Figure 2 of [6] the authors expect the deepest penetrations of LISM into the heliosheath to occur on time scales of 50 - 180 years.

We suggest that if it were possible to run comparable simulations with volume elements and time steps smaller than those that were necessary to complete these simulations with the available computational resources (which were provided by some of the largest supercomputer facilities in the world at the time) the fractal structure would be observed to continue down to the smallest simulatable scales, as is the case for many other instabilities in various types of fluids, so that the Voyager observations would be understood as resulting from smaller and faster fluctuations in the HP caused by the same instabilities as were studied by these researchers. Since neither V1 nor V2 had a trajectory close to the nose, where the Rayleigh-Taylor instability is expected to predominate [5], they probably both crossed the HP in flank regions dominated by the Kelvin-Helmholtz instability [5].

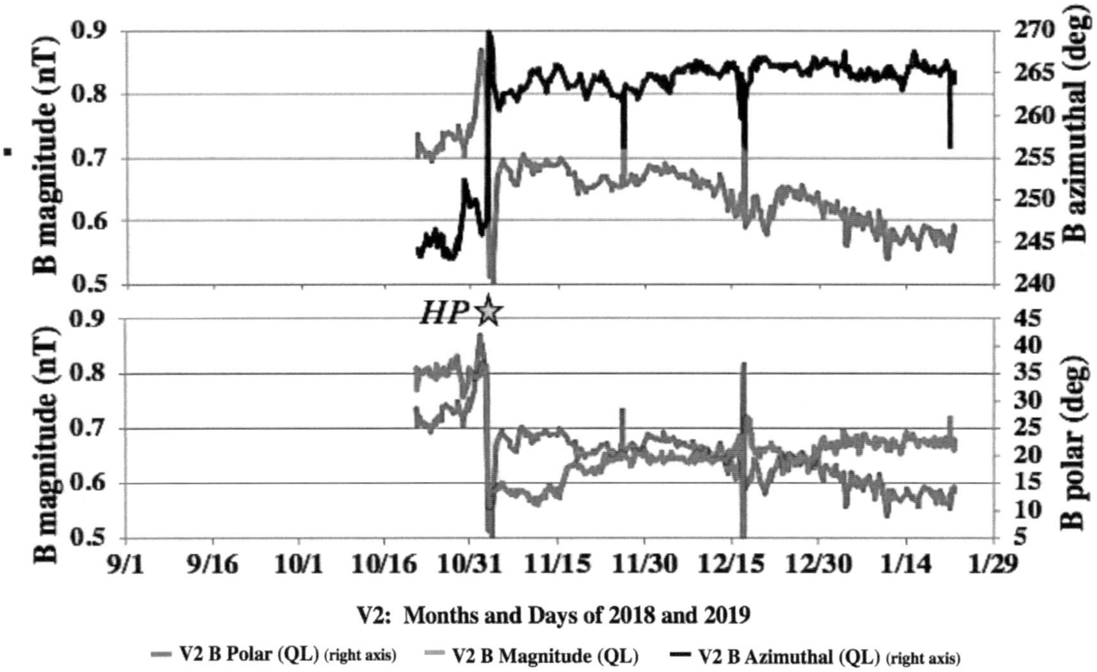

Figure 4. V2 HP-epoch magnetic public quicklook (QL) data that may be revised after calibration.

Combining these considerations with the results of [4] implies that the fluctuations around the trajectory of V1 were significantly larger than those around the trajectory of V2, since even though the HP probably was more nearly stationary overall near V1 in 2012, it fluctuated enough for V1 to observe multiple crossings, but in late 2018 and early 2019, even though the HP was probably moving outward at a substantial fraction of V2's speed, the temporary increases in the SSCR count rate at V2 in Figure 2(b) near December 1 and after December 16 suggest that for periods of several days after the crossing the HP moved closer to the spacecraft but never caught up with it. This contrast between the crossings may merely have resulted from fortuitous differences in the fractal fluctuations, or it may be connected with the crossings' being separated by about 120 AU and half a solar cycle.

4. Comparison with Simulations: Plasma Flow Effects on Cosmic Rays

The zigzag SSCR history in Figure 2(b) is taken as evidence for irregular HP motion because both 2(a) and 2(b) show that the SSCR count rates decline steeply beyond the HP and become undetectable only a few tenths of an AU away, thus raising the question of why the decline occurs. No solid bodies are present to absorb them, so in a static situation SSCR particles might be expected still to be present at greater distances from the HP, even if they had spent months or years diffusing that far. Of course, the situation is not static, as depicted in Figure 7.

Figure 7 is reproduced from Figure 2 of Opher, et al. [7], and depicts a meridional slice of results from another huge simulation (done on a system with about 2000 CPUs), depicting plasma flows inside and outside the HP, with colors representing the field magnitudes. Figure 7 extends 200 AU latitudinally and 140 AU radially around the nose of the heliosphere and expands part of Figure 1 of [7], in which the axes give these dimensions.

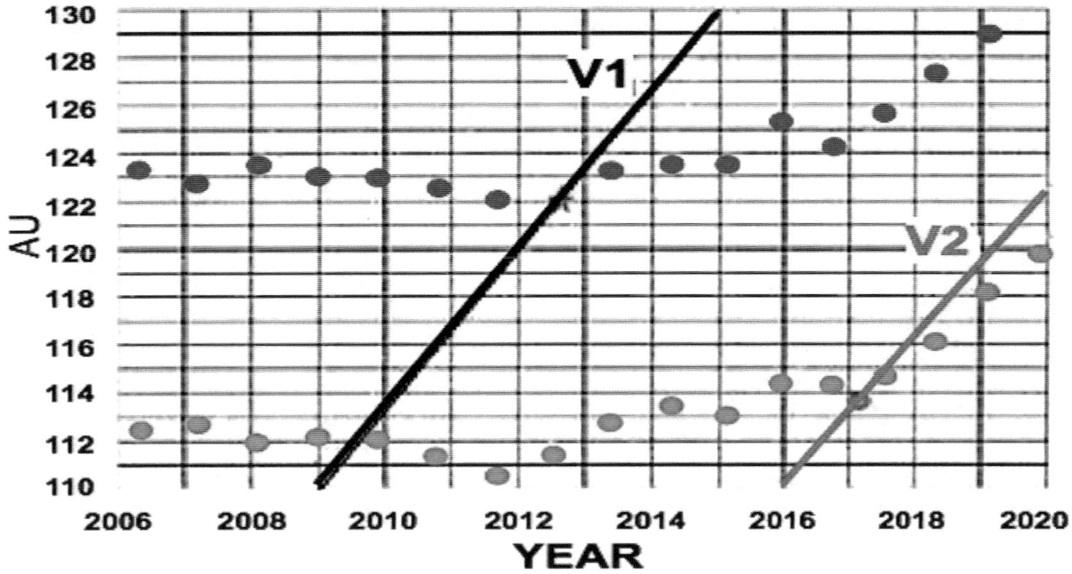

Figure 5. V1 and V2 spacecraft trajectories and estimated HP locations (Washimi, et al. [4])

Figure 7 shows motions only in the radial-normal plane, but the divergence of the flows from the stagnation point also produces tangential components of motion. In the website in [8] the PLS experiment group has posted daily average V2 plasma observations from within the heliosphere up to a few days before the crossing in early November, 2018, so Figure 8 shows strong deflections of the flow from the radial in both azimuthal and polar angle that are discussed in much more detail by Richardson, et al. in [9] and many other publications. The main Faraday cup cluster of the PLS is not favorably located for observing the flows beyond the HP, so plasma data from this region are not available yet, but in [10] Richardson described how the side-facing Faraday cup D of the instrument might be able to observe the LISM plasma, and if so then this will observationally confirm the expected high degree of deflection of the LISM plasma near the HP.

Combining the LISM field magnitudes in Figures 2 - 4 with the V1 PWS plasma density estimates and with estimated pickup ion temperatures implies β values around 1 beyond the HP, quantifying the expectation of a high β in [10]. Specifically, for $|B| = 0.4 - 0.7$ nT (= 4 - 7 μG, thus including field magnitudes higher than considered in [5] and [11]), as in Figures 2 - 4, the magnetic field pressure is $0.6 - 1.9 \times 10^{-13}$ pascal. Correspondingly, for LISM plasma densities of 0.08 - 0.12 cm^{-3} from V1 PWS observations [12] and a mean plasma thermal energy of 3 eV, which includes pickup ions formed by the charge-exchange collisions described below in Section 5, the plasma thermal pressure would be $0.4 - 0.6 \times 10^{-13}$ pascals. Comparing these ranges gives a β range of 0.25 - 1.0.

Thus, the magnetic field is sufficiently “frozen” into the plasma that the flows beyond the HP depicted in Figure 7 and perhaps observed by the PLS can be expected to transport the field with them. Then over times that are short compared to the typical time scale for cross-field diffusion the cosmic rays are carried along with the field, according to the conservation of the first adiabatic invariant. This is analogous to our description in [15] of how shocks locally stretch the LISM magnetic field so that conservation of the first adiabatic invariant causes the V-shaped GCR count rate profiles seen at V1 over weeks or months around the 3 kHz plasma wave events caused by the shocks [16].

All of this implies that as SSCRs emerge through the HP they are rapidly swept away by the LISM plasma as it flows around the HP, as shown in Figure 7, thus explaining the rapid decline in SSCR count rate with distance from the HP. Moreover, the few that have been able to diffuse to larger distances, such as those in Figure 2(a) that were observed early in September, 2012, more than a week after the August 25 crossing, must be understood as having emerged close to the nose so that they have had time to diffuse to the distances where they were observed by the time they reached the latitude and longitude of V1. Corresponding reasoning applies to the particles observed at the low but nonzero count rates at V2 in December, 2018. (Thus, combining these considerations with those in Section 3 implies that a simulation that accounted for the SSCR fluctuations observed by V1 and V2 would combine the flow deflections of [7] with the instabilities of [5,6]. This may not yet be feasible.)

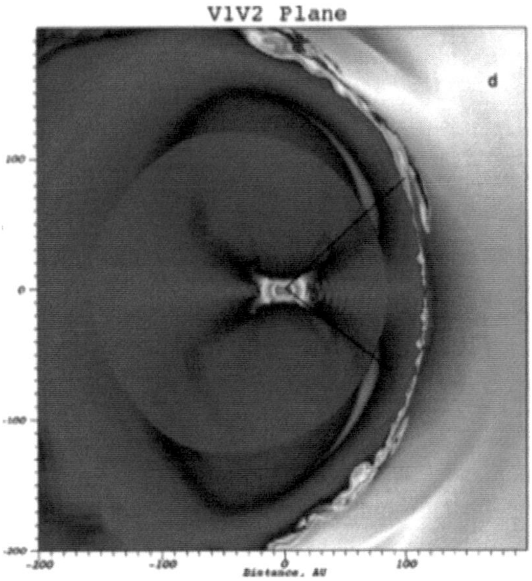

Figure 6. Estimated HP irregularities from instabilities, 2012 [5].

Figure 7. Schematic depiction of LISM and HS plasma flows near the HP [7].

The same reasoning implies that the flows within the heliosheath depicted on the right side of Figure 7 and recorded in Figure 8 should have a sweeping effect on GCRs crossing the HP from outside. This would explain at least part of the GCR count rate declines in Figure 2 with distance from the HP, (observed in reverse as the spacecraft approached the HP), mirroring the declines of the SSCR count rates beyond the HP. Figure 2 shows that the dependences were also far weaker than for the SSCRs, due at least in part to the much higher energies and consequently larger gyroradii of the GCRs, which at V1 were farther enlarged by the lower magnetic field strength in the heliosheath. On the other hand, despite the weaker dependence the higher energies of the GCRs may have contributed to slowing the plasma as it approached the HP, in a way possibly analogous to the deceleration when V2 approached the TS discussed by Florinski, et al. [17], since the fit data from [8] show that from early May to late October, 2018, the plasma speed declined irregularly from around 130 km/s to around 95 km/s. The slowing plasma speed also may be due to the piling up of plasma as in a snowplow effect.

5. Interactions in the LISM between Plasma and Neutral Atoms near the HP

The plasma flows depicted in Figure 7 also lead to a possible explanation for anomalous Lyman α observations reported by Katushkina, et al. [18]. These authors explain that, in an effort to observe the ultraviolet environment in the outer solar system, from 2003 until the instrument apparently began to malfunction in 2014 the V1 ultraviolet spectrometer (UVS) was left pointing in the same direction with respect to the spacecraft, and therefore watched a very small region of the sky upwind of the heliosphere as the spacecraft's orientation periodically changed slightly to maintain radio contact with Earth. During these years the observed Lyman α brightness consistently was significantly higher than what could be explained by the authors' model of Lyman α propagation in the solar system, which had been well confirmed in other contexts, and the presence of an excess over a broader area of the upwind sky was shown by observations from the Alice UV spectrograph on the New Horizons spacecraft.

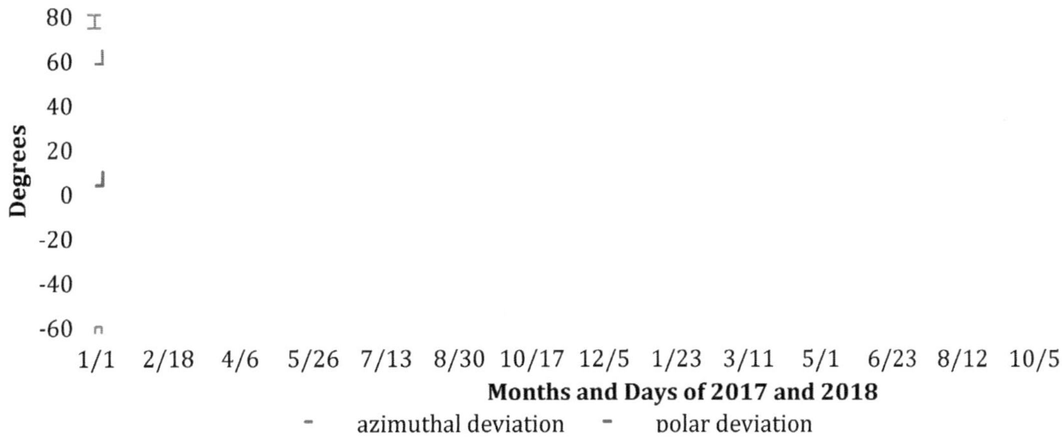

Figure 8. Azimuthal and polar deviations from radial flow from daily average V2 PLS velocities [8].

Figure 9 reproduces Figure 10 of [18] and compares the model to the Alice Lyman α data from the second yearly spacecraft roll at 11.3 AU from the Sun. The continuous line shows the observations, with spikes from stars that happened to be in the field of view, but between the spikes it is clear that from an ecliptic longitude of about 170 degrees all the way to 360 and wrapping around to about longitude 20 there is a consistent excess of about 20 Rayleighs (R) above the small diamonds of the model predictions. Compared to their earlier work, adding a previously unpredicted 15 - 20 R to the model also turned out to improve the agreement with the V1 UVS data from 1993 - 2003, and in a more thorough study of the Alice data Gladstone, et al. [19] also found an excess, estimating it at 40 R.

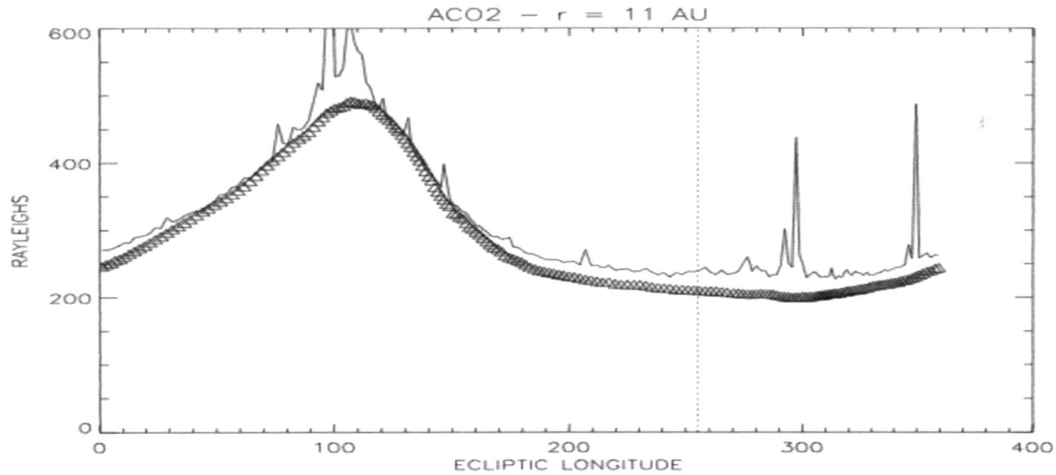

Figure 9. New Horizons Alice ultraviolet spectrograph Lyman α observations during a roll at 11.3 AU from the Sun (continuous line) and model predictions (small diamonds) [17].

Katushkina, et al. described two possible ways by which an excess could be produced. Backscatter of solar Lyman α from a thick, dense hydrogen "wall" in front of the heliosphere was one and unexpectedly high Lyman α emission from distant galaxies in these directions was the other. They probably recognized that these are not plausible mechanisms, since the Lyman α absorption studies described in Section 3 of [11] would have seen such a thick and dense hydrogen "wall", and UV light from distant galaxies is highly unlikely to correlate with the heliosphere's motion through the LISM. However, a natural explanation that they evidently did not consider could account for at least part of the excess. Charge-exchange interactions between the plasma ions and the neutral atoms of the LISM in the region near the HP could produce hydrogen atoms in the n=2 excited state, from which they would fall to the ground state by emitting Lyman α photons, and combining several considerations implies that sufficiently energetic charge-exchange collisions between ions and neutral atoms should be frequent enough in this region to produce significant Lyman α emission.

First, considering Figure 7 from the viewpoint of the rest frame of the LISM implies that in this context the heliosphere is plowing leftward through the LISM at 26 km/s, as determined by reconciling the IBEX and Ulysses ENA measurements [20]. Then the LISM plasma is accelerated by this interaction to flow in the way shown at left in Figure 7, but the LISM neutral atoms are unaffected by all the electric and magnetic field activity, and at a distance of roughly 120 AU the effect of the Sun's gravity is negligible, so in this frame the neutral gas has little if any bulk motion.

Nevertheless, the thermal motion of the neutral atoms is not negligible. In [20] the IBEX and Ulysses measurements were reconciled by estimating the temperature of the neutral gas in the range 7000 - 9500 K. Applying Boltzmann's Constant shows that a thermal energy of 0.75 eV corresponds to a temperature of just under 9000 K, so assuming 0.75 eV for convenience of calculation implies that a proton or a hydrogen atom with this energy is moving at about 12 km/s.

Further consideration of Figure 7 shows that the flow speed of the plasma ions following the streamlines on the left side is 26 sec θ, where θ is the angle between the streamline and the horizontal. Thus, for example, if θ = 45 degrees then sec θ = sqrt(2) or approximately 1.4, for a speed of about 37 km/s. (Figure 7 is reproduced in its correct proportions, showing many streamlines on the left side at angles around 45 degrees.) The energy of a Lyman α photon is 10.2 eV, corresponding to a proton or hydrogen atom speed of about 44.3 km/s, so a head-on collision between a proton going 37 km/s and a neutral atom going 12 km/s would give a total collision speed of 49 km/s, or more than 10.2 eV for a charge-exchange collision to convert the proton into an n=2 excited atom. Of course not all collisions would be head-on, but the thermal distribution includes energies above the mean of 0.75 eV, so many slightly oblique collisions would be sufficiently energetic.

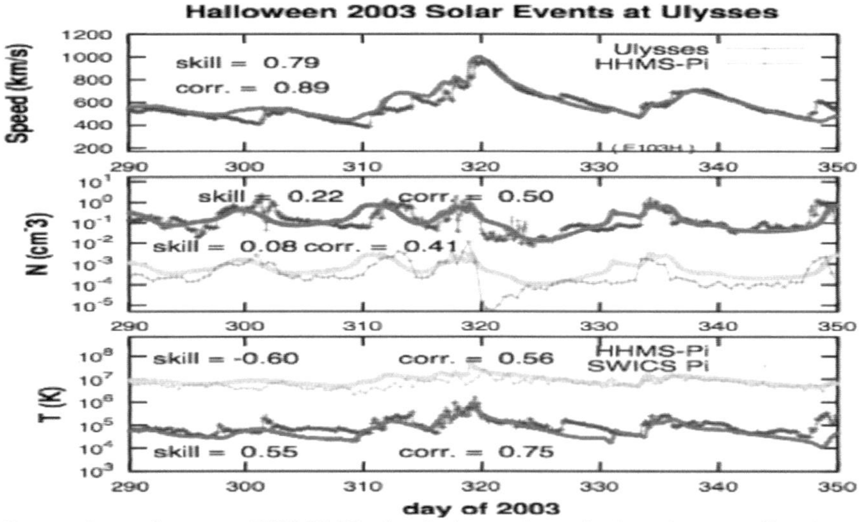

Figure 10. Comparisons between HHMS-PI simulation values (red and green lines) and Ulysses SWOOPS solar wind observations (blue lines) and SWICS PI observations (faint gray lines) [14].

Continuing this reasoning suggests that some collisions would be energetic enough to produce atoms excited to higher energy states that would then emit higher spectral lines in the Lyman series. However, Bransden and McDowell [21] report that significant numbers of atoms excited to the n=3 state are not produced until the collision energy is around 1 MeV, so this does not happen near the HP.

It is important to observe that in the aftermath of a charge-exchange collision the ion newly formed from a neutral atom of the interstellar gas would be picked up by the LISM magnetic field as it is pushed along by the advance of the heliosphere at 26 km/s with respect to the rest frame of the LISM neutral gas. Thus, it would begin gyrating around the field at a speed of 26 km/s, plus or minus the perpendicular component of whatever thermal motion it had had at the time of the collision. Thus, as noted above in Section 4, many of these pickup ions would be expected to have gyration speeds of 26 to >30 km/s, thus contributing disproportionately to the plasma thermal pressure just beyond the HP.

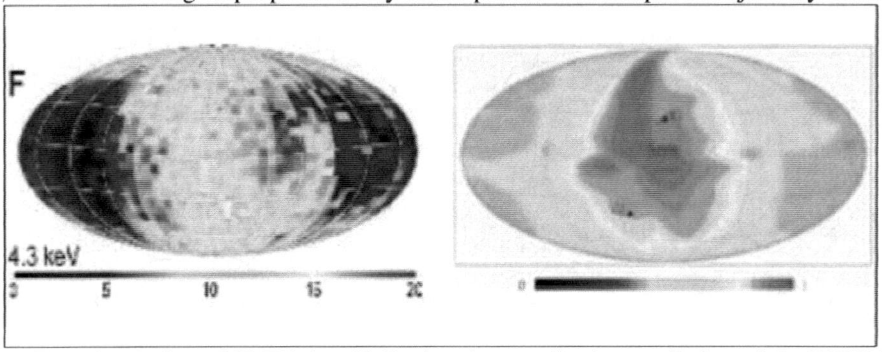

Figure 11. HAFSS simulation of IBEX data [24], showing qualitative prediction of the IBEX Ribbon.

In [22] by identifying corresponding events at different spacecraft and analyzing the delays between them we calculated that the shocks observed in the 3 kHz channel by the V1 plasma wave system were probably propagating through the LISM at a speed of about 69 km/s with respect to the rest frame of the LISM. Furthermore, the shock propagation speeds of around 15 AU/yr through the LISM depicted in Figures 5 and 6 of [23] correspond to speeds of around 71 km/s, agreeing remarkably well with the estimate from the delays (an observation that was omitted from the reference in [22] commending the realism of the modeling in [23]). As charge-exchange collisions between ions and neutral atoms at around 70 km/s evidently are well above the threshold for producing excited atoms, this raises the possibility, which may not have been recognized before, that shock propagation in the LISM may be associated with Lyman α emissions, and that it may be possible to observe them under sufficiently favorable conditions, in addition to the emissions from the LISM near the HP.

6. Modeling

Although we noted in [22] that the modeling in [23] was more realistic and detailed than many other studies, we believe that detailed understanding of longitudinally and latitudinally varying flows and shocks far from the Sun requires modeling from the comprehensive plasma outflow near the Sun, as provided by the Wang-Sheeley-Arge (WSA) maps of a source surface at 2.5 solar radii (below all Parker probe perihelia) instead of spacecraft data from points $\geq$ 1 AU. Our Hybrid Heliospheric Modeling System with Pickup Ions (HHMS-PI) is an MHD model of solar wind flow and shock propagation, and the Hakamada-Akasofu-Fry Source Surface model (HAFSS) is a simpler kinematic model.

Exemplifying our studies, Figure 10 reproduces Figure 3b of [14] (from [24]), showing the generally good agreement between a time dependent HHMS-PI simulation and Ulysses data from a 150-day period in 2003 that includes merged disturbances from the great flares that erupted around Halloween of 2003. The simulation did miss the very small SWICS PUP densities near Day 320, which may have slowed the shocks. Likewise, Figure 11 depicts a HAFSS simulation of IBEX ENA data in the 4.3 keV channel from [25], showing that HAFSS qualitatively predicted the existence of the IBEX Ribbon, although it overestimated the ENA count rates. We hope to improve on these results.

7. Concluding Remarks

Now that V2 has at last crossed the HP, it has answered some questions, but raised many others. For example, if we simply assume that the TS and HP remained where they were respectively observed by V1 and V2 then the delay calculations of Intriligator, et al. in [22] and [14] imply that a shock from the major flares in mid-September 2017 is likely to reach V2 late in August, 2019, and V1 between May and December, 2023. However, assuming constant HS thicknesses along the respective trajectories of V1 and V2 but that the TS and HP locations changed as in Figure 4 merely shifts the arrival at V2 to early September, 2019, but the time for V1 ranges from November of 2021 to April, 2022. (This would be consistent with how the shock predicted for early 2018 in [22] actually arrived at V1 in August 2017). All of these possible arrival times are within the expected mission lifetime until 2025, so we hope to see whether each spacecraft observes a shock from the September 2017 flares.

In this context, the discussions of pressure waves in the HS by Richardson et al. in [8] and [9] evidently provide insight into how slowly the thickness of the HS may change. It also appears worth noting that the Cassini spacecraft's INCA team reported in [26] that ENA observations indicated that the HS was about 30 AU thick along the V1 trajectory and 40 - 70 AU thick along the V2 trajectory, so that the actual HS distance traversed by each spacecraft was about 90% of the respective INCA estimate. More understanding of the heliosheath may come from further study of the Voyagers' data from this region, including whether Opher and others are right about the heliosphere being spherical.

If current hopes for the Voyagers to continue in operation and to be tracked until 2025 or longer are fulfilled then at least 6 or 7 years of informative observations of interstellar space lie ahead as the Voyagers move farther from the Sun and farther apart from each other.

8. References

[1] Stone E C, et al., 1977, *Space Sci. Rev.*, **21**, 355
[2] CRS homepage: https://voyager.gsfc.nasa.gov; rates: https://voyager.gsfc.nasa.gov/rates.html
[3] Burlaga L F et al., 2009, *Astrophys. J. Lett.*, **691,** L82
[4] Washimi H, et al., 2017, *Astrophys. J. Lett.*, **846,** L9
[5] Pogorelov N, et al., 2017, *Astrophys. J.*, **845**, 9
[6] Borovikov S, et al., 2014, *Astrophys. J. Lett.*, **783**, L16
[7] Opher M, et al, 2011, *Astrophys. J.*, **734**, 71
[8] Richardson J D, et al., 2017, *Astrophys. J.*, **834,** 190;
http://web.mit.edu/afs/athena/org/s/space/www/voyager/voyager_data/voyager_data.html
[9] Richardson J D et al, 2018, *17th AIAC, J. of Phys. Conf. Ser.,* **1100**, 012019
[10] Richardson J D, 2015, *14th AIAC, J. of Phys. Conf. Ser.,* **642**, 012022
[11] Zank G P, et al., 2013, *Astrophys. J.*, **763,** 20
[12] Kurth W S, et al., 2019, Voyager PWS Recent Observations,
http://www-pw.physics.uiowa.edu/~wsk/vgr/v1mos02-19.gif
[13] Zank G P, et al., 2010, *Astrophys. J.*, **708,** 1092
[14] Intriligator D S, et al., 2016, *15th AIAC, "The Science of Ed Stone Celebrating his 80th Birthday:" Journal of Physics Conference Series,* **761**, 012013
[15] Intriligator D S, et al., 2018, *17th AIAC, J. of Phys. Conf. Ser.,* **1100,** 012013
[16] Gurnett D A, et al., 2015, *Astrophys. J.*, **809**, 121
[17] Florinski, V et al., 2009, *Geophys. Res. Lett.*, **36,** L12101
[18] Katushkina O A, et al., 2017, *J. Geophys. Res.*, **122**, 10,921
[19] Gladstone, G R et al., 2018, *Geophys. Res. Lett.*, **45,** 8022
[20] McComas D J, et al., 2016, *Astrophys. J.*, **801**, 28
[21] Bransden B H, and McDowell M R C, 1992, *Charge Exchange and the Theory of Ion-Atom Collisions*, Oxford University Press
[22] Intriligator D S, et al., 2017, *16th AIAC, J. of Phys. Conf. Ser.,* **900,** 105
[23] Washimi H, et al., 2011, *Mon. Not. R. Astron. Soc.,* **416**, 1475
[24] Intriligator D S, et al., 2012, *J. Geophys. Res.*, **117**, A06104
[25] McComas D J, et al., 2011, *Geophys. Res. Lett.*, **38**, L18101
[26] Krimigis S, et al., 2016, AGU Fall Mtg., **SH23A-04**

Acknowledgments

We gratefully thank the Voyager projects for the vast information we have received, including CRS and mag data [2], PLS data [8], and PWS data [12]. We also are grateful for Ulysses SWOOPS from McComas and SWICS from the Gloecklers. We acknowledge the support of Carmel Research Center, Inc. We acknowledge Prof. Gary Zank and Adele Corona for these excellent conferences and papers.

Nanodust in the Heliosphere

Wing-Huen Ip [a], **Ian-Lin Lai** [b] **and Fang Shen** [c]

[a] Institute of Astronomy, National Central University, Taiwan
[b] Physikalisches Institut, University of Bern, Switzerland
[c] National Space Science Center, Chinease Academy of Sciences, China

wingipastro3@yahoo.com

Abstract. The NASA Parker Solar Probe and the ESA Solar Orbiter will explore the source region of the solar wind within 20 solar radii. Their unprecedented in-situ measurements are also expected to shed light on the nature of the F-corona and the existence of a halo of nanodust. Such a dust complex might play an important role in the generation of high-speed nanodust grains and the inner-source pickup ions. A brief summary of previous works on this topic is given here to be followed by a sketch on a plan to integrate MHD simulation of solar wind flow dynamics, orbital motion of charged nanodust and the production of energetic neutral atoms (ENAs) in the interplanetary space.

1. Introduction

Recent analysis of the HI-1 imaging data from the STEREO A solar telescope indicated the existence of a circumsolar dust belt in the vicinity of Mercury's orbit [1]. The source of such a dust belt could be associated with some small bodies in co-orbital motion with Mercury. The same author team [2] examined the white-light brightness distribution of the F corona inside of this newly discovered dust belt. Clearly, the in-situ measurements of the Parker Solar Probe would provide the first-hand information not just on the heating and acceleration of the solar wind, but also the near-solar dust cloud complex never directly probed by spacecraft. In retrospect, we should remember that there have been a number of intriguing space measurements providing a glimpse of the physical processes involving dust-plasma interaction in the source region of the solar wind. The first one has to do with the detection of the electromagnetic signals generated by high-speed impacts of nano-dust particles on the antennas of the plasma wave and radio science experiment on the STEREO spacecraft [3, 4]. A convenient explanation of the source mechanism of these tiny solid particles of 1-10 nm size has to do with the collisional erosion or breakup of the dust grains in the F corona [5,6,7]. However, an additional source of potential importance is comets – especially the so-called sun-grazing comets - with perihelion distances very close to the sun [8,9].

2. Nanodust dynamics

Because of photoemission, interplanetary dust particles would usually be charged to an electrostatic potential of a few Volts. Depending on the size and mass of individual dust grains, the corresponding charge-to-mass ratio in unit of electron charge (e) to the proton mass (m) can have a wide range of possible values, namely, from $e/m \sim 10^{-7}$ to $e/m \sim 10^{-5}$. As a result, the Lorentz force coupled with the solar gravitational force could generate different types of particle trajectories according to the e/m values and the injection locations of the nanodust. Computation of the particle trajectories is not as simple as it seems. This is because the details depend on the structures of the solar wind and the interplanetary magnetic fields (IMF) that are quite variable. For most cases of the MHD simulations of

Content from this work may be used under the terms of the Creative Commons Attribution 3.0 licence. Any further distribution of this work must maintain attribution to the author(s) and the title of the work, journal citation and DOI.
Published under licence by IOP Publishing Ltd

Journal of Physics: Conference Series **1332** (2019) 012007 doi:10.1088/1742-6596/1332/1/012007

the space weather effects, the inner boundary of the simulation box is larger than 20 $R_\odot$, thus missing the main part of the nanodust source region. To explore the dynamics of the charged nanodust, some special cases with the inner boundary set to be just a few solar radii would have to be run. The advantage is that with the initial condition of the solar wind and solar coronal magnetic field mapped to the solar surface, a more realistic picture of the solar wind structure and interplanetary magnetic field configuration would be available to the corresponding trajectory calculations. In the absence of such a more comprehensive description, the alternative is to use a simple kind of steady state Parker wind model with a prescribed, spherically symmetric, solar wind velocity profile and the resultant 3D Parker spiral configuration of the interplanetary magnetic field. For example, the solar wind speed can be assumed to be constant with $V_{sw} = 400$ km s^{-1}, and the magnetic field $B = 4x10^{-4}$ G at r = 1 AU. Even in this case, the dust motion is still quite variable. Figure1 illustrates the orbital structures of the nanodust with e/m between 10^{-4} and 10^{-6} and injection positions (r_i) at three different heliocentric distances from 10 $R_\odot$ to 25 $R_\odot$. The nanodust particles are assumed to be initially in circular Keplerian motion in the ecliptic plane before the switch-on of the Lorentz force. Some characteristic features to be noted.

(a) r_i= 10 $R_\odot$

Nanodust of different e/m ratios would all be deflected to inward trajectories. At the boundary of 5 $R_\odot$, the dust grains reach a radial speed of nearly 200 km s^{-1}. This means these fast-moving nanodust could be very efficient in eroding the F-corona dust cloud. The accelerated particles are generally confined to a flat disk with the vertical distance z < 1 $R_\odot$.

(b) r_i= 15 $R_\odot$

The trajectories are still generally inward except for e/m =10^{-5} in which case a large fraction of the nanodust is temporarily trapped. The dust population is still confined in a flat disk.

(c) r_i = 20 $R_\odot$

Depending on the magnetic field polarity, the highly charged nanodust with e/m~10^{-4} could be either deflected inward or outward. For the outward-moving component, their vertical distances could reach z ~ 50 $R_\odot$. The above-mentioned trend is followed by the dust grains with e/m~10^{-5}. Interestingly, the radial and vertical distances could be larger than 100-200 $R_\odot$ thus filling the whole heliosphere. For those dust particles with e/m ~10^{-6}, they would be trapped in oscillatory motion in an ellipsoidal-shaped zone.

(d) r_i = 25 $R_\odot$

At the outer edge of the nanodust source zone, the dust particles with large e/m values would be injected either inward or outward depending on the polarity of the IMF. For those reaching the orbital distance of the Earth at r ~ 200 $R_\odot$, the speed could be as high as 400 km s^{-1}. However, they also have rather large vertical distance (z~70 $R_\odot$). They hence do not necessarily account for the dust impact effect detected by the STEREO experiment (Bougerete et al., 2008; Meyer-Vernet et al., 2009). For e/m~10^{-5}, all the dust grains would be ejected outward and upward to r > 200 $R_\odot$ and z >100 $R_\odot$. The gravitational force becomes more significant at e/m~10^{-6} so that the charged nanodust would be trapped in orbital motion forming a kind of "knitting ball" surrounding the sun.

These numerical results are qualitative similar to the results of Mann, Murad and Czechowski [10] who also showed how charged nanodust in certain e/m range (~$5x10^{-6}$) could be trapped in different orbital configurations according to the their original injection positions near the sun. It is therefore expected that the Parker Solar Probe and the Solar Orbiter of ESA would enter this nanodust zone as they move inside the orbital region with r < 0.2 $R_\odot$.

Journal of Physics: Conference Series **1332** (2019) 012007 doi:10.1088/1742-6596/1332/1/012007

The fluxes of the high-speed nanodust are quite variable according to Meyet-Vernet et al. [4]. There could be a number of physical reasons. The first one, more exotic in nature, has to do with the occasional passages of sun-grazing comets with perihelion distances of a few solar radii [8, 11]. The charged cometary nanodust originally in near-parabolic orbits could be accelerated to very high speed in escaping trajectories. On the other hand, changes in the solar wind structures or solar activity could also be an important factor. Czechowski and Kleimann [6] using a MHD model of a coronal mass ejection from the solar surface to a radial distance of 0.14 $R_{\odot}$ to compute the trajectories of charged nanodust. These authors found that a speed as high as 1000 km s^{-1} could be achieved by some dust particles.

3. Discussion

In the above, a very brief summary is given to the topic of the existence and dynamics of nanodust in the near-solar region overlapping the F-corona. In-situ measurements and remote-sensing observations of the Parker Solar Probe and the Solar Orbiter related to dust impacts and dust cloud disribution might provide first-hand information on the near-sun dust environment. It can be seen that a lot remains to be done in this relatively unexplored area. Our plan is therefore to systematically assemble a set of numerical data from 3D MHD model calculations of (1) solar wind flow during quite-sun condition (cf. [12]), (2) the time evolution of the corotating interaction region of the high-speed solar wind and the low-speed solar wind (cf. [13]), and (3) expansion of the coronal mass ejection events from the solar corona to the interplanetary space (cf. [14]). These data sets providing detailed information on the solar wind plasma velocity, temperature, number density and interplanetary magnetic field in different regions of the heliosphere would allow us to simulate the orbital motion of nanodust with different e/m values.

An interesting potential byproduct of the planned study is the examination of the process of solar wind charge exchange with the interstellar neutral gas. By the same token, the dust-related source mechanism of the so-called inner source pickup ions [15-18] might be investigated in more detail once the orbital distribution and dynamics of the charged nanodust is better understood.

Journal of Physics: Conference Series **1332** (2019) 012007 doi:10.1088/1742-6596/1332/1/012007

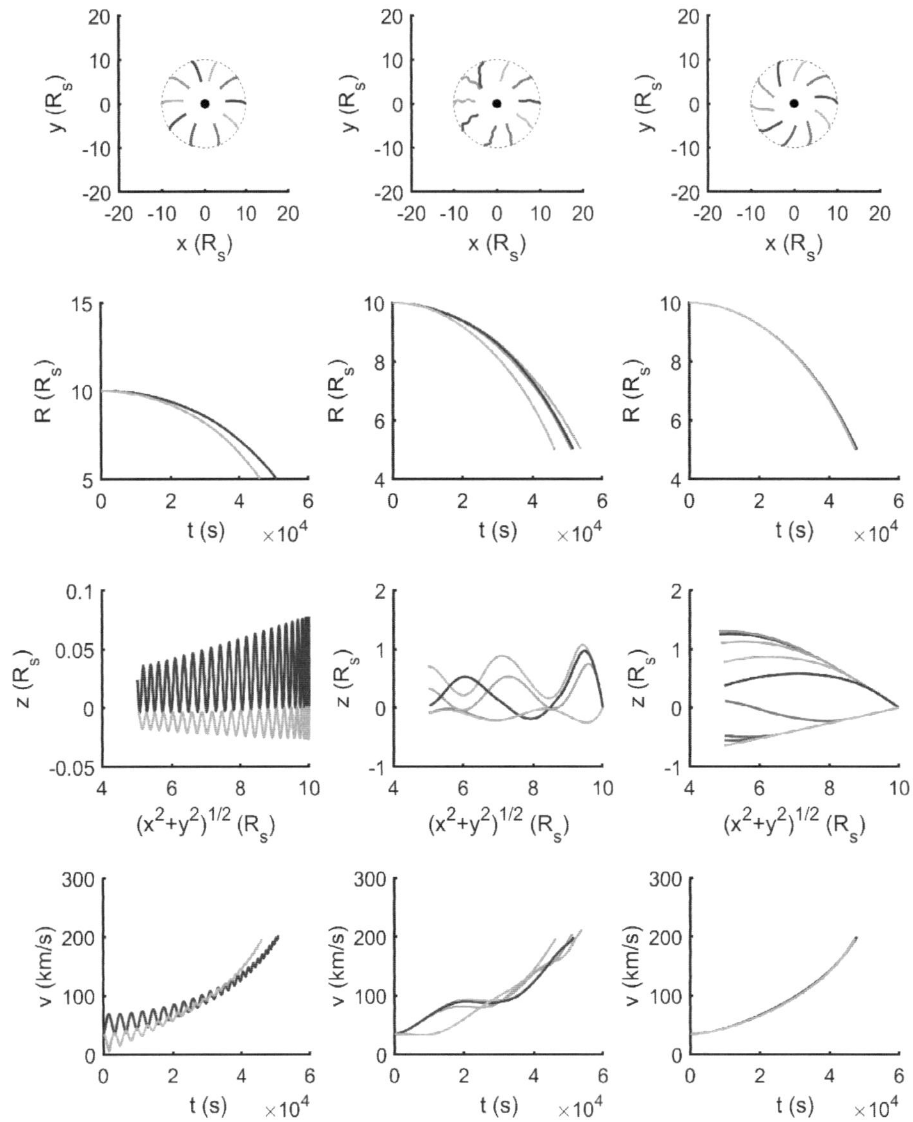

Figure 1. Orbital dynamics of charged nanodust with r_i = 10 $R_{\odot}$ and three different values of e/m (e.g., 10^{-4}, 10^{-5} and 10^{-6}). First row: 2D projections of the particle trajectories on the ecliptic plane; second row: the changes of the heliocentric distance; third row: the changes in the vertical distances; fourth row: the time profiles of the particle speeds after ejection.

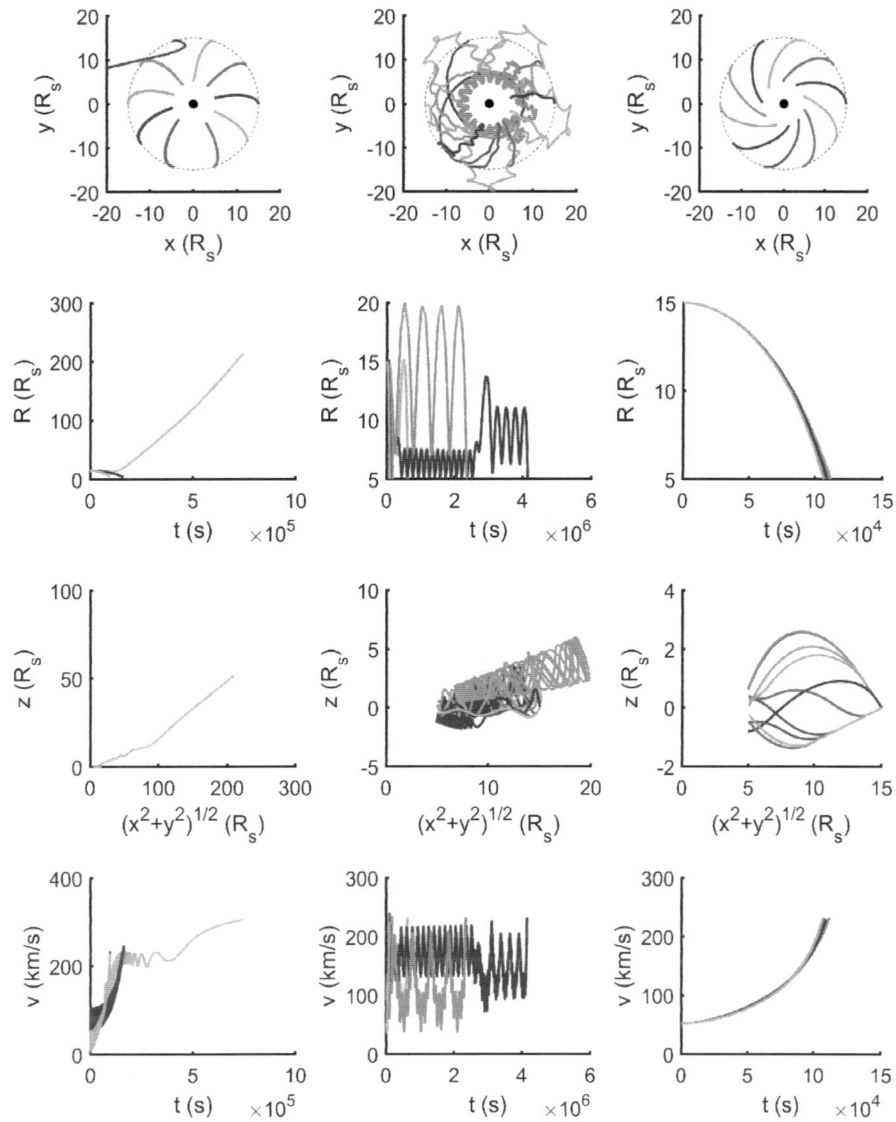

Figure 2. Same as Figure 1 but for $r_i = 15\ R_{\odot}$.

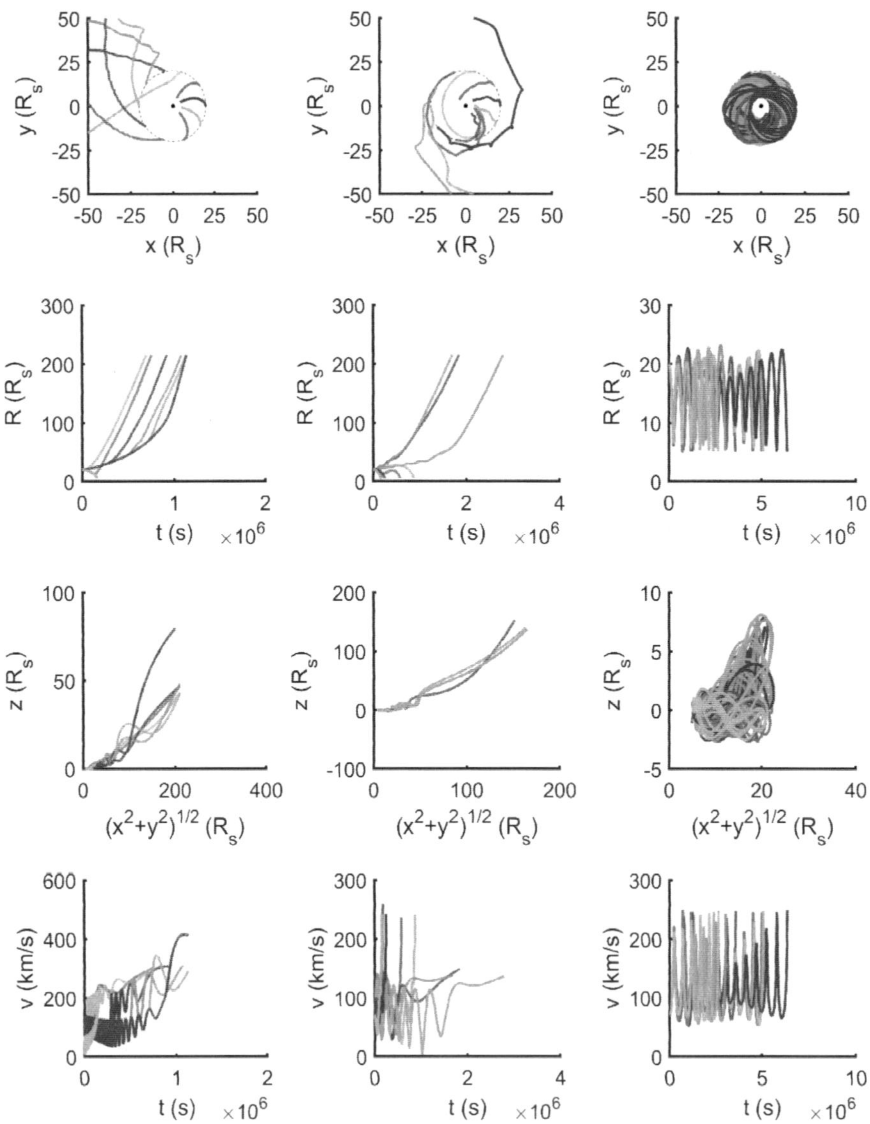

Figure 3. Same as Figure 1 but for $r_i = 20\ R_\odot$.

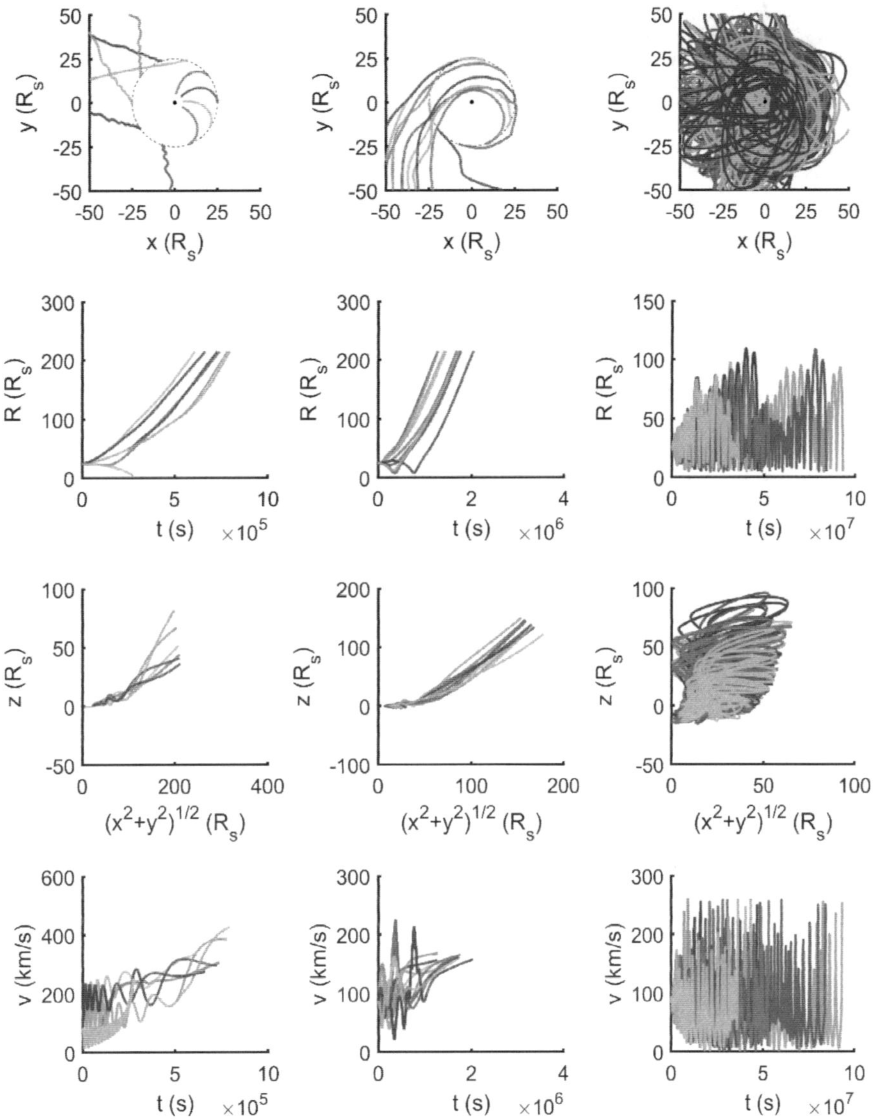

Figure 4. Same as Figure 1 but for $r_i = 25\ R_\odot$.

Acknowledgment.
This work was supported in part by grant No. 107-2119-M-008-012 of MOST, Taiwan.

References

[1] Stenborg, G., Stauffer, J.R., and Howard, R.A., ApJ, 868:74 (2018)
[2] Stenborg, G., Howard, R.A. and Styauffer, J.R., ApJ, 862:168 (2018)
[3] Bougeret, J.-L., et al., Space Sci. Rev., 136, 487 (2008)
[4] Meyer-Vernet, N., Lecacheux, A., Kaiser, M.L., and Gurnett, D.A., Solar Phys. (2009)
[5] Mann, I. and Murad, E., ApJ, 624:L125 (2005)
[6] Czechowski, A. and Mann, I., ApJ, 714:89 (2010)
[7] Quinn, P.R., et al., ApJ, 861:98 (2018)
[8] Ip, W.-H., and Yan, T.-H., AIP Conf.Ser., 1436, 30.(2012)
[9] Mann, I., Phil. Trans. R. Soc. A375:20160254 (2017).
[10] Mann, I., Murad, E., and Czechowski, A., Planet. Space Sci., 55, 1000 (2007)
[11] Czechowski< A. and Mann, I., A&A, 617,A43 (2018)
[12] Shen, F., et al., ApJ, 866:18 (2018)
[13] Wei, W. et al., J. Atm. Solar Terr. Phys., 182, 155 (2019)
[14] Shen, F., et al., Jour. Geophys. Res., 119, 7128 (2014)
[15] Gloeckler, G. and Geiss, J., Space Sci. ev., 86, 27 (1998)
[16] Bochsler, P., Moebius, E., Wimmer-Schweingruber, R.F., Geophys. Res. Lett., 33, L06102 (2006)
[17] Schwadron, N. and McComas, D.J., ApJ, 712, L157 (2010)

Journal of Physics: Conference Series **1332** (2019) 012008 doi:10.1088/1742-6596/1332/1/012008

An empirically-based model of the upstream heliopause and outer heliosheath - Current status

Philip A. Isenberg and Harald Kucharek

Institute for the Study of Earth, Oceans and Space and Department of Physics
University of New Hampshire, Durham NH 03824, USA

phil.isenberg@unh.edu

Abstract. We describe a simplified model of the outer heliosheath upstream of the heliopause in an attempt to extract the heliosheath properties from the IBEX-Lo neutral He observations. The distribution of incident He on the sky shows a primary beam surrounded by an irregular cloud of secondary particles. These secondaries are generated by charge-exchange interactions between the neutral primary beam, carrying the properties of the distant interstellar medium, and the nearby interstellar plasma that is deflected around the obstacle posed by the heliosphere. We construct an analytical model for the plasma flow around an arbitrarily-shaped obstacle and compute the resulting distribution of neutral He at IBEX, to compare with the observed He fluxes. Here, we report on the current status of this modelling effort.

1. Introduction

The IBEX spacecraft was launched in 2008 in an orbit around the Earth to measure neutral atoms from the surrounding interplanetary and interstellar medium [1]. Neutral particles, streaming unaffected by electromagnetic fields, can be collected to form remote images of the particle sources and provide information on the plasma processes at large distances from the observer. This capability is illustrated by Figure 1, which shows a sky-map of neutral Helium originating in the interstellar medium (ISM) as detected by the IBEX-Lo instrument [2]. The data shown here compiles the total sputter-corrected fluxes of He in the lowest three energy bins of IBEX-Lo (11 – 77 eV), as observed during "super good times" in the Spring ram-direction periods of 2009 - 2011 (see e.g. [3-5] for explanation of these terms). The data are shown in a Mercator projection that preserves the size of the $6^\circ \times 6^\circ$ pixels in J2000 ecliptic coordinates.

In this paper, we focus on the He component of the interstellar neutral gas since it is the second-most abundant species in the ISM, and since the motion of He through the inner heliosheath and supersonic solar wind has only a simple effect on the intensities of these particles. The sky-map of Figure 1 shows the superposition of a primary beam of ISM particles and an irregular cloud of secondary He resulting from the charge-exchange interaction, $He^+ + He \rightarrow He + He^+$ [6], in the outer heliosheath beyond the heliopause. The distribution of this flux on the sky carries information on the global structure of the heliosphere. We are interested in extracting this information from the observations, and this paper will describe our attempts to construct a simple, empirically-based, model of the upstream heliopause and outer heliosheath consistent with these data.

2. Simple Model of the Upstream Heliosheath

Content from this work may be used under the terms of the Creative Commons Attribution 3.0 licence. Any further distribution of this work must maintain attribution to the author(s) and the title of the work, journal citation and DOI.
Published under licence by IOP Publishing Ltd

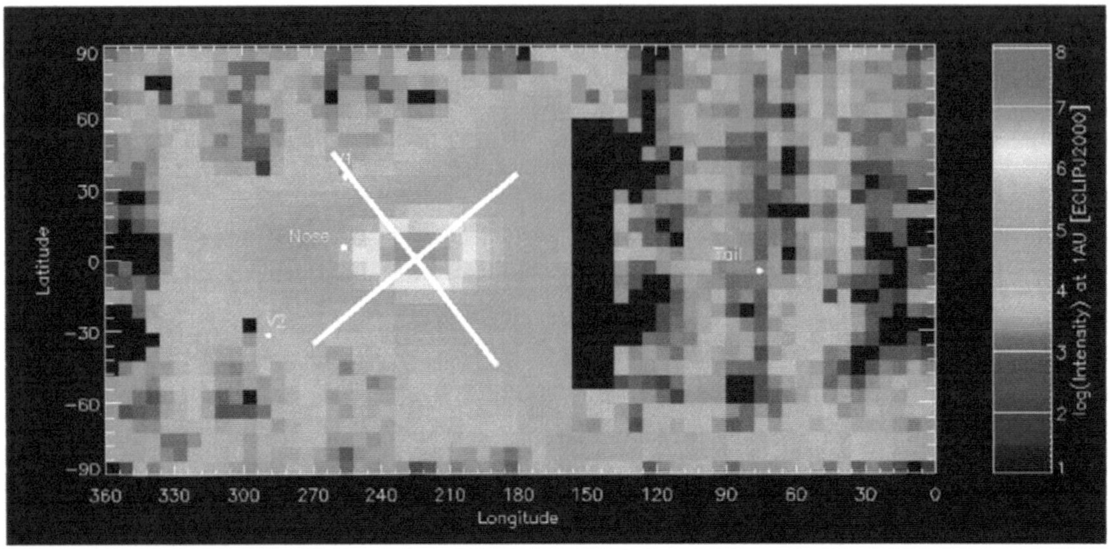

Figure 1. Sky map of neutral He, taken by IBEX-Lo during the super good times of 2009 – 2011, in Mercator projection. Fluxes are the sum of the lowest three energy steps (11 – 77 eV), and have been sputter-corrected. Straight lines in white are the cutlines defined in the text from which we take the data used for comparison with the model.

2.1. *Overview*

The ISM at large distance from the Sun is a partially ionized plasma, flowing uniformly at ~ 26 km/s, with a temperature of ~ 8000°K [4, 7]. We assume that, far from the Sun, both the neutral and ionized components of this gas are in equilibrium. For the He component, this means that although charge-exchange is taking place as elsewhere, the relative densities and bulk properties of each component are not changing. However, as a parcel of ISM moves closer to the Sun, the presence of the heliosphere begins to affect the magnetized plasma, which deflects to stream around this obstacle in the flow. In this essentially collisionless system, the neutral particles continue to travel along their original paths. A non-zero relative bulk speed develops between the He^+ plasma and the neutral He gas, and now the charge-exchange interactions begin to affect the properties of these components. When an atom in the He primary beam charge-exchanges with a deflected He^+ ion, the new neutral particle possesses the flow properties of the deflected plasma. Some of these secondary He atoms are detected at IBEX, and the distribution of these fluxes is connected to the flow characteristics of the deflected plasma and the global shape of the heliopause.

We seek a simple model of this interaction in order to deduce this heliospheric shape. To include all the relevant physics in a heliospheric model, one requires a global simulation that computes the MHD flow field of the magnetized interstellar plasma, coupled with kinetic computations describing the neutral particle flow, as well as an accurate representation of the internal pressures due to the shocked solar wind and inner heliosheath whose pressure balance with the external plasma and field defines the heliopause [8-10]. Such simulations take weeks to run for each choice of physical parameters, and so cannot be easily adjusted to fit observations. The simple model presented here approximates the physical situation in several ways, while retaining the essential properties that determine the flux of He at IBEX. This model will not be completely rigorous or self-consistent, but it will allow consideration of a wide variety of heliospheric shapes for comparison to the observations.

For this comparison, we will initially consider a subset of the data, chosen to exhibit the largest variations and asymmetries. The two cutlines shown in Figure 1 are set normal to each other, intersecting at the flux maximum. The lines are oriented such that one line points in the direction of the maximum secondary flux. This line roughly corresponds to the offset of the "warm breeze"

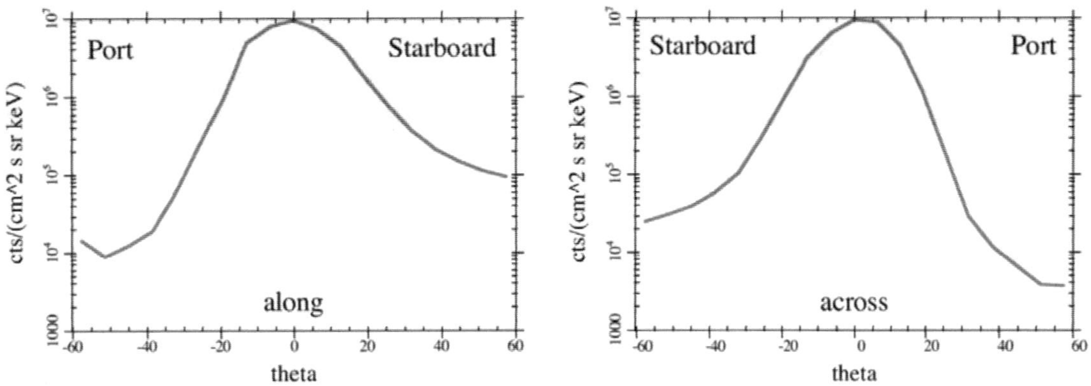

Figure 2. Interpolated fluxes on the two cutlines through the sky map shown in Figure 1, as functions of the angular position from the maximum where the lines cross. (left panel) The "along" line travels from the lower left to the upper right (South-Port to North-Starboard) in the figure. (right panel) The "across" line travels from lower right to upper left (South-Starboard to North-Port).

characterization of Kubiak et al. [11], so it also lies essentially parallel to the direction of the interstellar magnetic field as defined by the center of the IBEX ribbon [12, 13]. Consequently, we refer to the observations on this line as the "along" fluxes, and call the values on the other line the "across" fluxes. The fluxes along these cutlines are shown in Figure 2. The specific ecliptic coordinates of the cutlines in longitude φ and latitude α are

$$\begin{aligned} &\text{along line:} \quad \varphi = -121.3^\circ - \theta\cos(0.6747), \quad \alpha = 4^\circ + \theta\sin(0.6747) \\ &\text{across line:} \quad \varphi = -121.3^\circ - \theta\cos(0.896), \quad \alpha = 4^\circ + \theta\sin(0.896) \end{aligned} \quad (1)$$

where θ is the angular distance away from the maximum flux position along each line.

2.2. *Upstream Flow Field*

The basic idea behind our simple model is to recognize that the interstellar plasma and magnetic field flowing around the heliopause is responding to the shape of that surface, while also acting to determine that shape. Rather than trying to compute that shape as a result of the complicated internal and external pressures, we simply treat this undetermined shape as an obstacle in the flow field of the ISM. We then ask what flow field would be consistent with a given obstacle shape, compute the charge-exchange fluxes that would be seen at IBEX from that flow field, and attempt to adjust the obstacle shape to give good agreement with the IBEX observations. The construction of this model is still a work in progress, and this paper will describe the current state of our work.

In order to define a flow field without resorting to a global MHD simulation, we approximate the effect of the magnetized interstellar plasma as that of a gas with an effective isotropic pressure equal to the sum of the true thermal pressure and the isotropic magnetic pressure. By allowing the shape of our model obstacle to be non-axisymmetric and distorted with respect to the flow, we are already including some of the effects of the magnetic tension force, which we otherwise neglect in our description of the plasma flow.

Having made this assumption, and recognizing that the flow at large distances is uniform and curl-free, we can then represent the upstream plasma number flux in steady state as the gradient of a potential function, $n\mathbf{v} = \nabla\varphi$. It then follows from the conservation of number flux, $\nabla \bullet (n\mathbf{v}) = 0$, that this potential is a solution of Laplace's equation, $\nabla^2\varphi = 0$. We note that we are here neglecting the possible presence of an upstream bow shock or bow wave [14].

This simplification allows the specification of an essentially infinite variety of obstacle shapes and analytical solutions to the upstream flow fields around them. To demonstrate this, we start with Parker's heliospheric model [15-18] as illustrated in Figure 3. The streamlines shown in this figure are obtained by superimposing the potential function due to a uniform flow from the right and that from a

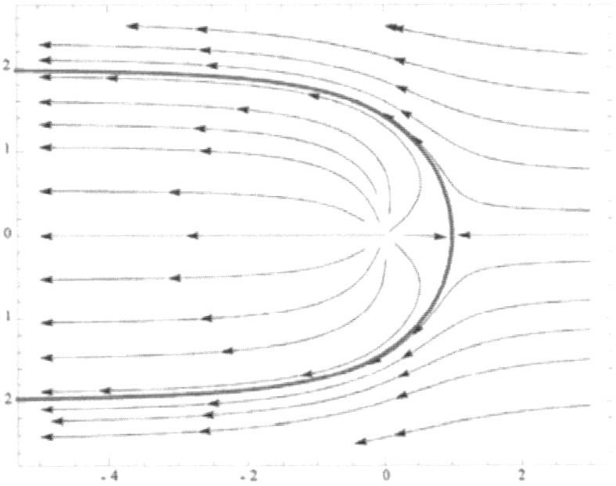

Figure 3. Streamlines for a Rankine half-body, an axisymmetric structure formed by the superposition of a point source and a uniform flow. The red curve separates the flows from the two sources, which can serve as a model heliopause.

point source placed at the origin. The red curve indicates the axisymmetric surface that separates the two flows, which can serve as the representation of a simple heliopause. This shape is well studied in hydrodynamics [19], and is typically called a "Rankine half-body". The superposition of sources is a standard technique for solving linear differential equations, such as Laplace's equation. The analytical character of the flow field is a simple consequence of the fact that when the sources and boundaries of a system can be expressed along single-coordinate surface in spherical coordinates, Laplace's equation is separable in three dimensions and admits analytical solutions.

There are a limited number of other coordinate systems for which Laplace's equation is separable in 3D [20], and we consider two further cases, both derived from the 2D elliptical coordinate systems shown in Figure 4. Elliptical coordinates are constructed from nested ellipses defining one coordinate, $0 \leq \xi \leq \infty$, and the normal set of hyperbolae defining the other coordinate. If we rotate the elliptical coordinate system about the vertical axis, as indicated in Figure 4a, we obtain the prolate spheroidal coordinate system, which can be used to describe a line source at $\xi = 0$. Placing a line source at an angle to a uniform flow gives the streamlines illustrated in Figure 5, where the blue curve separating the two flows gives an example of a non-axisymmetric heliopause. Another example is given in Figure 6, which shows three views of the boundary between a tilted line source and a uniform flow.

Rotating the elliptical coordinates about the horizontal axis, as in Figure 4b, gives oblate spheroidal coordinates and describes the flow field from a disc source at $\xi = 0$. Figure 7 shows three views of a surface separating a tilted disc source from a uniform flow. Since Laplace's equation is linear, the flow fields of any number of point, line, and disc sources of any size, strength and orientation may be superimposed on a uniform flow from infinity to give analytical flow fields representing heliopause surfaces of essentially any plausible shape. We will use this capability to test the distributions of secondary He produced by different outer heliosheath plasma flow fields around a variety of heliopause shapes.

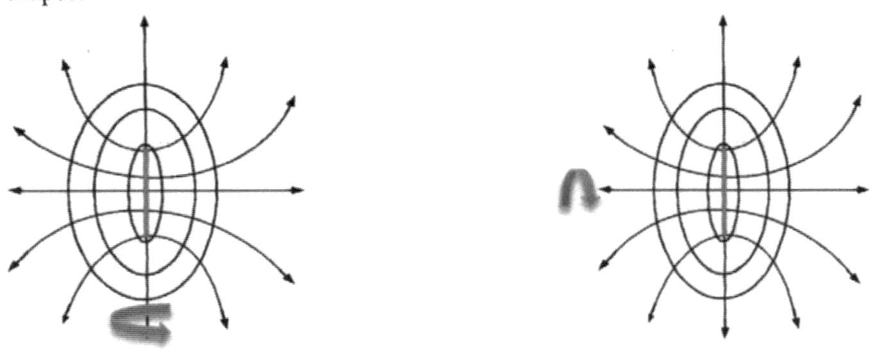

Figure 4. Three-dimensional coordinate systems derived from the 2D elliptical coordinates. (Left panel) Rotation about the vertical axis results in the prolate ellipsoidal coordinate system, containing a line source at the origin. (Right panel) Rotation about the horizontal axis results in the oblate ellipsoidal coordinate system, containing a disc source at the origin. Source objects are shown in red.

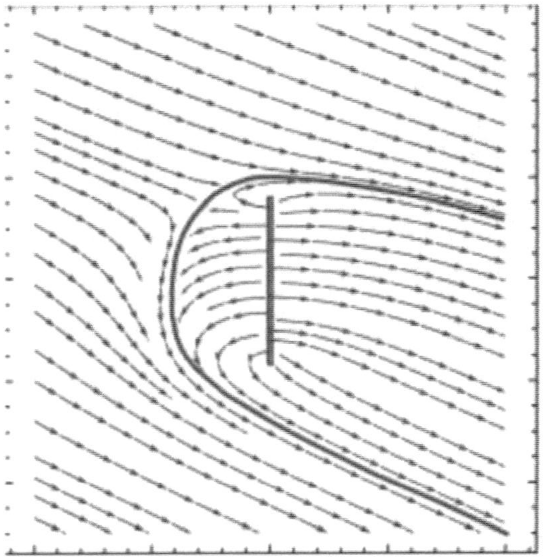

Figure 5. Plane slice through the non-axisymmetric superposition of a line source (red) and a uniform flow. The blue curve locates the separating surface, acting as a model heliopause in this system.

2.3. Plasma Properties

For a given obstacle shape and corresponding flow field, we then describe the plasma properties upstream of the obstacle resulting from a typical undisturbed ISM at infinity. We assume that the partially ionized plasma is composed of protons and He^+ ions, with a neutralizing population of electrons, along with neutral populations of H and He atoms. All the components of this plasma are in equilibrium with each other at large distance from the Sun, and their distributions are taken to be Maxwellian there. The proton number density at infinity is taken to be n_p = 0.04 cm^{-3}, the He^+ ions have a number density of 0.1 n_p, and the bulk speed of all the co-moving particles is V_o = 26 km s^{-1}. The kinetic plasma temperature there is taken to be T_o = 8000 °K, but we will treat the inflowing plasma as having an isotropic pressure with an effective sound speed equal to the MHD fast speed $c_f = \sqrt{c_s^{\,2} + V_A^{\,2}}$, such that kT_o/m = 3/5 c_f^2, where c_s is the true kinetic sound speed, V_A is the Alfvén speed, k is Boltzmann's constant, and m is the mass/particle in the plasma. An undisturbed interstellar magnetic field intensity of 4.4 μG gives an Alfvén speed of 40.6 km s^{-1}, so the fast speed at infinity is taken to be 41.6 km s^{-1}.

A given set of sources in the otherwise uniform flow yields the potential $\varphi(\mathbf{x})$, which provides the number flux as a function of position, $n_p\,\mathbf{v} = \nabla\varphi$. We assume that this interstellar plasma behaves adiabatically as it approaches the heliosphere. Taking the effective sound speed as c_f, the adiabatic equation of state can be manipulated to give $n_p\,c_f^{-3} \equiv N$ = constant. Similarly, Bernoulli's law yields $c_f^2 + v^2/3 \equiv C$ = constant. If we define $W = n_p^{\,2}v^2/3$, these equations can be combined to give the quartic

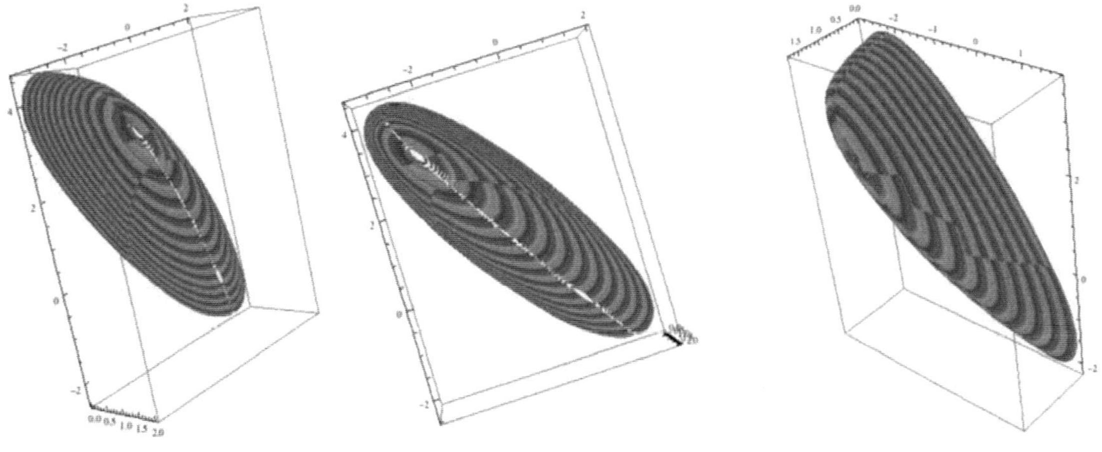

Figure 6. Three views of a model heliopause resulting from a tilted line source in a uniform flow, as seen from upstream of the surface.

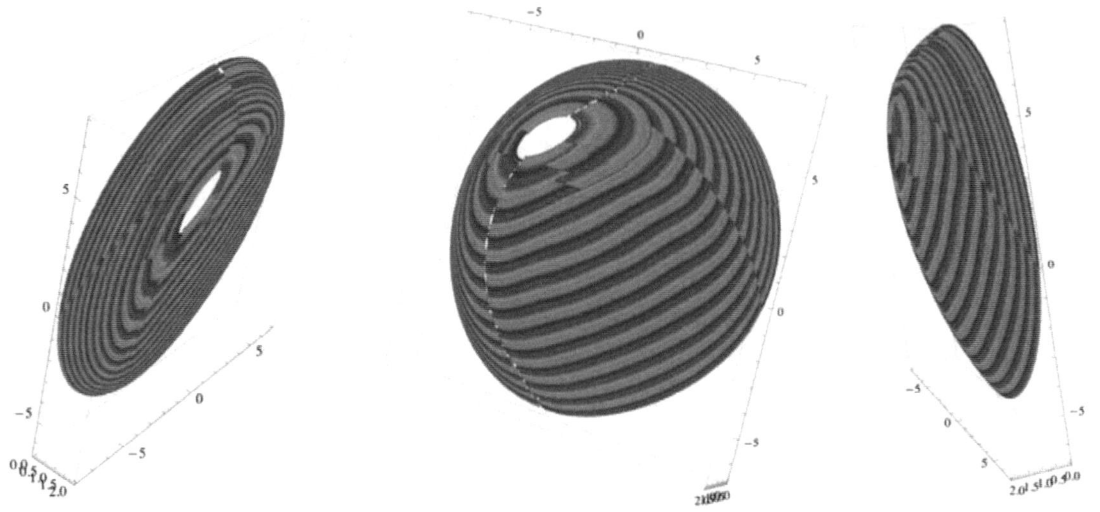

Figure 7. Three views of a model heliopause resulting from a tilted disc source in a uniform flow, as seen from upstream of the surface.

$$c_f{}^8 - C c_f{}^6 + \frac{W}{N^2} = 0 \,. \tag{2}$$

Both constants, C and N, are given by the ISM conditions far from the Sun, and W is known from the particular flow field, so the effective temperature is then determined from the appropriate solution of equation (2). For $0 < W/N^2 < 27\ C^4/256$, equation (2) has two real roots and the larger root is the physical one related to $c_f^2 = C$ when $v = 0$. As the plasma speeds up around the flanks of the obstacle, W/N^2 reaches $27\ C^4/256$ where the two roots coalesce, and the flow speed reaches the effective sound speed $v = c_f$. Beyond this point, equation (2) has no real solutions and the simple model constructed here is no longer valid.

The charge-exchange process that produces the He secondaries depends on the properties of the He^+ component of the upstream plasma, so these properties must be extracted from the model results. The density fraction of He^+ does not change, but the kinetic temperature of the ions is still an unknown fraction of the effective plasma temperature from the model. Here, we define a "partition function" ζ to relate the actual kinetic temperature of He^+ proportional to $c_s{}^2$, to the effective model temperature proportional to c_f^2, $\zeta = (c_s/c_f)^2$. The simplest assumption would be to take ζ = constant, but we know that the plasma is more compressible than the magnetic field so this would be a poor choice. Instead, we take a hint from the global simulation of Izmodenov & Alexashov [10], who plot the plasma density and pressures (kinetic and magnetic) along the stagnation line in their Figures 4a and 4b. We assume that the partition is a function of plasma density, and estimate from these figures the change in ζ between the two positions, far upstream and at the heliopause stagnation point. A linear dependence between these points, along with the statement that $\zeta = 0$ at zero density, gives the model partition function as

$$\zeta(n_p) = \begin{cases} 19.78 n_p - 0.826 & n_p \geq 0.044 \\ n_p & n_p < 0.044 \end{cases} \tag{3}$$

where the total ion density here is measured in cm^{-3}.

2.4. *Trajectories of He Incident at IBEX-Lo*

The sky-map in Figure 1 is composed of 6° × 6° pixels labeled according to the orientation of the IBEX-Lo instrument. The IBEX spacecraft spins about an axis nominally pointing toward the Sun, and the measurements are made in the spin plane perpendicular to the spin axis. For a given day-of-year, the spin plane nominally scans a strip of sky centered on a given ecliptic longitude, and fluxes are compiled as a function of spin angle corresponding to ecliptic latitude. Pixel position on the map simply refers to the direction of the instrument aperture when those specific fluxes were measured.

However, the measured atoms arrive at IBEX by traveling along Keplerian trajectories in the Sun's gravitational field. Thus, an incident atom has passed through regions of interstellar and interplanetary space on a curved path quite different from a traditional "line of sight". Furthermore, atoms arriving within a given pixel designation but at different speeds will have encountered different regions of space during their passage. Since we are dealing with energy-integrated measurements, the fluxes in the map are a superposition of many trajectories, each encountering different conditions and subject to different charge-exchange interactions.

To compute the model fluxes corresponding to a given pixel, we follow a straightforward, though cumbersome, procedure. We take IBEX to sit at 1 AU, at a given position in the Earth's orbit, with a given spin angle, and track the incident atoms backwards along their Keplerian trajectories [21-23]. We include the effects of both elliptical (closed) and hyperbolic (open) atom trajectories. Starting with an atom speed large enough to reach the heliopause, we first transform that incident velocity into the inertial frame of the Sun. In the inertial frame, both the particle total energy/mass, $E_o = w^2/2 - k_g/r$, and the angular momentum/mass, $\ell = r\, w_\theta$, are constants along the trajectory, where w is the particle speed in the inertial frame and k_g is the gravitational potential of the Sun at 1 AU. We then compute the He flux fraction from that trajectory by evaluating the charge-exchange gains and losses due to the model plasma properties along that path in the outer heliosheath, allowing for the contribution of primary interstellar He on the open trajectories [6]. Specifically, we solve the difference equation

$$\Delta n_{He} = \left[\beta_{\rm prod} - \beta_{\rm loss}\, n_{He} \right] \frac{\Delta s}{V_{traj}} \tag{4}$$

along the trajectory, where the production rate $\beta_{\rm prod}$ counts only those He atoms produced by charge-exchange from an He^+ ion already moving at the speed and direction appropriate to that trajectory at that position in space, and the loss rate $\beta_{\rm loss}$ removes any atom already on the trajectory which charge-exchanges with the He^+ plasma. At this point in our modeling, we take the charge-exchange cross-section for the process $He^+ + He \rightarrow He + He^+$ to be constant at $\sigma = 2. \times 10^{-15}$ cm^2. For closed trajectories, we start at the model heliopause, moving outward toward aphelion, with a boundary condition on equation (4) that $n_{He} = 0$ at the starting point. This assumes that the measured atoms reaching 1 AU will have been produced somewhere during their first traversal of the outer heliosheath. The boundary condition for the open trajectories sets the initial density at 40,000 AU upstream equal to the phase-space density of the Maxwellian distribution of primary interstellar particles.

The incident speed at IBEX is then incremented, the solution of equation (4) is obtained for the next trajectory, and the total flux for that pixel is eventually obtained as

$$j(\phi,\alpha) = G \int v_{\rm o}^3 \; n_{He}(v_{\rm o}) dv_{\rm o} \, , \tag{5}$$

where $v_{\rm o}$ is the incident speed, and G is a proportionality constant equivalent to the instrument geometric factor.

The map intensity distribution is affected by photoionization due to solar UV. In addition to an overall reduction at 1 AU, this process depletes the slower atoms more than the faster ones. We assume that the photoionization rate at 1 AU is a constant, $\beta_{\rm ph} = 10^{-7}$ s^{-1}, decreasing as r^{-2} [24]. Each incident flux value, $n_{He}(v_{\rm o})$ is then reduced by the factor

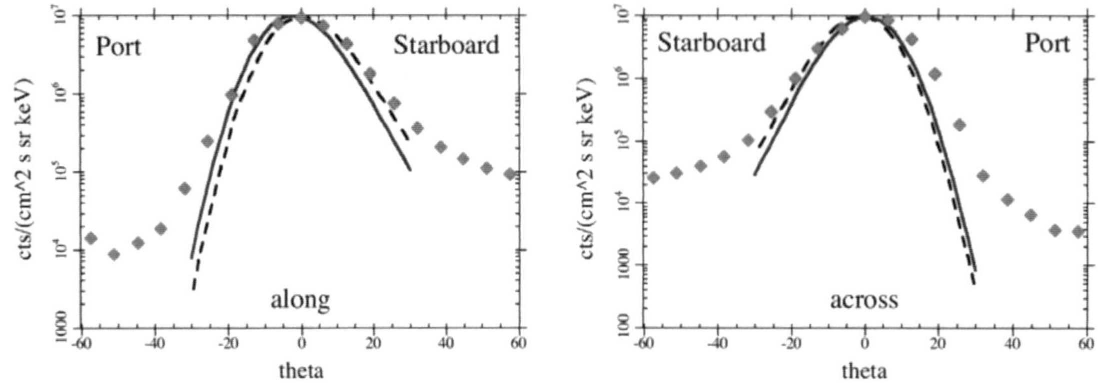

Figure 8. Comparison of model results for the interstellar primary neutral beam with the IBEX-Lo observations (red diamonds) on the two cutlines, along (left panel) and across (right panel). The blue curves show the model predictions for two pointing directions of the spacecraft spin axis, toward the Sun (dashed line) and offset 2° to the West (solid line).

$$\exp\left(-\beta_{ph} r_o^{\,2} \int_1^{\infty} \frac{dr}{r^2 \sqrt{E_o + k_g / r}}\right) \tag{6}$$

before adding to the integral (5).

Another effect is important when comparing models to the IBEX-Lo data. When spacecraft spin axis is pointed at the Sun, as is nominally described, the incident He atoms arrive at IBEX at the perihelion point of their trajectory. If the spin axis is shifted, the true particle trajectories will also be shifted, and the resulting fluxes will be different. It turns out that the spacecraft spin axis was often not pointed at the Sun during 2009 – 2011, but offset by several degrees at various times during the measurements shown in Figure 1 [22, 25]. The magnitude of this effect can be seen in Figure 8, which shows a comparison between the measured He fluxes given in Figure 2 (red diamonds) with two model curves representing the primary beam (that is, setting $\sigma = 0$ along the trajectories). The primary curves are normalized so their peak fluxes coincide with the peak of the data. The two panels correspond to the fluxes on the "along" and "across" cutlines of Figure 1, as functions of the angular distance along each cutline. The two curves in each panel show the modeled primary fluxes assuming no offset of the spin axis (dashed line) and a 2° offset of the axis to the West (solid line). The primary beam should dominate the measured fluxes above the 10% level, but discrepancies between either model and the data are evident.

It is apparently necessary to incorporate the actual position of the IBEX spin axis during the specific observation time of each pixel to truly model the measured fluxes. We are in the process of making these improvements to our models.

3. Current Status of the Model

The motivation for this modeling effort has always been to extract empirical information on the shape of the heliopause and plasma conditions of the outer heliosheath. We have explored the apparent effect of many trial obstacle shapes and flow patterns in our model. Generally, our findings seem to indicate that the heliopause is approximately symmetric about the *B-V* plane (see also [26]). At this time, we are still refining the model effects of spin axis positioning for the IBEX spacecraft. In the future, we will determine the type of obstacle shape that is most consistent with the observations. This information can be used to guide and validate the much more complicated global simulation efforts being undertaken by other groups.

Acknowledgments
The authors are grateful for valuable conversations with M. Bzowski, T. G. Forbes, V. Izmodenov, M. A. Lee, E. Möbius, P. Swaczyna, and B. J. Vasquez. Computations were performed on Trillian, a Cray XE6m-200 supercomputer at UNH supported by the NSF MRI program under grant PHY-1229408. This work was also supported in part by NASA grants 80NSSC18K0655, 80NNSSC18K1215, and by the Interstellar Boundary Explorer mission as a part of the NASA Explorer Program.

References
[1] McComas D J, *et al.* 2009 IBEX - Interstellar Boundary Explorer *Space Sci. Rev.* **146** 11
[2] Park J, Kucharek H, Möbius E, Galli A, Kubiak M A, Bzowski M and McComas D J 2016 *IBEX* observations of secondary interstellar helium and oxygen distributions *Astrophys. J.* **833** 130
[3] McComas D J, *et al.* 2012 The first three years of *IBEX* observations and our evolving heliosphere *Astrophys. J. Suppl.* **203** 1
[4] Möbius E, *et al.* 2012 Interstellar gas flow parameters derived from Interstellar Boundary Explorer-Lo observations in 2009 and 2010: Analytical analysis *Astrophys. J. Suppl.* **198** 11
[5] Swaczyna P, *et al.* 2018 Interstellar neutral Helium in the heliosphere from *IBEX* observations. V. Observations in *IBEX*-Lo ESA steps 1, 2, and 3 *Astrophys. J.* **854** 119
[6] Bzowski M, Kubiak M A, Czechowski A and Grygorczuk J 2017 The helium warm breeze in *IBEX* observations as a result of charge-exchange collisions in the outer heliosheath *Astrophys. J.* **845** 15
[7] Möbius E, *et al.* 2015 Interstellar flow and temperature determination with IBEX: Robustness and sensitivity to systematic effects *Astrophys. J. Suppl.* **220** 24
[8] Opher M, Drake J F, Zieger B and Gombosi T I 2015 Magnetized jets driven by the Sun: The structure of the heliosphere revisited *Astrophys. J.* **800** L28
[9] Pogorelov N V, Heerikhuisen J, Roytershteyn V, Burlaga L F, Gurnett D A and Kurth W S 2017 Three-dimensional features of the outer heliosphere due to coupling between the interstellar and heliospheric magnetic field. V. The bow wave, heliospheric boundary layer, instabilities and magnetic reconnection *Astrophys. J.* **845** 9
[10] Izmodenov V V Alexashov D B 2015 Three-dimensional kinetic-MHD model of the global heliosphere with the heliopause-surface fitting *Astrophys. J. Suppl.* **220** 32
[11] Kubiak M A, *et al.* 2014 Warm breeze from the starboard bow: A new population of neutral helium in the heliosphere *Astrophys. J. Suppl.* **213** 29
[12] Kubiak M A, *et al.* 2016 Interstellar neutral helium in the heliosphere from IBEX observations. IV. Flow vector, Mach number, and abundance of the warm breeze *Astrophys. J. Suppl.* **223** 25
[13] Zirnstein E J, Heerikhuisen J, Funsten H O, Livadiotis G, McComas D J and Pogorelov N V 2016 Local interstellar magnetic field determined from the *Interstellar Boundary Explorer* ribbon *Astrophys. J.* **818** L18
[14] McComas D J, *et al.* 2012 The heliosphere's interstellar interaction: No bow shock *Science* **336** 1291
[15] Isenberg P A, Forbes T G and Möbius E 2015 Draping of the interstellar magnetic field over the heliopause: A passive field model *Astrophys. J.* **805** 153
[16] Kleimann J, Röken C, Fichtner H and Heerikhuisen J 2016 Toward more realistic analytical models of the heliotail: Incorporating magnetic flattening vie distortion flows *Astrophys. J.* **816** 29
[17] Kleimann J, Röken C and Fichtner H 2017 An improved analytical model of the local interstellar magnetic field: The extension to compressibility *Astrophys. J.* **838** 75
[18] Parker E N 1961 The stellar-wind regions *Astrophys. J.* **134** 20
[19] Batchelor G K 1967 *An Introduction to Fluid Mechanics* (Cambridge: Cambridge Univ. Press)

[20] Morse P M and Feshbach H 1953 *Methods of Theoretical Physics* (New York: McGraw-Hill)

[21] Lee M A, Kucharek H, Möbius E, Wu X, Bzowski M and McComas D J 2012 An analytical model of interstellar gas in the heliosphere tailored to Interstellar Boundary Explorer observations *Astrophys. J. Suppl.* **198** 10

[22] Lee M A, Möbius E and Leonard T 2015 The analytical structure of the primary interstellar helium distribution function in the heliosphere *Astrophys. J. Suppl.* **220** 23

[23] Müller H R, Möbius E and Wood B E 2016 Modeling secondary neutral Helium in the heliosphere *J. Phys: Conf. Series* **767** 012019

[24] Rucinski D, Cummings A C, Gloeckler G, Lazarus A J, Möbius E and Witte M 1996 Ionization processes in the heliosphere - Rates and methods of their determination *Space Sci. Rev* **78** 73

[25] Leonard T, *et al.* 2015 Revisiting the ISN flow parameters, using a variable *IBEX* pointing strategy *Astrophys. J.* **804** 42

[26] Isenberg P A, Kucharek H and Park J 2017 Emperical estimate of the shape of the upstream heliopause from IBEX-Lo Helium measurements: Preliminary results arXiv:1711.09823

3D Turbulent Reconnection: 20 Years After

A. Lazarian[1], Grzegorz Kowal[2], Siyao Xu[1,3], and Amir Jafari[4]

[1] Department of Astronomy, University of Wisconsin, 475 North Charter Street, Madison, WI 53706, USA
[2] Escola de Artes, Ciências e Humanidades, Universidade de São Paulo, Av. Arlindo Béttio, 1000 - Vila Guaraciaba, São Paulo - SP, CEP: 03828-000, Brazil
[3] Hubble Fellow
[4] Department of Physics & Astronomy, Johns Hopkins University, 3400 N. Charles Street, Baltimore, MD 21218, USA

E-mail: `lazarian@astro.wisc.edu`

Abstract. A theory of 3D turbulent reconnection was published 20 years ago. At that time it was suggested that the fast reconnection requires X-point Petschek-type configuration of magnetic field in the reconnection region, as well as plasma effects that would stabilize this configuration. On the contrary, the turbulent reconnection theory (i) identified 3D Alfvénic turbulence as the cause of fast reconnection, (ii) postulated the ubiquitous nature of turbulent reconnection in astrophysical high-Reynolds number environments, and (iii) identified Y-point volume reconnection as the typical magnetic reconnection on length scales where the magnetohydrodynamic (MHD) approximation is valid. In this article we briefly summarize the status of the 3D turbulent reconnection theory. We show that other alternative models of reconnection have evolved toward supporting the essential features of the original turbulent reconnection theory. We also provide numerical and observational evidence supporting the turbulent reconnection and its consequences, e.g., the breakdown of classical magnetic flux freezing in highly conducting turbulent fluids.

1. Magnetic reconnection in astrophysics

Magnetic fields are ubiquitous in astrophysical systems and they critically affect the dynamics and properties of magnetized plasmas on all length scales, including those much larger than any relevant plasma scales, e.g., the ion inertial length. In fact, the extensively studied magnetic fields of Earth's magnetosphere are an exception in the astrophysical context in terms of the involved length scales.

It is widely believed in astrophysical studies, that magnetic fields in highly conducting plasmas are nearly perfectly frozen into the fluid and retain their topology all the time (1; 2). The concept of flux-freezing is at the heart of many astrophysical theories, e.g., the theory of star formation in magnetized interstellar medium (ISM).

The theory of 3D turbulent reconnection was introduced by (3, henceforth LV99). This model was followed by several subsequent theoretical studies, for instance, (4, henceforth ELV11), (5), and (6). The LV99 theory has been supported by numerical simulations in non-relativistic (see 7; 8; 9; 10; 11; 12), as well as relativistic settings (see 13; 14). In addition, different pieces of observational evidence support the predictions of the 3D turbulent reconnection theory of LV99.

2. Turbulence in astrophysical fluids

2.1. Observational evidence

Observations of the ISM strongly support its turbulent state (see 15; 16), see Figure 1, (see 17, for a review). This is also clear from the *in situ* measurements of the solar wind fluctuations (18) and via

Content from this work may be used under the terms of the Creative Commons Attribution 3.0 licence. Any further distribution of this work must maintain attribution to the author(s) and the title of the work, journal citation and DOI.
Published under licence by IOP Publishing Ltd

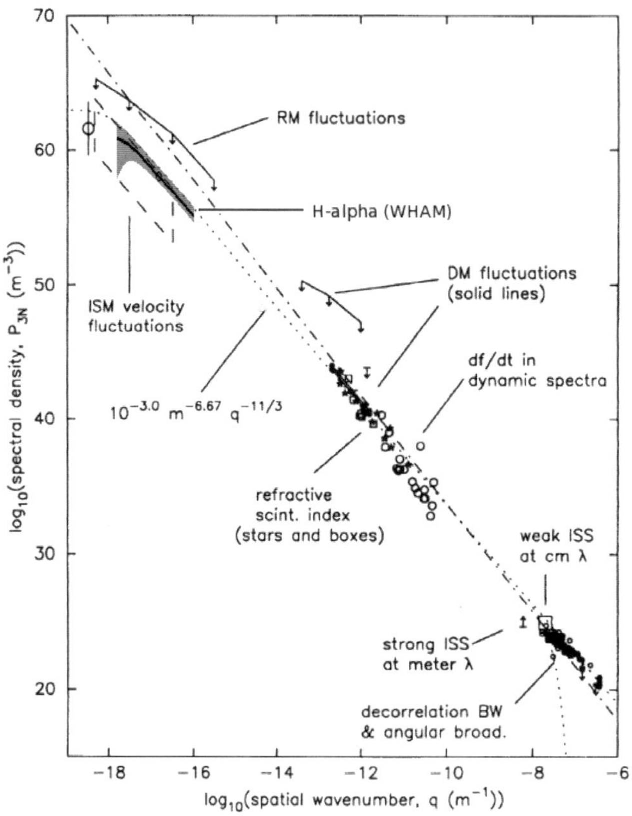

Figure 1. Big power law in the sky from (15) extended to length scales of tens of parsecs using WHAM data. From (16).

studies of non-thermal broadening of spectral lines as well as measures obtained by other techniques (see 19). Turbulence naturally arises from magnetized astrophysical plasmas generally with very large Reynolds numbers. Indeed, the length scales of the motions are large, while the motions of charged particles in the direction perpendicular to magnetic fields are of the order of the Larmor radius. As a result, laminar plasma flows are prey to numerous linear and finite-amplitude instabilities that induce turbulent motions.

Turbulence can be driven by an external energy source, such as supernovae in the ISM (20; 21), merger events and active galactic nuclei outflows in the intracluster medium (ICM) (22; 23; 24), and baroclinic forcing behind shock waves in interstellar clouds. In addition, magneto-rotational instability (MRI) in accretion disks (25; 26), kink instability of twisted flux tubes in the solar corona (27; 28), etc. can drive turbulence. In some cases, magnetic reconnection can also be the driver.

Observations show that turbulence is ubiquitous in all astrophysical plasmas. The spectrum of electron density fluctuations in Milky Way is presented in Figure 1, as one dramatic piece of evidence for cosmic turbulence. Similar spectra are observed in solar wind (18) and (29), molecular clouds (30), and the ICM (31). New techniques for studying turbulence, e.g. the Velocity Channel Analysis (VCA) and Velocity Coordinate Spectrum (VCS) techniques (32; 33; 34), have provided important insight into the velocity spectra of turbulence in molecular clouds (see 30; 35), Galactic and extragalactic atomic hydrogen gas (36; 37; 38, see also the review by (17), where a compilation of velocity and density spectra obtained with contemporary HI and CO data is presented).

2.2. *Alfvénic turbulence: eddies perpendicular to magnetic field*

The Goldreich-Sridhar (1995) model of MHD turbulence predicts the Kolmogorov spectrum of motions that are perpendicular to the direction of magnetic field. This dynamics suggests that the magnetic field does not resist such mixing motions. This is possible if the field lines do not get entangled by such motions, which, in its turn means that the reconnection of magnetic field lines belonging to the eddy happens over the eddy-turnover time. This is exactly the prediction of LV99 theory. An important consequence of LV99 picture is that magnetic eddies should be aligned not with the mean magnetic field but with the magnetic field of the eddies. Note that this notion of the local magnetic field, which is currently an accepted corner stone of the MHD turbulence theory, was not in the original GS95 formulation of the turbulence theory. The MHD turbulence theory in LV99 identified an important intrinsic connection between the MHD turbulence and turbulent magnetic reconnection.

Adopting the picture of eddies perpendicular to the local magnetic field of the eddies, in terms of perpendicular motions we expect to have the Komogorov scaling, i.e. $u_\ell \sim l_\perp^{1/3}$, and also obtain the relation between the parallel and perpendicular scales of the eddies. The so-called critical balance in the picture of eddies follows naturally. Indeed, as the eddy provides mixing of magnetic field lines in the perpendicular direction over the time scale $l_\perp/u_\ell$, it sends out an Alfvén wave with the same period. This is exactly

$$l_\parallel^{-1} V_A \sim \ell_\perp^{-1} u_l, \tag{1}$$

where $l_\parallel$ and $l_\perp$ are the eddy scales parallel and perpendicular to the *local* direction of magnetic field (LV99, 39; 40).

It is assumed in GS95 that the energy is injected at the scale L_i with the injection velocity equal to the Alfvén velocity V_A. If the injection velocity u_L is smaller than V_A, then the eddies on smaller length scales get more elongated. For $M_A = u_L/V_A < 1$, the eddy description is valid only for the strong turbulence on length scales smaller than $l_{trans} = LM_A^2$.

2.3. *MHD turbulence in compressible and partially ionized media*

The compressible MHD turbulence can be presented by the superposition of three cascades of basic MHD modes, namely, Alfvénic, slow, and fast. For magnetic reconnection one should concentrate on Alfvénic turbulence as this component is the most important for the theory of reconnection. In non-relativistic MHD turbulence, the backreaction of slow and fast magnetosonic modes (41; 42; 43) on Alfvénic cascade is insignificant (41; 44; 45), and therefore the Alfvénic turbulence is not much affected by the presence of compressible modes.

In partially ionized astrophysical plasmas in e.g., the early universe, interstellar medium, protoplanetary disks, solar atmosphere, MHD turbulence is subjected to the damping effects due to the presence of neutrals. The damping effects in a partially ionized medium mainly arise from the collisional friction between ions and neutrals, i.e., ion-neutral collisional damping and the viscosity in neutral fluid, i.e., neutral viscous damping (45; 46). When the damping effects are significant, MHD turbulence is suppressed. The damping scale where the MHD turbulence is dissipated can be determined by comparing the turbulent energy cascade rate with the damping rate. The general expressions that describe the damping scales in different conditions were obtained in (47).

The scaling relations of relativistic Alfvénic turbulence were first derived in (48), and in terms of the spectral slope and anisotropy they coincide with those in GS95 theory. These relations were numerically tested later in (49). For relativistic compressible turbulence theory, the earlier studies of relativistic MHD turbulence (see 50; 51; 52; 53; 54; 55) did not reveal significant differences between the non-relativistic and relativistic turbulence. However, more recent studies in (13; 56) provided the decomposition of relativistic turbulent motions into Alfvén, slow, and fast modes and found that the coupling between compressible and incompressible turbulence is stronger than that in the non-relativistic case.

Journal of Physics: Conference Series **1332** (2019) 012009 doi:10.1088/1742-6596/1332/1/012009

3. Analytical model of turbulent reconnection

The 3D model of turbulent reconnection proposed by LV99 generalizes the classical Sweet-Parker model (57; 58). There two regions with uniform *laminar* magnetic fields are separated by a thin current sheet. The magnetic fields are frozen in through the two regions, and the violation of flux freezing takes place over a thin slot Δ, determined by Ohmic diffusion. Thus the speed of reconnection is,

$$V_{rec1} \approx \eta/\Delta. \tag{2}$$

The Sweet-Parker model deals with the *steady state reconnection*, and thus the plasma in the diffusion region must be ejected from the edge of the current sheet at the Alfvén speed V_A. Therefore, the reconnection speed is limited by the conservation of mass condition

$$V_{rec2} \approx V_A \Delta/L_x, \tag{3}$$

where L_x is the lateral extend of the current sheet. Combining Eq. (refeq.1) and (3) one gets the Sweet-Parker reconnection speed, which is reduced from V_A by the square root of the Lundquist number, $S \equiv L_x V_A/\eta$, i.e.

$$V_{rec,SP} = V_A S^{-1/2}. \tag{4}$$

The Lundquist number in astrophysical conditions is very large, and therefore the reconnection rate is negligible. The disparity between L_x and Δ makes the Sweet-Parker model unable to explain most of the observed astrophysical reconnection events, e.g. Solar flares. A way to deal with this problem was suggested by (59), who proposed that in special conditions the reconnection configurations may have magnetic field lines converging to the reconnection zone at a sharp angle making L_x and Δ comparable. The corresponding configuration is known as X-point reconnection as opposed to the Sweet-Parker extended current sheet Y-type reconnection. Since the pioneering work of Petschek, the research exploring X-point reconnection was the dominant trend in the attempts of obtaining fast reconnection. However, as it was discussed in LV99 the major problem of X-point reconnection is how to preserve the X-point configuration over astrophysically large scales and in the presence of external driving of motions, e.g. turbulence.

Instead, LV99 identified fast reconnection in the presence of the Y-type magnetic configurations. The authors showed a way to make Δ macroscopic and even astrophysically large. The LV99 proposal was based on property of MHD turbulence that they also quantified, namely, the magnetic field line wandering. The corresponding model of magnetic reconnection is illustrated in Figure 2. In LV99 model, the outflow thickness Δ is determined by the large-scale magnetic field wandering that depends on the level of turbulence. Therefore, the rate of reconnection in LV99 model is determined by the level of MHD turbulence. To quantify the reconnection rate, the scaling relations of Alfvénic turbulence in § 2 should be used.∗

On length scales less than the injection scale of turbulence L_i, the mean squared magnetic field wandering is equal to:

$$\langle y^2 \rangle \approx \frac{27x^3}{L_i}\left(\frac{u_L}{V_A}\right)^4, x < L_i \tag{5}$$

which provides a superdiffusive dependence of $\langle y^2 \rangle \sim x^3$ with x as the measured distance along the magnetic field. If the scale of the current sheet is larger than L_i, i.e. $L_x \gg L_i$, the magnetic field wandering obeys the usual random walk scaling. The corresponding random walk step is L_x/L_i with the mean squared displacement per step equal to $L_i^2(u_L/V_A)^4$. Therefore,

$$\langle y^2 \rangle^{1/2} \approx (L_i x)^{1/2}(u_L/V_A)^2, \quad x > L_i, \tag{6}$$

∗The reconnection at low Alfvén Mach numbers is astrophysically important, and this motivated LV99 to generalize the MHD turbulence theory for arbitrary Alfvén Mach numbers.

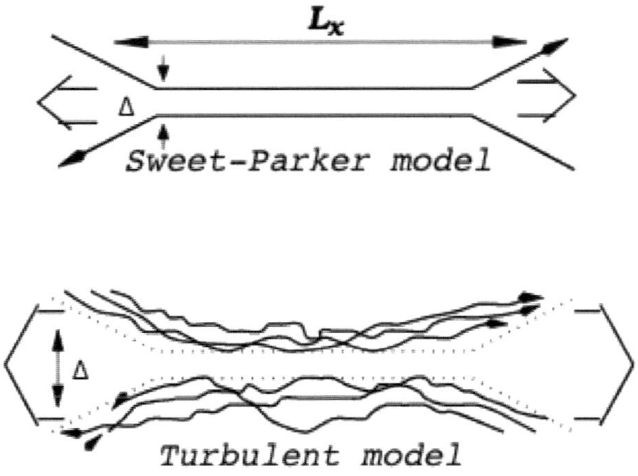

Figure 2. *Upper plot*: Sweet-Parker model of reconnection. The thickness of the outflow Δ is limited by Ohmic diffusivity that takes place on microscopic scales. The disparity between Δ and the astrophysically large scale L_x makes the reconnection slow. *Lower plot*: LV99 model of turbulent reconnection. The width of outflow Δ is determined by macroscopic field line wandering, and it can be comparable to L_x for trans-Alfvénic turbulence. (46).

which provides the other limit for the magnetic field wandering. Combining Eqs. (5) and (6), it is possible to derive the thickness of the outflow Δ in the aforementioned two regimes. Then, using the mass conservation Eq. (3), LV99 obtained

$$V_{rec} \approx V_A \min\left[\left(\frac{L_x}{L_i}\right)^{1/2}, \left(\frac{L_i}{L_x}\right)^{1/2}\right] M_A^2, \tag{7}$$

where $V_A M_A^2$ is proportional to the turbulent eddy speed. When $M_A \sim 1$, the reconnection rate reaches a large fraction of V_A. Importantly, when using Eq. (7), one should keep in mind that the value of the guide magnetic field, i.e. the field component shared by two reconnecting fluxes, does not affect the reconnection speed. Indeed, both turbulent velocity and magnetic field can be decomposed into components in the direction of the guide field and perpendicular to it. The field wondering in the direction perpendicular to the guide field is affected by the velocity perturbations in the same direction. The guide field does not affect the bending of magnetic field. The outflow velocity is, however, is determined only by V_A corresponding to the perpendicular component of magnetic field.

The LV99 theory of turbulent reconnection has provided a number of predictions. (1) First of all, the turbulent reconnection was identified as a generic process that takes place everywhere in magnetized turbulent fluids. A violation of flux freezing follows from that. (2) Second, LV99 showed that the speed of reconnection varies depending on the level of turbulence, which enables the theory to explain diverse observations. (3) Third, the independence on plasma parameters makes the turbulent reconnection a generic reconnection process in different astrophysical settings.

(1) Flux freezing violation: Richardson dispersion

The "flux freezing principle" (1) has become a powerful tool to study astrophysical problems (2; 60). For instance, if magnetic field were frozen into the fluid in astrophysical turbulent dynamos, the tangled structure of the generated magnetic field would quench its further growth.

The Richardson dispersion manifests a gross violation of the flux freezing in magnetized turbulent fluids. This numerically confirmed phenomenon (see §4.2) provides another way of re-deriving LV99 rate

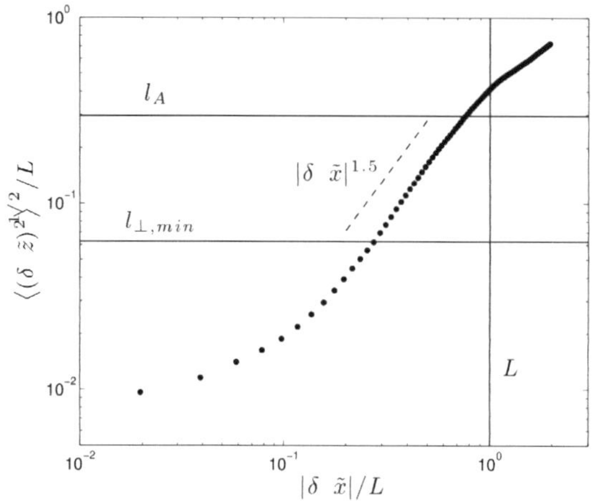

Figure 3. The separation of particle trajectories in MHD turbulence. From (65).

of turbulent reconnection. It was initially introduced in hydrodynamic turbulence (61). Using this theory one can predict that the separation between two particles obeys the equation $d/dt[l(t)] \sim v(l) \sim \alpha l^{1/3}$, where α is proportional to the cube-root of the energy cascading rate. The corresponding solution $l(t) = [l^{2/3} + \alpha(t - t_0)]^{1/3}$ describes the Richardson dispersion, i.e. the mean square separation between particles increasing in proportion to the cube of time, i.e. $\langle l^2 \rangle \sim t^3$.

In terms of motions perpendicular to the local direction of magnetic field, the scaling of velocities in GS95 picture corresponds to the Kolmogorov one. Therefore we expect that the plasma particles separate in accordance to the Richardson law over the inertial range of turbulence. The Richardson dispersion is a numerically proven phenomenon (see §4.2), and it was used in ELV11 to re-derive the LV99 rate of reconnection.

(2) Magnetic field wandering

For magnetized fluids one can trace magnetic fields and study their separation. This effect was discussed in the cosmic ray literature as the cause of the perpendicular diffusion of cosmic rays in Galactic magnetic field (62). Dealing with magnetic reconnection we have discussed the separation of magnetic field lines. Figure 3 illustrates the spread of magnetic field lines as traced by test particles in the perpendicular direction. Eq. (5) exhibits two remarkable results. First of all, the squared separation between magnetic field lines increases in proportion to the cube of the distance measured along the magnetic field lines. The other result is that the square of the separation between magnetic field lines increases with the fourth power of Alfvén Mach number, i.e. $\sim M_A^4$. The LV99 prediction of magnetic field line superdiffusion in space was successfully tested by (63) and later by (64) and (65) (see Figure 3). The latter study also confirmed the M_A^4 dependence.

We would like to stress again that magnetic wandering cannot be understood by assuming that magnetic field lines preserve their identify in turbulent flows. Magnetic field lines wondering in turbulent magnetized flow constantly reconnect, inducing the exchange of plasmas and magnetic field. It was decided in ELV11 to term this new type of dispersion, *the Richardson diffusion in space*, as opposed to the original Richardson dispersion that is a time-dependent process.

(3) Irrelevance of plasma effects in large-scale turbulent reconnection

While plasma effects play a crucial role in reconnection of magnetic reversals near electron- and ion-scales, they are totally negligible and irrelevant to reconnection of magnetic structures at scales $\ell \gg \rho_i$, where ρ_i is the ion gyroradius. This has been demonstrated in different ways, i.e. with either the generalized Ohm's law (5) or the full Vlasov-Maxwell-Landau kinetic equations (66). This conclusion does *not* mean that the large-scale turbulent reconnection in a weakly collisional plasma such as the

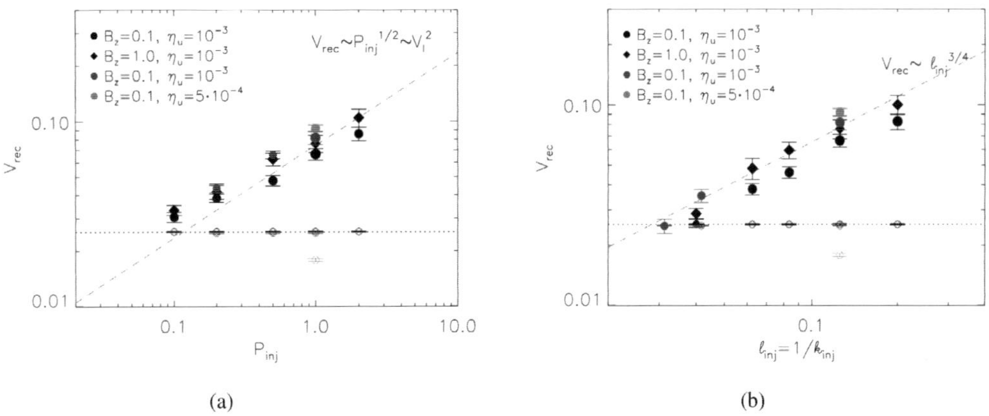

Figure 4. Dependence of the reconnection speed V_{rec} on the injection power P_{inj} (left) and the injection scale k_{inj} (right). Different colors correspond to different turbulence driving methods (black - Fourier driving in velocity, blue and red - real space driving in velocity and magnetic field, respectively). The dotted line corresponds to the Sweet-Parker reconnection rate for models with $\eta_u = 10^{-3}$. A unique red symbol shows the reconnection rates for model with driving in velocity performed with higher resolution (512x1024x512) and resistivity coefficient reduced to $\eta_u = 5 \cdot 10^{-4}$. From (8).

solar wind can be described by ideal MHD equations. Although an ideal Ohm's law holds, there are not enough collisions to make the ion distribution function close to a local Maxwellian and to make the ion pressure tensor an isotropic function of local density and temperature. The approximation of ideal MHD should, however, is a suitable model to investigate some of the basic effects of turbulence at inertial-range scales. As the incompressible shear-Alfvén description is applicable on large scales, quantitative measures of field-line stochasticity and turbulent Richardson dispersion should carry over with little change to this collisionless plasma environment.

4. Testing of turbulent reconnection and flux freezing violation

4.1. Testing LV99 theory with MHD codes

Numerical two-dimensional studies of turbulent reconnection in a periodic box were done by (67; 68). The results indicated a strong effect of turbulence on the reconnection process. Unfortunately, the numerical setup precluded the calculation of long-term reconnection speed, and there were no studies on how the properties of turbulence affect reconnection. More recently, (69) studied the effects of small-scale turbulence on two-dimensional reconnection. However, no significant effects of turbulence on reconnection were observed. In what follows we consider realistic 3D magnetic reconnection. Indeed, any 2D configuration significantly restricts the magnetic field freedom, making impossible, for example, field line wandering. Moreover, magnetized turbulence is very different in 2D and 3D †.

The LV99 model is intrinsically 3D, requiring significant computational resources and sufficient resolution to resolve all length scales from large ones where turbulence is injected to dissipation ones where magnetic resistivity operates. The first extensive testing of the theoretical LV99 model with numerical simulations was performed by (7), a decade after the LV99 theory was published. The tests were carried out using MHD compressible code in 3D domain with open boundary conditions in the subAlfvénic turbulence regime. Some additional tests in terms of different turbulence driving was published in (8). The authors performed numerical experiments by changing three parameters, the power

†In fact, no 2D magnetized turbulence model was proposed until today, while for the 3D turbulence we could list a number of existing theoretical models (see, e.g. 70; 71; 44).

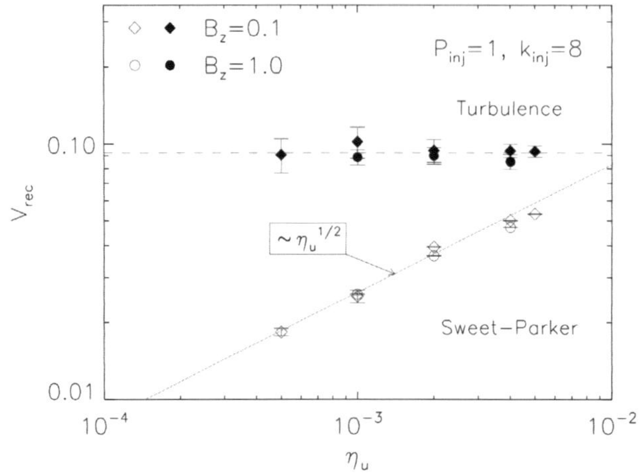

Figure 5. Dependence of the reconnection speed V_{rec} on the resistivity η for models with weak (diamonds) and strong (circles) guide field. From (7).

P_{inj} of turbulence driving, length scale k_{inj} of driving, and resistivity coefficient η. Figure 4 presents the measured dependence of the reconnection speed V_{rec} on the turbulence parameters P_{inj} and k_{inj}. Even though relatively low resolution was used, the simulations confirmed the theoretical predictions by LV99 on the dependence of reconnection rate on the turbulence parameters. More importantly, the numerical studies by (7) proved that the reconnection is indeed fast in the presence of turbulence, since it does not depend on the resistivity η, as shown in Figure 5. Moreover, the same dependence was observed in different conditions with weak and strong guide fields B_z. In (7), they also tested the role of the anomalous resistivity intended to mimic plasma effects and found that there is no essential dependence of the reconnection rate on anomalous plasma effects as well.

As expected in the LV99 model, the current sheet is broad with individual currents distributed widely within a 3D volume, and the turbulence within the reconnection region is similar to the turbulence within a statistically homogeneous volume. The structure of the reconnection region was analyzed in more details by (72) based on the numerical work by (7). The results supported LV99 picture about the reconnection region being broad, the magnetic shear being more or less coincident with the outflow zone, and the turbulence within it being similar to that in a homogeneous system.

4.2. Flux freezing violation test

As we discussed earlier, the LV99 theory of turbulent reconnection is closely related to the gross violation of flux freezing in turbulent environments. A direct test of the temporal Richardson dispersion of magnetic field lines was performed recently by (9). In this experiment, stochastic fluid trajectories were tracked backward in time to determine which field lines at earlier times would arrive to the point of interest. In the flux-freezing situation, the one to one mapping is expected for different times, but the Richardson dispersion introduces a very different behavior. To test the idea, many time frames of an MHD simulation were stored so that the equations for the trajectories could be integrated backward. The results of this study are presented in Figure 6. It shows the trajectories of the arriving magnetic field lines, which are widely dispersed backward in time resembling a spreading plume of smoke and are very different from a single "frozen-in" line. The quantitative results of the time-dependent separation are consistent with the temporal Richardson dispersion behavior.

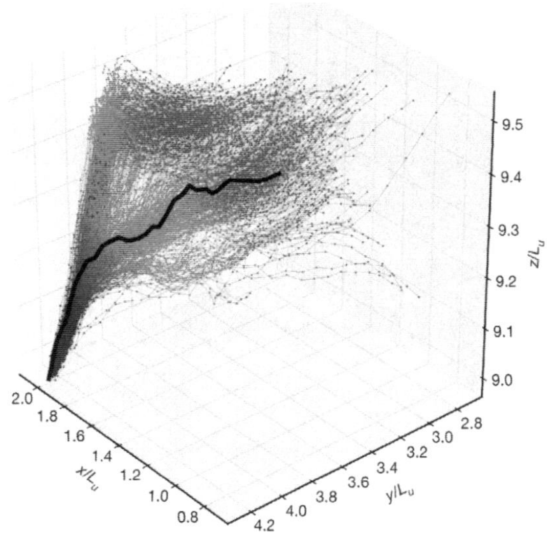

Figure 6. Stochastic trajectories that arrive at a fixed point in the archived MHD flow, color-coded red, green, and blue from earlier to later times. From (9).

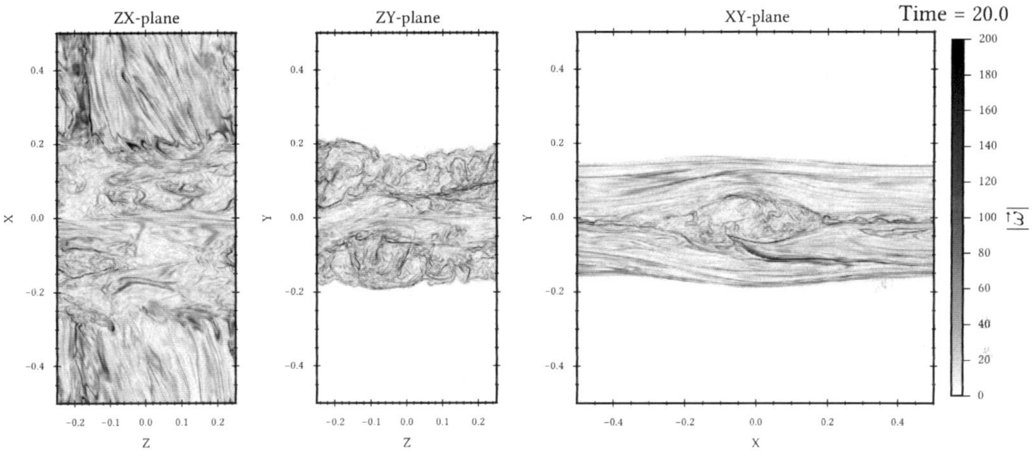

Figure 7. Slices along three main planes, ZX (left), ZY (center), and XY (right), of the amplitude of vorticity $\omega = \nabla \times \vec{v}$ at time $t = 20.0V_A$ for a model with sound speed $a = 1.0$ and grid size $h = 1/1024$. The turbulent region develops near the initial current sheet and expands in the Y direction. From (12).

4.3. Self-driven turbulent reconnection

The LV99 theory assumes that the level of turbulence is given. In fact, turbulence can also be produced by the outflow from a reconnection zone, as well as the instabilities of current sheets. This is the case of the self-driven reconnection. It was first addressed in LV99 and further discussed in (73). Turbulence in self-driven reconnection is generated by reconnection itself. The developed turbulence, in return, influences the reconnection process, resulting in its enhanced rate.

The problem of self-driven reconnection was approached by numerical simulations within last several years.The first numerical simulations of turbulence generated in spontaneous magnetic reconnection were presented in (10) with an incompressible MHD code. Later works on numerical simulations of stochastic reconnection were done by (11) and (74), both using compressible MHD codes. The most

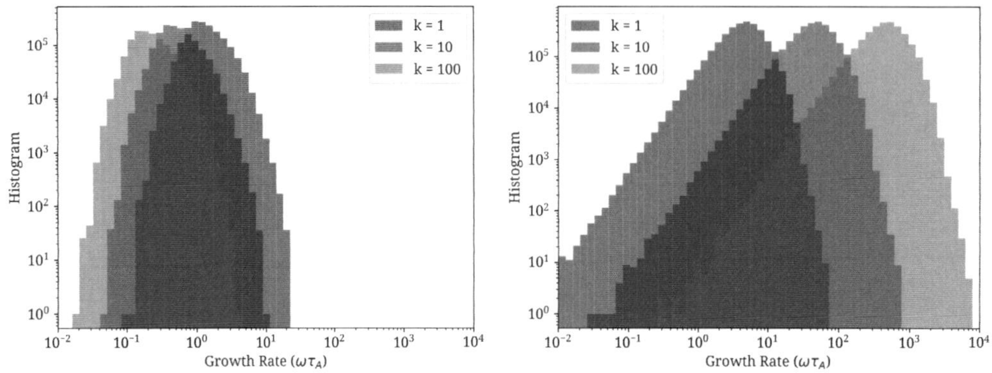

Figure 8. Distribution of the growth rate of the tearing mode (left) and Kelvin-Helmholtz instability (right) observed in reconnection-driven turbulence simulations by (75). Distributions with different colors correspond to three different perturbation wave numbers (see legend). From (75).

extensive studies of the reconnection-driven turbulence were carried out by (12) (see Fig. 7). They investigated the dependence of the properties of developed turbulence on the plasma β parameter. They reported Kolmogorov power spectra of generated velocity fluctuations, which also exhibit scale-dependent Goldreich-Sridhar anisotropy at later times in large-β models. Their results demonstrated that the turbulence statistics are similar to that of strong MHD turbulence.

(75) addressed the issue whether the tearing instability is important for generating turbulence. They considered two instabilities, tearing mode and Kelvin-Helmholtz instability. For each analyzed model they determined the magnetic (in the case of tearing mode) and velocity (in the case of Kelvin-Helmholtz instability) shear location. The obtained results indicated that the tearing mode growth is significantly suppressed at all studied wave numbers. In Figure 8, the statistical distribution of the growth rates of tearing mode (left panel) and Kelvin-Helmholtz instability (right panel) at three different perturbation wave numbers are shown. Strong suppression of tearing mode can be explained by the fact, that its growth rate actually decreases with the wave number k, while for Kelvin-Helmholtz instability it increases. It could also be attributed to the presence of the transverse component of magnetic field, which is generated by turbulence. The study demonstrates that a significant number of cells within current sheets are stable due to the strong transverse magnetic field component.

5. Observations of turbulent reconnection

Turbulent reconnection provides a natural explanation for the variations of solar activity. Indeed, the reconnection speed that can be inferred varies dramatically. This is difficult to explain on the basis of any plasma reconnection process. On the contrary, as we discussed earlier, the LV99 predictions suggest that the rate of magnetic reconnection varies with the level of turbulence, and this level of turbulence can be affected by the reconnection itself. Using the observational data, (76) provided the observational evidence in favor of LV99 model, namely, thick current sheets consistent with the theoretical predictions. Their observational results were reanalyzed later by taking into account the nature of turbulent driving (see (9; 77)) and even better correspondence to the LV99 predictions was reported. A very distinct prediction in LV99 is the spread of magnetic reconnection from one active reconnection region to adjacent reconnection regions. Accordingly, the turbulence induced by the reconnection in one flaring region is expected to induce reconnection in adjacent regions. In (78) and (79), the authors reported this effect.

The events of turbulent reconnection were also compared with the in situ measurements of the solar wind. There are extensive observations of strong narrow current sheets in the solar wind (80).

However, within the LV99 picture these are small-scale events representing the microphysics, but not much affecting the large-scale reconnection. The theory deals with very large-scale reconnection events in the solar wind, often associated with interplanetary coronal mass ejections and magnetic clouds or occasionally magnetic disconnection events at the heliospheric current sheet (81; 80). The outflows associated with these events have reconnection widths up to nearly 10^5 of the ion gyroradius ρ_i and exhibit a prolonged, quasi-stationary regime with reconnection lasting for several hours. The "current sheet" producing such large-scale reconnection contains many small-scale intense current sheets embedded in a diffuse turbulence background of weaker current. A successful comparison between the reconnection events in High-Speed Solar Wind and those in MHD turbulence simulations was performed in (82).

The violation of flux freezing required by the LV99 model of turbulent reconnection can be demonstrated in the case of the heliospheric current sheet (HCS). By analyzing the HCS data, (5) showed its turbulent nature and argued that the LV99 magnetic reconnection can explain the properties of the HCS. In addition, the Parker's spiral model (83) of the interplanetary magnetic field is a classical example used to demonstrate the concept of magnetic flux freezing. However, the deviations from this model are notable. (84) interpreted the observed deviations arising from a quasi-continuous magnetic reconnection. Their interpretation is consistent with the results of (5). The magnetic slippage through the fluid due to the pervasive turbulence in the near-ecliptic solar wind leads to a less tightly wound spiral and a stronger radial field strength than those in the Parker model, in agreement with the observations in (84). This is closely related to the reconnection diffusion proposed by (63) and (85).

6. Astrophysical implications of turbulent reconnection

6.1. Nonlinear turbulent dynamo, reconnection diffusion and star formation

The turbulent dynamo is the most promising mechanism that accounts for the cosmic magnetism (86). For the so-called "small-scale" turbulent dynamo, which operates on length scales smaller than the driving scale of turbulence, the nonlinear stage is the most astrophysically important. It gives rise to both the amplification of magnetic field strength and the increase of the correlation length. Numerical studies show that the nonlinear dynamo has a universal linear-in-time growth of magnetic energy with a low growth rate as a small fraction of the constant turbulent energy cascading rate (87; 88; 89; 90). In fact, the dynamo efficiency is determined by the interaction between the turbulent dynamo and the diffusion of magnetic fields. In earlier theoretical studies, only the microscopic diffusion of magnetic fields due to plasma effects, e.g., resistive diffusion, ambipolar diffusion, was considered (91; 92). However, their theoretical predictions cannot explain the numerical findings on the low dynamo efficiency and the spectral shape of magnetic fluctuations.

As demonstrated by the analytical study by (93), the reconnection diffusion of magnetic fields facilitated by the turbulent reconnection dominates over microscopic diffusion processes in the nonlinear stage of dynamo. As discussed in (63; 94; 95), the Richardson dispersion describes the magnetic field dynamics based on the Lagrangian approach, while the Eulerian approach in describing this dynamics is known as the "reconnection diffusion". The name stresses the importance of turbulent reconnection that enables this diffusion process. After incorporating the reconnection diffusion, (93) derived a small energy growth rate as $3/38$ of the turbulent energy cascading rate for the nonlinear dynamo. This result is in a good agreement with earlier numerical measurements (88; 90). As a result of reconnection diffusion, the growing magnetic energy spectrum peaks at the dynamo driving scale, where the growing magnetic energy reaches equipartition with the turbulent energy. A comparison between the analytical prediction by (93) on the magnetic energy spectrum and the numerical measurements by (86) confirms that the reconnection diffusion plays a crucial role in the nonlinear turbulent dynamo (see Fig. 9).

The nonlinear dynamo theory incorporating reconnection diffusion has important astrophysical implications for studying the role of magnetic fields for the first star formation (93) and the acceleration of Galactic cosmic rays (47). (47) found that the low efficiency of nonlinear dynamo due to reconnection diffusion well explains the magnetic field evolution and structure in shock simulations (96) and X-ray

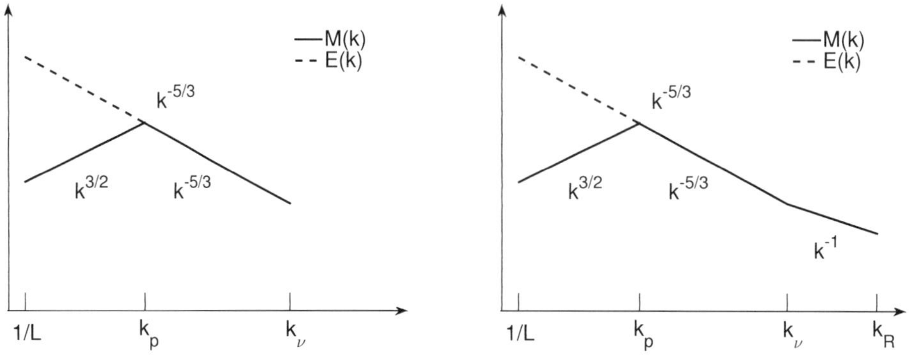

(a) $P_m = 1$ (b) $P_m > 1$

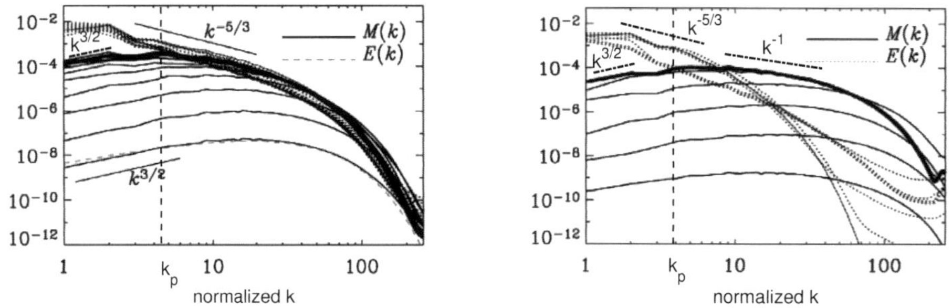

(c) $P_m = 1$, figure. 5.1 in (86) (d) $P_m = 50$, figure. 5.2 in (86)

Figure 9. Upper panel: illustration of the magnetic (solid line) and turbulent (dashed line) energy spectra for the nonlinear dynamo based on the theory by (93). Lower panel: numerical measurements by (86). From (93).

observations of supernova remnants (97).

The reconnection diffusion is also an important effect changing the classical theory of star formation. The standard theory is based on the assumption that magnetic field lines preserve their identify and the diffusion of charged particles perpendicular to magnetic field lines is restricted. Therefore, the mass loading of magnetic field lines does not change. However, the LV99 theory suggests that this is not true for turbulent magnetized fluids. MHD turbulence is expected to induce the diffusion of matter perpendicular to magnetic field. Magnetic field lines reconnect all the time in a turbulent flow. The process of reconnection connects magnetic fields with different mass loading and plasma pressures. As a result, plasmas stream along magnetic field lines to equalize the pressure. In the process, the portions of magnetic flux tubes with higher magnetic pressure expand as plasma pressure increases due to the flow of plasma along magnetic field lines. In other words, the process of diffusion driven by turbulent reconnection does not depend on the degree of ionization of the matter. In the absence of gravity, the outcome of reconnection diffusion is to make magnetic fields and plasmas more homogeneously distributed through the turbulent volume. Figure 10 presents a toy model of two adjacent magnetized eddies. The magnetic flux tubes moving with the eddies reconnect and exchange plasmas and magnetic fields. This induces turbulent diffusion of both magnetic fields and plasmas perpendicular to the mean magnetic field. In real turbulence, the above process takes place at every length scale. The concept of reconnection diffusion has been applied to different aspects of star formation, from the loss of magnetic support of molecular clouds to the problem of magnetic braking of accretion disks (see 98; 99).

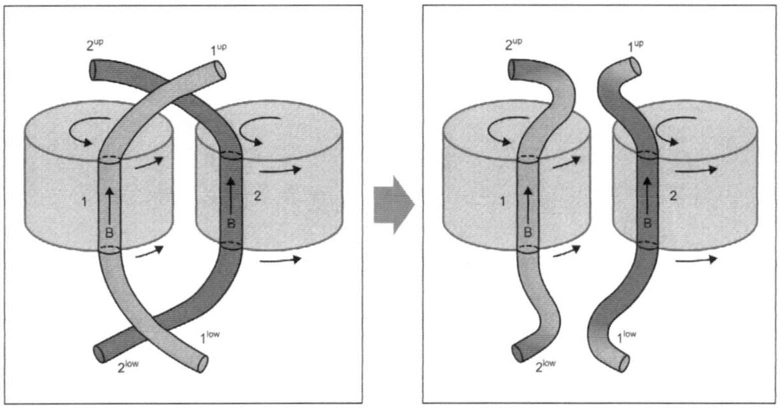

Figure 10. Reconnection diffusion: exchange of magnetic flux with entrained matter. Illustration of the mixing of matter and magnetic fields due to reconnection as two magnetic flux tubes of different eddies interact. In real turbulent cascade, such interactions proceed at every length scale of turbulent motions. From (95).

6.2. Energetic particle acceleration, flares and bursts of reconnection

Particle acceleration induced by magnetic reconnection was discussed before the appearance of LV99 model (e.g., (100)). The problem was that the Sweet-Parker reconnection is too slow to transfer an appreciable amount of energy to particles. The Petschek model faces the same difficulty as only a tiny part of magnetic field lines undergo reconnection. In this respect, the LV99 model with its volume reconnection offers a great opportunity to explore particle acceleration. In turbulent reconnection regions magnetic reconnection results in shrinking of magnetic loops and the charged particles entrained on magnetic loops get accelerated. It is important that the process takes place even in incompressible fluids, as the volume available for particles moving with magnetic field lines decreases as the magnetic field lines shrink. This acceleration process in the turbulent reconnection region was introduced in (101) (henceforth GL05). The process is illustrated in the left panel of Figure 11 that is taken from (63). The process of acceleration is the first-order Fermi acceleration, and it increases the parallel component of particle momentum as illustrated in the right panel of Figure 11. A similar acceleration processes was discussed later in the case of 2D tearing reconnection in (102). In 2D tearing reconnection, one has to consider magnetic islands, and the shrinking of magnetic field lines in 2D islands is limited by them getting circular. In 3D turbulent reconnection, the magnetic loops can shrink all the way with no constraints arising from the extraneous dimensional constraints.

The new mechanism of particle acceleration by reconnection has been invoked in many studies (see (104; 105; 106; 107; 108; 109; 110; 111), see also (112) for a review). In particular, it was discussed in the context of acceleration of energetic particles in pulsars (e.g. (107; 108; 109; 113)) and relativistic jets of active galactic nuclei (AGNs) (e.g. (114)). Most of the studies referred to the 2D tearing reconnection. However, the turbulent reconnection is the necessary feature of realistic astrophysical settings, and therefore the corresponding 3D turbulent reconnection is more relevant for particle acceleration in astrophysical environments. The particle acceleration in a large-scale turbulent current sheet was numerically studied by (115) with 3D MHD simulations and the injection of 10,000 test particles. The results are presented in Figure 12. The exponential growth of particle energy is shown in the figure. During this stage the acceleration rate scales as $\sim E^{-0.4}$ (116). We note that these results are difficult to reproduce with PIC simulations, as most of PIC simulations still do not have enough particles to reproduce the turbulent behavior of fluid, which is the key physical ingredient of astrophysical magnetic reconnection.

The fast turbulent reconnection acceleration is a widespread phenomenon. In particular, it has been

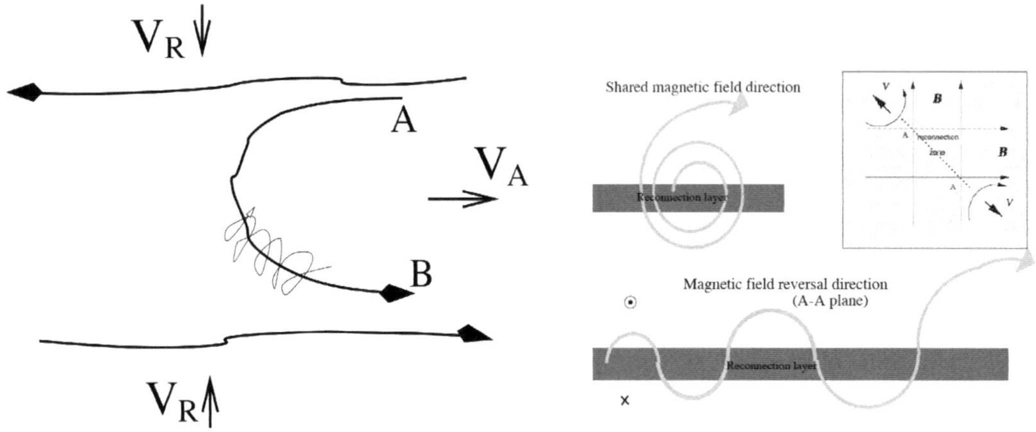

Figure 11. *Left*: Shrinking magnetic loop within the turbulent reconnection region with a particle gaining higher $p_{\|}$ as it follows the magnetic field. From (63). *Right*: A particle with a large Larmor radius interacts with converging magnetized flows and its $p_{\perp}$ increases as a result. From (103).

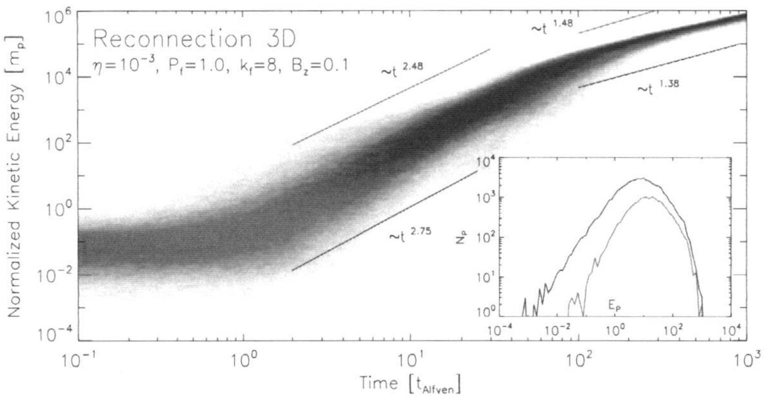

Figure 12. Particle kinetic energy distributions for 10,000 protons injected in the simulated turbulent reconnection region. The color indicates the velocity component that is accelerated (red or blue for parallel or perpendicular, respectively). The energy is normalized by the rest proton mass. From (115).

discussed in (117) as a source of anomalous cosmic rays observed by Voyagers and in (118) as a source of the observed cosmic ray anisotropies. It was also discussed for the environments around black holes (GL05, (119; 120; 121; 116; 122)).

In a magnetically dominated medium, the release of magnetic energy via reconnection results in the outflow that induces turbulence in astrophysical high-Reynolds number plasma. This in turn inevitably increases the rate of reconnection and magnetic energy dissipation, leading to a reconnection instability (73). The applications of the theory range from solar flares to gamma ray bursts (GRBs) (see 123). It was the basis of the Internal-Collision-induced MAgnetic Reconnection and Turbulence (ICMART) model of GRBs by (124). They showed that the model can overcome several difficulties of the traditional internal shock model (125; 126; 127; 128; 129) and well interpret the lightcurves and spectra of GRBs (130; 131; 132; 133). (134) used the theoretical advances in relativistic turbulent reconnection to improve the ICMART model. They also modified the ICMART model by identifying the kink instability as the most plausible mechanism of creating the magnetic configurations prone to reconnection and triggering the turbulent reconnection. In addition, the magnetic energy release and particle acceleration arising from turbulent magnetic reconnection associated with AGNs and galactic black hole binaries (GHBs) has also been explored using the model in GL05. (120) revisited the aforementioned model and extended the study to investigating the gamma-ray flare emission of these sources. They confirmed the predictions in GL05. The study by (121) suggests that the results in (120) are robust and depend on properties of turbulent reconnection rather than the detailed accretion physics.

7. Turbulent and tearing reconnection

7.1. Comparison of turbulent and tearing reconnection models

Tearing reconnection was discussed in the literature prior to the recent surge of interest in the process (see 135). It can act as a trigger for turbulent reconnection process in the settings where the reconnecting magnetic fields are initially laminar. Most studies of tearing reconnection were carried out in 2D (136; 137; 138; 139; 140; 141; 142; 143; 144; 145; 146; 147; 148; 74; 149; 150; 151; 152; 153; 154; 155), and this is a serious theoretical deficiency. The reason that the 2D tearing reconnection model is adopted by the reconnection community is that in 3D the particle-in-cell (PIC) simulations do not yet have enough particles to reproduce fluid/turbulent behavior. Therefore, for most PIC simulations the regime of MHD-like turbulent reconnection is very difficult to achieve.

In 2D MHD numerical simulations, the transfer from Sweet-Parker to tearing reconnection was observed in the 2D current sheet starting with a particular Lundquist number larger than $S \approx 10^4$. The resulting reconnection does not depend on the fluid resistivity. The study of tearing reconnection momentarily eclipsed the earlier mainstream research based on the Petschek-type X point reconnection (156; 157; 102). Conceptually, the tearing reconnection, which invokes a cascade of mergers that increases the size of plasmoids, is closer to the turbulent reconnection compared with the Petschek X-point reconnection. Tearing introduces stochasticity into reconnection and brings the mainstream models of magnetic reconnection closer to the LV99 turbulent reconnection model. In particular, similar to the LV99 model, the tearing reconnection presents volume-filling Y-type reconnection with a broad outflow. This allows some ideas, e.g. first-order Fermi acceleration of cosmic rays, developed within the framework of the LV99 theory to be transferred to the case of tearing reconnection.

LV99 model may not be applicable to 2D configurations (4), and we do not expect to see the turbulent reconnection dominating the 2D reconnection process. Numerical tests in a number of recent studies (8; 9; 11; 10; 12; 14) show that the transfer to turbulent reconnection is inevitable in 3D. This transfer is also inevitable in 3D and large-Lundquist number astrophysical systems. Indeed, the mass conservation constraint requires that one has to increase the outflow region thickness Δ in proportion to L_x. The Reynolds number Re of the outflow, approximately $\Delta V_A/\nu$ with ν as the viscosity, grows with the Lundquist number S. As Re increases the flow gets turbulent.

Recently, an extensive literature has developed suggesting that reconnection in MHD turbulence is necessarily induced by tearing instability (see 158; 159; 160; 161; 162; 163; 164; 165). The authors explored the possible modifications of the properties of MHD turbulence through this instability. They believed that for high Reynolds and Lundquist numbers, the microscale current sheets become tearing unstable, and this instability affects the turbulence on length scales smaller than some critical scale but larger than the dissipation scale of turbulence. This idea is based on a dubious assumption that the properties of laminar tearing instability can be directly applied to turbulent environments. The modification of turbulence spectrum does not affect the LV99 picture, which does not directly depend on the particular GS95 scaling and also holds for an arbitrary spectrum of MHD turbulence (see LV99, Appendix D). However, we feel that it is a serious misconception to talk about the "reconnection-mediated turbulence" or a modification of MHD turbulence "due to reconnection" at scales smaller than some critical scale. It implies that the fast magnetic reconnection does not occur in the inertial range on scales larger than this critical length. For example, (158) regarded it as a surprise that in MHD turbulence "the anisotropic, current-sheetlike eddies become the sites of magnetic reconnection *before* the formal Kolmogorov dissipation scale is reached" and they argued that such fast reconnection sets in only because "Sweet-Parker current sheets above a certain critical aspect ratio ... are violently unstable to the formation of multiple magnetic islands, or plasmoids." This contradicts to what we know about turbulent reconnection and related processes, i.e. Richardson dispersion, flux freezing violation, that are discussed above. Indeed, fast reconnection *on all length scales* is an essential part of the MHD turbulent cascade, which is required for the dynamical consistency of MHD turbulence.

7.2. Reconnection diffusion, turbulent resistivity, and turbulent ambipolar diffusion

LV99 predicted dramatic changes of the dynamics of magnetic fields in turbulent fluids compared to their laminar counterparts. These changes may be erroneously associated with the concept of "turbulent resistivity". The description of the phenomena in MHD turbulence based on the concept of "turbulent resistivity" can be very problematic. If we know the reconnection rate, e.g. from LV99, then an eddy-resistivity can always be adjusted to achieve the required rate. Although the manually tuned reconnection rate is numerically satisfactory, the model itself is unphysical. For instance, the required large-eddy resistivity is expected to smooth out all magnetic structures on length scales below the injection scale. Obviously, this contradicts to the physical reality, where turbulent magnetic fluctuations are observed on all length scales down to the dissipation micro-scale. All the relevant MHD turbulence physics, including the Richardson diffusion, magnetic helicity conservation, is also lost within the turbulent resistivity approach.

In partially ionized gas, the concept "turbulent ambipolar diffusion" (166; 167) was developed to explain the magnetic flux removal from molecular clouds. The concept is based on the idea that turbulence can create gradients in neutrals, and those can accelerate the overall pace of ambipolar diffusion. The simulations in (168) were claimed as the numerical support for the "turbulent ambipolar diffusion". However, in 3D if there is no fast turbulent reconnection, the turbulent mixing and thus the "turbulent ambipolar diffusion" is not possible. Moreover, (168) reported no dependence of the "turbulent ambipolar diffusion" on the ambipolar diffusion rate. Both facts taken together suggest that in turbulent partially ionized gas the reconnection diffusion dominates over the ambipolar diffusion for the magnetic field diffusion process.

8. Past, present, and future of turbulent reconnection theory

It has been 20 years since the introduction of the 3D turbulent reconnection model. The tearing reconnection model shares many common features with the original LV99 model of turbulent reconnection, and the two models can complement each other in some instances. For example, the tearing reconnection can be the initial transient stage of magnetic reconnection before turbulence is developed. Numerical simulations testify that the steady-state 3D reconnection is generally turbulent and astrophysically more relevant.

The turbulent reconnection is an essential part of the magnetic field dynamics in turbulent fluids. The turbulent reconnection theory is a part of the theories of MHD turbulence, turbulent dynamo, magnetic field stochasticity, etc. It has induced significant changes in understanding fundamental astrophysical processes and interpreting observational phenomena involving magnetic fields, including cosmic ray acceleration, star formation, and non-thermal radiation associated with high-energy astrophysical systems. Similar to other turbulence induced processes, it is important to note that turbulent reconnection does not depend on plasma microphysics, and this justifies the MHD treatment of turbulent fluids in numerical studies of reconnection.

Acknowledgments

AL acknowledges the support of NSF AST grant 1715754 and ACI grant 1713782. GK acknowledges the support of the Brazilian agency CNPq (no. 304891/2016-9). SX acknowledges the support for Program number HST-HF2-51400.001-A provided by NASA through a grant from the Space Telescope Science Institute, which is operated by the Association of Universities for Research in Astronomy, Incorporated, under NASA contract NAS5-26555.

References

[1] Alfvén H 1943 *Arkiv for Matematik, Astronomi och Fysik* **29** 1–7

[2] Parker E N 1979 *Cosmical magnetic fields. Their origin and their activity* International series of monographs on physics (Clarendon Press) ISBN 9780198512905 URL `https://books.google.com.br/books?id=Msp-AAAAIAAJ`

[3] Lazarian A and Vishniac E T 1999 *The Astrophysical Journal* **517** 700–718

[4] Eyink G L, Lazarian A and Vishniac E T 2011 *The Astrophysical Journal* **743** 51

[5] Eyink G L 2015 *The Astrophysical Journal* **807** 137

[6] Jafari A, Vishniac E T, Kowal G and Lazarian A 2018 *The Astrophysical Journal* **860** 52

[7] Kowal G, Lazarian A, Vishniac E T and Otmianowska-Mazur K 2009 *The Astrophysical Journal* **700** 63–85

[8] Kowal G, Lazarian A, Vishniac E T and Otmianowska-Mazur K 2012 *Nonlinear Processes in Geophysics* **19** 297–314

[9] Eyink G, Vishniac E, Lalescu C, Aluie H, Kanov K, Bürger K, Burns R, Meneveau C and Szalay A 2013 *Nature* **497** 466–469

[10] Beresnyak A 2017 *The Astrophysical Journal* **834** 47

[11] Oishi J S, Mac Low M M, Collins D C and Tamura M 2015 *ArXiv e-prints*

[12] Kowal G, Falceta-Gonçalves D A, Lazarian A and Vishniac E T 2017 *The Astrophysical Journal* **838** 91

[13] Takamoto M and Lazarian A 2016 *The Astrophysical Journal Letters* **831** L11

[14] Takamoto M 2018 *Monthly Notices of the Royal Astronomical Society* **476** 4263–4271

[15] Armstrong J W, Rickett B J and Spangler S R 1995 *The Astrophysical Journal* **443** 209–221

[16] Chepurnov A and Lazarian A 2010 *The Astrophysical Journal* **710** 853–858

[17] Lazarian A 2009 *Space Science Reviews* **143** 357–385

[18] Leamon R J, Smith C W, Ness N F, Matthaeus W H and Wong H K 1998 *Journal of Geophysical Research* **103** 4775

[19] Burkhart B, Stanimirović S, Lazarian A and Kowal G 2010 *The Astrophysical Journal* **708** 1204–1220

[20] Norman C A and Ferrara A 1996 *The Astrophysical Journal* **467** 280

[21] Ferrière K M 2001 *Reviews of Modern Physics* **73** 1031–1066

[22] Subramanian K, Shukurov A and Haugen N E L 2006 *Monthly Notices of the Royal Astronomical Society* **366** 1437–1454

[23] Enßlin T A and Vogt C 2006 *Astronomy & Astrophysics* **453** 447–458

[24] Chandran B D G 2005 *The Astrophysical Journal* **632** 809–820

[25] Balbus S A and Hawley J F 1998 *Reviews of Modern Physics* **70** 1–53

[26] Jafari A and Vishniac E T 2018 *The Astrophysical Journal* **854** 2

[27] Galsgaard K and Nordlund Å 1997 *Journal of Geophysical Research* **102** 231–248

[28] Gerrard C L and Hood A W 2003 *Solar Physics* **214** 151–169

[29] Bale S D, Kellogg P J, Mozer F S, Horbury T S and Reme H 2005 *Physical Review Letters* **94** 215002

[30] Padoan P, Juvela M, Kritsuk A and Norman M L 2006 *The Astrophysical Journal Letters* **653** L125–L128

[31] Vogt C and Enßlin T A 2005 *Astronomy & Astrophysics* **434** 67–76

[32] Lazarian A and Pogosyan D 2000 *The Astrophysical Journal* **537** 720–748

[33] Lazarian A and Pogosyan D 2004 *The Astrophysical Journal* **616** 943–965

[34] Lazarian A and Pogosyan D 2006 *The Astrophysical Journal* **652** 1348–1365

[35] Padoan P, Kritsuk A G, Lunttila T, Juvela M, Nordlund A, Norman M L and Ustyugov S D 2010 MHD Turbulence In Star-Forming Clouds *American Institute of Physics Conference Series* (*American Institute of Physics Conference Series* vol 1242) ed Bertin G, de Luca F, Lodato G, Pozzoli R and Romé M pp 219–230

[36] Stanimirović S and Lazarian A 2001 *The Astrophysical Journal Letters* **551** L53–L56

[37] Chepurnov A, Lazarian A, Stanimirović S, Heiles C and Peek J E G 2010 *The Astrophysical Journal* **714** 1398–1406

[38] Chepurnov A, Burkhart B, Lazarian A and Stanimirovic S 2015 *ArXiv e-prints*

[39] Cho J and Vishniac E T 2000 *The Astrophysical Journal* **539** 273–282

[40] Maron J and Goldreich P 2001 *The Astrophysical Journal* **554** 1175–1196

[41] Cho J and Lazarian A 2002 *Physical Review Letters* **88** 245001

[42] Cho J and Lazarian A 2003 *Monthly Notices of the Royal Astronomical Society* **345** 325–339

[43] Kowal G and Lazarian A 2010 *The Astrophysical Journal* **720** 742–756

[44] Goldreich P and Sridhar S 1995 *The Astrophysical Journal* **438** 763–775

[45] Lithwick Y and Goldreich P 2001 *The Astrophysical Journal* **562** 279–296

[46] Lazarian A, Vishniac E T and Cho J 2004 *The Astrophysical Journal* **603** 180–197

[47] Xu S and Lazarian A 2017 *New Journal of Physics* **19** 065005

[48] Thompson C and Blaes O 1998 *Physical Review D* **57** 3219–3234

[49] Cho J 2005 *The Astrophysical Journal* **621** 324–327

[50] Zhang W, MacFadyen A and Wang P 2009 *The Astrophysical Journal Letters* **692** L40–L44

[51] Inoue T, Asano K and Ioka K 2011 *The Astrophysical Journal* **734** 77

[52] Beckwith K and Stone J M 2011 *The Astrophysical Journal Supplement Series* **193** 6

[53] Zrake J and MacFadyen A I 2012 *The Astrophysical Journal* **744** 32

[54] Zrake J and MacFadyen A I 2013 *The Astrophysical Journal Letters* **763** L12

[55] Garrison D and Nguyen P 2015 *ArXiv e-prints*

[56] Takamoto M and Lazarian A 2017 *Monthly Notices of the Royal Astronomical Society* **472** 4542–4550

[57] Parker E N 1957 *Journal of Geophysical Research* **62** 509–520

[58] Sweet P A 1958 *The Observatory* **78** 30–32

[59] Petschek H E 1964 *NASA Special Publication* **50** 425

[60] Kulsrud R M 2005 *Plasma physics for astrophysics* Plasma Physics for Astrophysics (Princeton University Press) ISBN 9780691120737 URL `https://books.google.com.br/books?id=kuk9DwAAQBAJ`

[61] Richardson L F 1926 *Proceedings of the Royal Society of London. Series A, Containing Papers of a Mathematical and Physical Character* **110** 709–737

[62] Jokipii J R 1973 *The Astrophysical Journal* **183** 1029–1036

[63] Lazarian A 2005 Astrophysical Implications of Turbulent Reconnection: from cosmic rays to star formation *Magnetic Fields in the Universe: From Laboratory and Stars to Primordial Structures.* (*American Institute of Physics Conference Series* vol 784) ed de Gouveia dal Pino E M, Lugones G and Lazarian A pp 42–53

[64] Beresnyak A 2013 *The Astrophysical Journal Letters* **767** L39

[65] Xu S and Yan H 2013 *The Astrophysical Journal* **779** 140

[66] Eyink G L and Drivas T D 2018 *Physical Review X* **8** 011022

[67] Matthaeus W H and Lamkin S L 1985 *Physics of Fluids* **28** 303–307

[68] Matthaeus W H and Lamkin S L 1986 *Physics of Fluids* **29** 2513–2534

[69] Watson P G, Oughton S and Craig I J D 2007 *Physics of Plasmas* **14** 032301

[70] Iroshnikov P S 1964 *Soviet Astronomy* **7** 566

[71] Kraichnan R H 1965 *Physics of Fluids* **8** 1385–1387

[72] Vishniac E T, Pillsworth S, Eyink G, Kowal G, Lazarian A and Murray S 2012 *Nonlinear Processes in Geophysics* **19** 605–610

[73] Lazarian A and Vishniac E T 2009 Model of Reconnection of Weakly Stochastic Magnetic Field and its Implications *Revista Mexicana de Astronomia y Astrofisica Conference Series* (*Revista Mexicana de Astronomia y Astrofisica, vol. 27* vol 36) pp 81–88

[74] Huang Y M and Bhattacharjee A 2016 *The Astrophysical Journal* **818** 20

[75] Kowal G, Falceta-Gonçalves D A, Lazarian A and Vishniac E T 2019 *in preparation*

[76] Ciaravella A and Raymond J C 2008 *The Astrophysical Journal* **686** 1372–1382

[77] Lazarian A, Eyink G L, Vishniac E T and Kowal G 2015 *ArXiv e-prints*

[78] Sych R, Nakariakov V M, Karlicky M and Anfinogentov S 2009 *Astronomy & Astrophysics* **505** 791–799

[79] Sych R, Karlický M, Altyntsev A, Dudík J and Kashapova L 2015 *Astronomy & Astrophysics* **577** A43

[80] Gosling J 2012 *Space science reviews* **172** 187–200

[81] Phan T D, Gosling J T and Davis M S 2009 *Geophysical Research Letters* **36** L09108

[82] Lalescu C C, Shi Y K, Eyink G L, Drivas T D, Vishniac E T and Lazarian A 2015 *Physical Review Letters* **115** 025001

[83] Parker E N 1958 *The Astrophysical Journal* **128** 664

[84] Khabarova O and Obridko V 2012 *The Astrophysical Journal* **761** 82

[85] Santos-Lima R, Lazarian A, de Gouveia Dal Pino E M and Cho J 2010 *The Astrophysical Journal* **714** 442–461

[86] Brandenburg A and Subramanian K 2005 *Physics Reports* **417** 1–209

[87] Ryu D, Kang H, Cho J and Das S 2008 *Science* **320** 909–

[88] Cho J, Vishniac E T, Beresnyak A, Lazarian A and Ryu D 2009 *The Astrophysical Journal* **693** 1449–1461

[89] Beresnyak A, Jones T W and Lazarian A 2009 *The Astrophysical Journal* **707** 1541–1549

[90] Beresnyak A 2012 *Physical Review Letters* **108** 035002

[91] Schekochihin A A, Cowley S C, Hammett G W, Maron J L and McWilliams J C 2002 *New Journal of Physics* **4** 84

[92] Kulsrud R M and Anderson S W 1992 *The Astrophysical Journal* **396** 606–630

[93] Xu S and Lazarian A 2016 *The Astrophysical Journal* **833** 215

[94] Lazarian A and Yan H 2014 *The Astrophysical Journal* **784** 38

[95] Lazarian A, Esquivel A and Crutcher R 2012 *The Astrophysical Journal* **757** 154

[96] Inoue T, Yamazaki R and Inutsuka S i 2009 *The Astrophysical Journal* **695** 825–833

[97] Uchiyama Y and Aharonian F A 2008 *The Astrophysical Journal Letters* **677** L105

[98] Lazarian A 2014 *Space Science Reviews* **181** 1–59
[99] González-Casanova D F, Lazarian A and Santos-Lima R 2016 *The Astrophysical Journal* **819** 96
[100] Litvinenko Y E 1996 *The Astrophysical Journal* **462** 997
[101] de Gouveia Dal Pino E M and Lazarian A 2005 *Astronomy & Astrophysics* **441** 845–853
[102] Drake J F, Swisdak M, Che H and Shay M A 2006 *Nature* **443** 553–556
[103] Lazarian A, Vlahos L, Kowal G, Yan H, Beresnyak A and de Gouveia Dal Pino E M 2012 *Space Science Reviews* 86
[104] Jaroschek C H, Lesch H and Treumann R A 2004 *The Astrophysical Journal Letters* **605** L9–L12
[105] Zenitani S and Hoshino M 2008 *The Astrophysical Journal* **677** 530–544
[106] Zenitani S, Hesse M and Klimas A 2009 *The Astrophysical Journal* **696** 1385–1401
[107] Cerutti B, Werner G R, Uzdensky D A and Begelman M C 2013 *The Astrophysical Journal* **770** 147
[108] Cerutti B, Werner G R, Uzdensky D A and Begelman M C 2014 *The Astrophysical Journal* **782** 104
[109] Sironi L and Spitkovsky A 2014 *The Astrophysical Journal Letters* **783** L21
[110] Werner G R, Uzdensky D A, Cerutti B, Nalewajko K and Begelman M C 2014 *ArXiv e-prints*
[111] Guo F and Giacalone J 2015 *The Astrophysical Journal* **802** 97
[112] Hoshino M and Lyubarsky Y 2012 *Space Science Reviews* **173** 521–533
[113] Uzdensky D A and Spitkovsky A 2014 *The Astrophysical Journal* **780** 3
[114] Giannios D 2010 *Monthly Notices of the Royal Astronomical Society* **408** L46–L50
[115] Kowal G, de Gouveia Dal Pino E M and Lazarian A 2012 *Physical Review Letters* **108** 241102
[116] Khiali B, de Gouveia Dal Pino E M and del Valle M V 2015 *Monthly Notices of the Royal Astronomical Society* **449** 34–48
[117] Lazarian A and Opher M 2009 *The Astrophysical Journal* **703** 8–21
[118] Lazarian A and Desiati P 2010 *The Astrophysical Journal* **722** 188–196
[119] de Gouveia Dal Pino E M, Kowal G, Kadowaki L H S, Piovezan P and Lazarian A 2010 *International Journal of Modern Physics D* **19** 729–739
[120] Kadowaki L H S, de Gouveia Dal Pino E M and Singh C B 2015 *The Astrophysical Journal* **802** 113
[121] Singh C B, de Gouveia Dal Pino E M and Kadowaki L H S 2015 *The Astrophysical Journal Letters* **799** L20
[122] Khiali B and de Gouveia Dal Pino E M 2015 *ArXiv e-prints*
[123] Lazarian A, Petrosian V, Yan H and Cho J 2003 *ArXiv Astrophysics e-prints*
[124] Zhang B and Yan H 2011 *The Astrophysical Journal* **726** 90
[125] Rees M J and Meszaros P 1994 *The Astrophysical Journal Letters* **430** L93–L96
[126] Kobayashi S, Piran T and Sari R 1997 *The Astrophysical Journal* **490** 92
[127] Daigne F and Mochkovitch R 1998 *Monthly Notices of the Royal Astronomical Society* **296** 275–286
[128] Ghisellini G, Celotti A and Lazzati D 2000 *Monthly Notices of the Royal Astronomical Society* **313** L1–L5
[129] Kumar P and McMahon E 2008 *Monthly Notices of the Royal Astronomical Society* **384** 33–63
[130] Zhang B and Zhang B 2014 *The Astrophysical Journal* **782** 92
[131] Uhm Z L and Zhang B 2014 *Nature Physics* **10** 351–356

[132] Xu S and Zhang B 2017 *The Astrophysical Journal Letters* **846** L28
[133] Xu S, Yang Y P and Zhang B 2018 *The Astrophysical Journal* **853** 43
[134] Lazarian A, Zhang B and Xu S 2018 *arXiv e-prints*
[135] Syrovatskii S I 1981 *Annual Review of Astronomy & Astrophysics* **19** 163–229
[136] Biskamp D 1986 *Physics of Fluids* **29** 1520–1531
[137] Shibata K and Tanuma S 2001 *Earth, Planets, and Space* **53** 473–482
[138] Daughton W, Scudder J and Karimabadi H 2006 *Physics of Plasmas* **13** 072101
[139] Daughton W, Roytershteyn V, Albright B J, Karimabadi H, Yin L and Bowers K J 2009 *Physical Review Letters* **103** 065004
[140] Daughton W, Roytershteyn V, Karimabadi H, Yin L, Albright B J, Bergen B and Bowers K J 2011 *Nature Physics* **7** 539–542
[141] Daughton W S, Akcay C, Billey Z, Finn J, Zweibel E and Gekelman W N 2014 *AGU Fall Meeting Abstracts* SH22A-04
[142] Fermo R L, Drake J F and Swisdak M 2012 *Physical Review Letters* **108** 255005
[143] Loureiro N F, Schekochihin A A and Cowley S C 2007 *Physics of Plasmas* **14** 100703
[144] Loureiro N F, Samtaney R, Schekochihin A A and Uzdensky D A 2012 *Physics of Plasmas* **19** 042303
[145] Lapenta G 2008 *Physical Review Letters* **100** 235001
[146] Bhattacharjee A, Huang Y M, Yang H and Rogers B 2009 *Physics of Plasmas* **16** 112102
[147] Cassak P A, Shay M A and Drake J F 2009 *Physics of Plasmas* **16** 120702
[148] Huang Y M and Bhattacharjee A 2010 *Physics of Plasmas* **17** 062104
[149] Shepherd L S and Cassak P A 2010 *Physical Review Letters* **105** 015004
[150] Uzdensky D A, Loureiro N F and Schekochihin A A 2010 *Physical Review Letters* **105** 235002
[151] Huang Y M, Bhattacharjee A and Sullivan B P 2011 *Physics of Plasmas* **18** 072109–072109
[152] Bárta M, Büchner J, Karlický M and Skála J 2011 *The Astrophysical Journal* **737** 24
[153] Shen C, Lin J and Murphy N A 2011 *The Astrophysical Journal* **737** 14
[154] Takamoto M 2013 *The Astrophysical Journal* **775** 50
[155] Wyper P F and Pontin D I 2014 *Physics of Plasmas* **21** 102102
[156] Shay M A and Drake J F 1998 *Geophysics Research Letters* **25** 3759–3762
[157] Drake J F 2001 *Nature* **410** 525–526
[158] Loureiro N F and Boldyrev S 2017 *Physical Review Letters* **118** 245101
[159] Boldyrev S and Loureiro N F 2017 *The Astrophysical Journal* **844** 125
[160] Loureiro N F and Boldyrev S 2017 *The Astrophysical Journal* **850** 182
[161] Walker J, Boldyrev S and Loureiro N F 2018 *Physical Review E* **98** 033209
[162] Mallet A, Schekochihin A A and Chandran B D G 2017 *Journal of Plasma Physics* **83** 905830609
[163] Mallet A, Schekochihin A A and Chandran B D G 2017 *Monthly Notices of the Royal Astronomical Society* **468** 4862–4871
[164] Comisso L, Huang Y M, Lingam M, Hirvijoki E and Bhattacharjee A 2018 *The Astrophysical Journal* **854** 103
[165] Vech D, Mallet A, Klein K G and Kasper J C 2018 *The Astrophysical Journal Letters* **855** L27
[166] Fatuzzo M and Adams F C 2002 *The Astrophysical Journal* **570** 210–221
[167] Zweibel E G 2002 *The Astrophysical Journal* **567** 962–970
[168] Heitsch F, Zweibel E G, Slyz A D and Devriendt J E G 2004 *The Astrophysical Journal* **603** 165–179

On the origin of the polytropic behavior in space plasmas

George Livadiotis

Southwest Research Institute, USA
glivadiotis@swri.edu

Abstract. The paper addresses the connection between the polytropic behavior – the specific power-law relationship among the thermal plasma moments– and the functional form of the distribution of particle velocities and energies. Surprisingly, the polytropic behavior requires the statistical mechanics of the plasma particles to turn to the framework of kappa distributions. While it was already known that kappa distributions can lead to the polytropic relationship, the new result shows that the reverse derivation is also true; thus, the polytropic behavior has the role of a mechanism generating kappa distributions. Therefore, when observations confirm the existence of a polytropic behavior in plasma particles, then the framework of kappa distributions for describing particle velocities and energies can be indirectly confirmed.

1. Introduction

Space plasmas are collisionless and weakly coupled particle systems, characterized by a collective behavior that induces correlations among their particles. The distributions of particle velocities and energies are frequently described by a specific type of non-Maxwellian distributions, the so-called, kappa distributions (e.g., see the book of kappa distributions [1]). These distributions generalize the Maxwell-Boltzmann distribution that applies in ideal systems characterized by no correlations among their particles.

Kappa distributions were first introduced to describe magnetospheric electron data, by Olbert and its Ph.D. students and colleagues [2-4]; Binsack (1966) [2] was the first to publish the usage of kappa distributions, but he acknowledged that the kappa function was actually "introduced by Prof. Olbert of MIT in his studies of IMP-1." Since then, single types of kappa distributions are frequently used to model the space plasma populations (e.g., [5-7]); however, more complicated models of kappa distributions have been employed to describe rare features (e.g., anisotropy, superposition) (e.g., [8-10]). A kappa distribution is primarily formulated to describe a Hamiltonian, i.e., the particle kinetic and potential energy ([11,12]; [1], Chapter 3).

Various mechanisms responsible to generate kappa distributions in space plasmas have been examined and verified. Some examples are: macroscopic extensivity of entropy [13], thermodynamics [14], superstatistics [15-18], effect of shock waves [19], weak turbulence [20], turbulence with a diffusion coefficient inversely proportional to velocity [21], effect of pickup ions [22], pump acceleration mechanism [23], polytropic behavior [24-26]; (see also: [1], Chapters 5,6,8,10,15,16). Also, common processes, characteristic of space plasmas, such as the Debye shielding and magnetic coupling, have an important role in the generation of kappa distributions in plasmas [27]. In general, long-range interactions or other causes of local particle correlations implicate the particle system in the statistical framework of kappa distributions [28]. Such an example is the state of a charged test particle in a constant temperature heat bath of a second species of charged particles, modeled by [29]. The time dependence of the distribution function of the test particle is given by a Fokker-Planck

Content from this work may be used under the terms of the Creative Commons Attribution 3.0 licence. Any further distribution of this work must maintain attribution to the author(s) and the title of the work, journal citation and DOI.
Published under licence by IOP Publishing Ltd

equation for Coulomb collisions and wave-particle interactions; the stationary state of this equation can be described by kappa distributions (for certain choices of the involved parameters). (For more examples, see: [30-37].)

The kappa distributions have become increasingly widespread across the physics of space plasma processes, describing particles in the heliosphere, from the solar wind and planetary magnetospheres to the heliosheath and beyond, the interstellar and intergalactic plasmas: *inner heliosphere*, including solar wind (e.g., [13,20,26,38-48]), solar spectra (e.g., [49,50]), solar corona (e.g., [51-54]), solar energetic particles (e.g., [55,56]), corotating interaction regions (e.g., [57]), and solar flares related (e.g., [58,33,21,59]); *planetary magnetospheres*, including magnetosheath (e.g., [60,61]), magnetopause (e.g., [62]), magnetotail (e.g., [63]), ring current (e.g., [64]), plasma sheet (e.g., [65-67]), magnetospheric substorms (e.g., [68]), Aurora (e.g., [69]), magnetospheres of giant planets, such as Jovian (e.g., [70-73]), Saturnian (e.g., [74-76]), Uranian (e.g., [77]), Neptunian (e.g., [78]), magnetospheres of planetary moons, such as Io (e.g., [79]) and Enceladus (e.g., [80]), cometary magnetospheres (e.g., [81,82]); *outer heliosphere and the inner heliosheath* (e.g., [83-91,22,92-97,13]); (iv) *beyond the heliosphere*, including HII regions (e.g., [98]), planetary nebula (e.g., [99,100]), and supernova magnetospheres (e.g., [101]); and in cosmological scales (e.g., [102]). The kappa distributions have also been applied in general plasma-related analyses (e.g., [103-115,92,116-121).

A breakthrough in the field came with the connection of kappa distributions with statistical mechanics and thermodynamics, accomplished by the following findings: (1) kappa distributions maximize entropy under the constraints of canonical ensemble [122]; (2) particle systems exchanging heat and reaching thermodynamic equilibrium are stabilized always into a kappa distribution [14]; and (3) polytropes - Systems characterized by a polytropic behavior - are uniquely consistent to kappa distributions and their statistical formalism [24]. The latter is the topic presented in this paper, that is, the equivalence between the polytropic behavior and the formalism of kappa distributions.

2. General profiles of thermal observables

The phase-space distribution function of plasma particles is constructed by considering the distribution of particle energy, $p(E)$, and substituting the Hamiltonian function instead of the energy, $f(\vec{r},\vec{u}) = p\left[H(\vec{r},\vec{u})\right]$. We consider the 1-particle Hamiltonian function, $H(\vec{r},\vec{u}) = \varepsilon_{\mathrm{K}}(\vec{u}) + \Phi(\vec{r})$, for an ideal-gas type of plasma, where the inter-particle dynamical terms are negligible and the potential energy depends only on the position vector. The kinetic energy is expressed in the reference frame of the co-moving system, i.e., $\varepsilon_{\mathrm{K}}(\vec{u}) = \frac{1}{2}m\cdot(\vec{u}-\vec{u}_{\mathrm{b}})^2$, with $\vec{u}_{\mathrm{b}}$ noting the bulk plasma velocity.

The integration of $f(\vec{r},\vec{u})$ over both the positions and velocities gives the normalization of the probability distribution,

$$1 = \int_{-\infty}^{+\infty}\int_{-\infty}^{+\infty} f(\vec{r},\vec{u})\,d\vec{r}\,d\vec{u} \ . \tag{1}$$

The velocity distribution comes from the integration of $f(\vec{r},\vec{u})$ over the positions $\vec{r}$, while the positional distribution is derived from the integration over the velocities $\vec{u}$ (Figure 1).

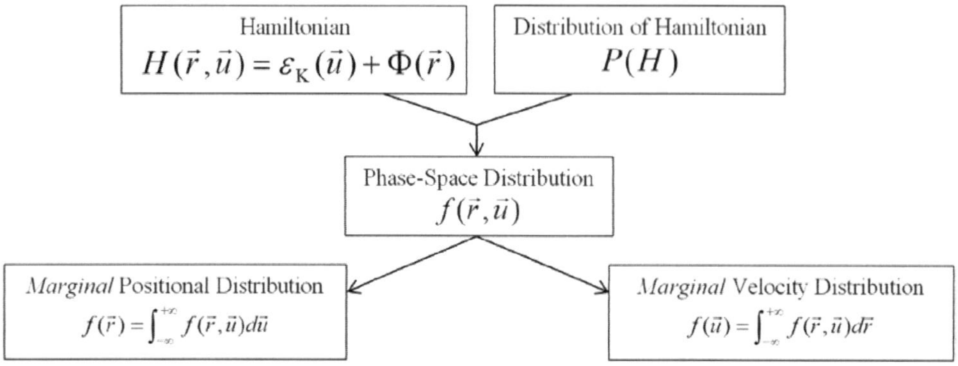

Figure 1. Phase-space and marginal distributions

The statistical moments of velocity space are defined in the co-moving system. The global moments are given by:

$$\mu_m=\int_{-\infty}^{+\infty}\int_{-\infty}^{+\infty} f(\vec{r},\vec{u})d\vec{r}\left|\vec{u}-\vec{u}_b\right|^m d\vec{u} \ , \tag{2}$$

while the unnormalized and normalized local moments are defined by:

$$\tilde{\mu}_m(\vec{r})=\int_{-\infty}^{+\infty} f(\vec{r},\vec{u})\left|\vec{u}-\vec{u}_b\right|^m d\vec{u} \ , \tag{3}$$

and

$$\mu_m(\vec{r})=\frac{\int_{-\infty}^{+\infty} f(\vec{r},\vec{u})\left|\vec{u}-\vec{u}_b\right|^m d\vec{u}}{\int_{-\infty}^{+\infty} f(\vec{r},\vec{u})d\vec{u}} \ . \tag{4}$$

The density profile is proportional to the positional distribution, that is, the unnormalized local distribution for m=0, $f(\vec{r})=\tilde{\mu}_0(\vec{r})$. The proportionality constant, n_∞, provides the density at a position where the effect of the potential energy is negligible; here, we consider that $\vec{r}_\infty\to 0$. Hence,

$$n(\vec{r})=n_\infty\cdot f(\vec{r})=n_\infty\cdot\int_{-\infty}^{+\infty} f(\vec{r},\vec{u})d\vec{u} \ . \tag{5}$$

The global temperature T is defined by the global 2nd moment, μ_2,

$$\tfrac{1}{2}d_K k_B T\equiv\left\langle\varepsilon_K\right\rangle_{(\vec{r},\vec{u})}=\tfrac{1}{2}m\mu_2=\tfrac{1}{2}m\int_{-\infty}^{+\infty}\int_{-\infty}^{+\infty} f(\vec{r},\vec{u})d\vec{r}\left|\vec{u}-\vec{u}_b\right|^2 d\vec{u} \ . \tag{6}$$

(Note: The notions $\langle X\rangle_{(\vec{r},\vec{u})}$, $\langle X\rangle_{\vec{r}}$, and $\langle X\rangle_{\vec{u}}$ mean averaging the argument $X(\vec{r},\vec{u})$ over the whole phase space, the positional space, and the velocity space, respectively.)

The temperature profile $T(\vec{r})$ is given by the normalized local 2nd moment, $\mu_2(\vec{r})$, i.e.,

$$\tfrac{1}{2}d_K k_B T(\vec{r})\equiv\left\langle\varepsilon_K\right\rangle_{\vec{u}}=\frac{\int_{-\infty}^{+\infty} f(\vec{r},\vec{u})\varepsilon_K(\vec{u})d\vec{u}}{\int_{-\infty}^{+\infty} f(\vec{r},\vec{u})d\vec{u}}=\frac{\int_{-\infty}^{+\infty} n_\infty f(\vec{r},\vec{u})\varepsilon_K(\vec{u})d\vec{u}}{n(\vec{r})} \ . \tag{7}$$

Therefore, the thermal pressure is given by

$$p(\vec{r})=n(\vec{r})k_B T(\vec{r})=\frac{2}{d_K}\cdot\int_{-\infty}^{+\infty} n_\infty f(\vec{r},\vec{u})\varepsilon_K(\vec{u})d\vec{u} \ . \tag{8}$$

It is worth noting that the average temperature profile coincides with the global temperature.

$$\langle T(\vec{r})\rangle_{\vec{r}} = \frac{\int_{-\infty}^{+\infty} n(\vec{r})T(\vec{r})d\vec{r}}{\int_{-\infty}^{+\infty} n(\vec{r})d\vec{r}} = \frac{1}{k_B n_\infty}\cdot\int_{-\infty}^{+\infty} p(\vec{r})d\vec{r}$$
$$= \frac{1}{\frac{1}{2}d_K k_B}\cdot\int_{-\infty}^{+\infty}\int_{-\infty}^{+\infty} f(\vec{r},\vec{u})\varepsilon_K(\vec{u})d\vec{r}d\vec{u} = \frac{1}{\frac{1}{2}d_K k_B}\cdot\langle\varepsilon_K\rangle_{(\vec{r},\vec{u})} = T\,. \quad (9)$$

The specific relationship, $p=g(n)$, constructed by the graph curve between the pressure $p(\vec{r})$ and density $n(\vec{r})$ for all the positions $\vec{r}$, defines the generalized polytropic behavior of the space plasma. Next, we will examine how this relationship is reduced to a power-law in the case of kappa distributions.

3. Profiles of thermal observables in the case of kappa distributions

The kappa phase-space distribution function [1,11,12,123,124] gives the probability distribution of a particle having its position and velocity in the infinitesimal intervals $[\vec{r},\vec{r}+d\vec{r}]$ and $[\vec{u},\vec{u}+d\vec{u}]$, respectively,

$$f(\vec{r},\vec{u};\kappa,T) \propto \left[1+\frac{1}{\kappa}\cdot\frac{H(\vec{r},\vec{u})-\langle H\rangle_{(\vec{r},\vec{u})}}{k_B T}\right]^{-\kappa-1}$$
$$= \left[1+\frac{1}{\kappa}\cdot\frac{\varepsilon_K(\vec{u})+\Phi(\vec{r})-\left(\langle\varepsilon_K\rangle_{(\vec{r},\vec{u})}+\langle\Phi\rangle_{(\vec{r},\vec{u})}\right)}{k_B T}\right]^{-\kappa-1}, \quad (10)$$

which can be rewritten as:

$$f(\vec{r},\vec{u}) \propto \left[1-\frac{1}{\kappa}\cdot\frac{\langle H\rangle_{(\vec{r},\vec{u})}}{k_B T}+\frac{1}{\kappa}\cdot\frac{H(\vec{r},\vec{u})}{k_B T}\right]^{-\kappa-1}$$
$$\propto \left[1+\frac{1}{\kappa-\frac{\langle H\rangle_{(\vec{r},\vec{u})}}{k_B T}}\cdot\frac{H(\vec{r},\vec{u})}{k_B T}\right]^{-\kappa-1} \propto \left[1+\frac{1}{\kappa-\frac{1}{2}d}\cdot\frac{H(\vec{r},\vec{u})}{k_B T}\right]^{-\kappa-1}. \quad (11)$$

The mean Hamiltonian defines the total degrees of freedom or dimensionality, d, summing the kinetic and potential degrees of freedom,

$$\tfrac{1}{2}d = \frac{\langle H\rangle_{(\vec{r},\vec{u})}}{k_B T} = \tfrac{1}{2}d_K + \tfrac{1}{2}d_\Phi\,, \quad (12)$$

where the potential degrees of freedom are defined similar to the kinetic ones,

$$\tfrac{1}{2}d_K = \frac{\langle\varepsilon_K\rangle_{(\vec{r},\vec{u})}}{k_B T}\,,\ \tfrac{1}{2}d_\Phi \equiv \frac{\langle\Phi\rangle_{(\vec{r},\vec{u})}}{k_B T}\,. \quad (13)$$

(Note that d_Φ can be either positive or negative.)

The kappa index depends on the dimensionality as $\kappa=\kappa(d)=\text{constant}+\frac{1}{2}d$, so that the difference $\kappa(d)-\frac{1}{2}d$ remains invariant under changes of the dimensionality d. Hence, the invariant kappa index κ_0 is defined by $\kappa_0 \equiv \kappa-\frac{1}{2}d$; hence, $\kappa(d)=\kappa_0+\frac{1}{2}d$. The physical meaning of the thermodynamic parameter kappa is better carried by its invariant value κ_0, because this is independent of the degrees of freedom [1,10,11,12,28). Throughout this analysis, we use the notion of the invariant kappa index κ_0, but the typical 3-dimensional index can be easily retrieved, $\kappa_3=\kappa_0+\frac{3}{2}$. Then, the phase-space distribution (11) is rewritten as,

$$f(\vec{r},\vec{u};\kappa_0,T) \propto \left[1+\frac{1}{\kappa_0}\cdot\frac{\varepsilon_{\rm K}(\vec{u})+\Phi(\vec{r})}{k_{\rm B}T}\right]^{-\kappa_0-1-\frac{1}{2}d} . \quad (14)$$

The integration of the phase space kappa distribution over the positions leads to the standard kappa distribution of velocities. This is true for quite a large set of potential energy forms, which they can be analytically expressed and expanded (at a position $\vec{r}_0$), so that $\Phi(\vec{r}) \cong \Phi(\vec{r}_0)+\frac{1}{b}k\cdot|\vec{r}-\vec{r}_0|^b$ (e.g., [123]).

If we integrate the phase space kappa distribution over the velocity space, then we derive the positional kappa distribution:

$$f(\vec{r};\kappa_0,T) \propto \int_{-\infty}^{+\infty}\left[1+\frac{1}{\kappa_0}\cdot\frac{\varepsilon_{\rm K}(\vec{u})+\Phi(\vec{r})}{k_{\rm B}T}\right]^{-\kappa_0-1-\frac{1}{2}d} d\vec{u} \propto \left[1+\frac{1}{\kappa_0}\cdot\frac{\Phi(\vec{r})}{k_{\rm B}T}\right]^{-\kappa_0-1-\frac{1}{2}d_\Phi} , \quad (15)$$

where the density profile is proportional to the positional probability distribution, i.e.,

$$n(\vec{r}) = n_\infty\left[1+\frac{1}{\kappa_0}\cdot\frac{\Phi(\vec{r})}{k_{\rm B}T}\right]^{-\kappa_0-1-\frac{1}{2}d_\Phi} . \quad (16)$$

Again, the density n_∞ refers to the position where the potential energy becomes zero or practically negligible.

The temperature profile is determined by the local mean kinetic energy,

$$\tfrac{1}{2}d_{\rm K}k_{\rm B}T(\vec{r}) \equiv \langle\varepsilon_{\rm K}\rangle_{\vec{u}} = \frac{\int_{-\infty}^{+\infty} f(\vec{r},\vec{u};\kappa_0,T)\varepsilon_{\rm K}(\vec{u})d\vec{u}}{\int_{-\infty}^{+\infty} f(\vec{r},\vec{u};\kappa_0,T)d\vec{u}} , \quad (17)$$

hence,

$$T(\vec{r}) = T_\infty\cdot\left[1+\frac{1}{\kappa_0}\cdot\frac{\Phi(\vec{r})}{k_{\rm B}T}\right] , \text{ with } T_\infty = T\cdot\frac{\kappa_0}{\kappa_0+\frac{1}{2}d_\Phi} . \quad (18)$$

Then, the thermal pressure, $p(\vec{r}) = n(\vec{r})k_{\rm B}T(\vec{r})$, becomes

$$p(\vec{r}) = p_\infty\cdot\left[1+\frac{1}{\kappa_0}\cdot\frac{\Phi(\vec{r})}{k_{\rm B}T}\right]^{-\kappa_0-\frac{1}{2}d_\Phi} , \text{ with } p_\infty \equiv n_\infty k_{\rm B}T_\infty . \quad (19)$$

The polytropic behavior is defined by the power-law relationship between thermal observables:

$$p(\vec{r}) \propto n(\vec{r})^\gamma \Leftrightarrow n(\vec{r}) \propto T(\vec{r})^\nu , \quad (20)$$

where the exponent γ denotes the polytropic index (e.g., [24-26,123,125-128]):

$$\gamma = 1+1/\nu \Leftrightarrow \nu = 1/(\gamma-1) . \quad (21)$$

(Note that both the exponents γ and ν have been called as the polytropic index.)

In the classical case of Maxwell-Boltzmann distributions, the integration of the exponential of Hamiltonian over the velocity space gives the exponential distribution of the potential energy,

$$f(\vec{r};\kappa_0\to\infty,T) = \int_{-\infty}^{+\infty} f(\vec{r},\vec{u};\kappa_0\to\infty,T)\,d\vec{u} \propto \int_{-\infty}^{+\infty} e^{-\frac{\varepsilon_{\rm K}(\vec{u})+\Phi(\vec{r})}{k_{\rm B}T}} d\vec{u} \propto e^{-\frac{\Phi(\vec{r})}{k_{\rm B}T}} , \quad (22)$$

and

$$\tfrac{1}{2}d_{\rm K}k_{\rm B}T(\vec{r}) = \frac{\int_{-\infty}^{+\infty} e^{-\frac{\varepsilon_{\rm K}(\vec{u})+\Phi(\vec{r})}{k_{\rm B}T}}\varepsilon_{\rm K}(\vec{u})d\vec{u}}{\int_{-\infty}^{+\infty} e^{-\frac{\varepsilon_{\rm K}(\vec{u})+\Phi(\vec{r})}{k_{\rm B}T}} d\vec{u}} = \frac{\int_{-\infty}^{+\infty} e^{-\frac{\varepsilon_{\rm K}(\vec{u})}{k_{\rm B}T}}\varepsilon_{\rm K}(\vec{u})d\vec{u}}{\int_{-\infty}^{+\infty} e^{-\frac{\varepsilon_{\rm K}(\vec{u})}{k_{\rm B}T}} d\vec{u}} = \tfrac{1}{2}d_{\rm K}k_{\rm B}T , \quad (23)$$

thus, we obtain

$$n(\vec{r}) = n_\infty\cdot e^{-\frac{\Phi(\vec{r})}{k_{\rm B}T}} , \; T(\vec{r}) = T , \; p(\vec{r}) = n_\infty k_{\rm B}T\cdot e^{-\frac{\Phi(\vec{r})}{k_{\rm B}T}} . \quad (24)$$

Therefore, in the case of Maxwell-Boltzmann distributions, we obtain $p(\vec{r}) \propto n(\vec{r})$, hence γ=1 (or $\nu \to \pm\infty$), corresponding to a single thermodynamic process, the isothermal one.

In the case of kappa distributions, we obtain

$$p(\vec{r}) \propto n(\vec{r})^{\frac{\kappa_0+\frac{1}{2}d_\Phi}{\kappa_0+\frac{1}{2}d_\Phi+1}} \text{ or } n(\vec{r}) \propto T(\vec{r})^{-(\kappa_0+\frac{1}{2}d_\Phi+1)} , \tag{25}$$

leading to the relationship between kappa and polytropic indices:

$$\gamma = \frac{\kappa_0 + \frac{1}{2}d_\Phi}{\kappa_0 + \frac{1}{2}d_\Phi + 1} \text{ or } -\nu = \kappa_0 + \tfrac{1}{2}d_\Phi + 1 , \tag{26}$$

from which we can verify that the classical case of Maxwell-Boltzmann that corresponds to kappa index $\kappa_0 \to \infty$, leads to the isothermal case, i.e., $\gamma = 1$ or $\nu = \pm\infty$.

4. Connection between kappa distributions and polytropic behavior

In the previous section, we have seen how the kappa distributions can lead to the polytropic behavior. Here, we will show that the opposite statement is also true. Namely, stationary velocity and/or energy distributions associated with polytropic behavior are always connected with the framework of kappa distributions. This recently shown result [24,123,127] is briefly presented below.

The Navier-Stokes momentum equation in a conservative external field F is

$$m \cdot n \cdot \left[\frac{\partial \vec{u}}{\partial t} + (\vec{u} \cdot \vec{\nabla})\vec{u} \right] = -\vec{\nabla} p + n \cdot \vec{F} + \vec{\nabla} \cdot \mathrm{R} . \tag{27}$$

Furthermore, considering: (i) the viscosity tensor R can be neglected; (ii) the velocity vector field is independent of the position vector, i.e., the convective acceleration term vanishes, $(\vec{u} \cdot \vec{\nabla})\vec{u} = 0$; and (iii) the velocity field is stationary, $\partial \vec{u} / \partial t = 0$, then, the Navier-Stokes momentum equation above is reduced to the Euler's equation:

$$\vec{\nabla} p(\vec{r}) = n(\vec{r}) \cdot \vec{F} = -n(\vec{r}) \cdot \vec{\nabla}\Phi(\vec{r}) . \tag{28}$$

It is trivial to show that both Maxwell-Boltzmann and kappa distributions are consistent with the polytropic relationship and the Euler's equation. Then, we may ask: *What is the most general distribution function of particle velocities/energies, consistent with an inviscid and polytropic particle flow under a conservative field?*

In order to answer this question, we use the polytropic relationship, $n \propto T^{\nu}$, and express the density, temperature, and pressure, as arbitrary functions f, f^{ν}, and $f^{\nu+1}$,, respectively, of the ratio $\Phi(\vec{r}) / (k_B T)$:

$$n(\vec{r}) = n_\infty \cdot \left\{ f\left[\frac{\Phi(\vec{r})}{k_B T} \right] \right\}^{\nu} , \; T(\vec{r}) = T_\infty \cdot f\left[\frac{\Phi(\vec{r})}{k_B T} \right] , \; p(\vec{r}) = p_\infty \cdot \left\{ f\left[\frac{\Phi(\vec{r})}{k_B T} \right] \right\}^{\nu+1} , \tag{29}$$

and setting that $f(0)$=1. Then, we substitute $n(\vec{r})$ and $p(\vec{r})$ in Euler's equation, Eq.(28), and we find:

$$\left\{ 1 + (\nu+1)\frac{T_\infty}{T} \cdot f'\left[\frac{\Phi(\vec{r})}{k_B T} \right] \right\} \cdot n(\vec{r}) \cdot \vec{\nabla}\Phi(\vec{r}) = 0 , \tag{30}$$

leading to

$$f'\left[\frac{\Phi(\vec{r})}{k_B T} \right] = -\frac{1}{(\nu+1)\frac{T_\infty}{T}} = \text{constant} \equiv 1/C , \tag{31}$$

and given that $f(0)$=1, Eq.(31) is solved to

$$f\left[\frac{\Phi(\vec{r})}{k_B T} \right] = 1 + \frac{1}{-(\nu+1)\frac{T_\infty}{T}} \cdot \frac{\Phi(\vec{r})}{k_B T} , \text{ or} \tag{32}$$

$$f\left[\frac{\Phi(\vec{r})}{k_{\mathrm{B}}T}\right]=1+C\cdot\frac{\Phi(\vec{r})}{k_{\mathrm{B}}T}\text{ , with } C=-(\nu+1)\tfrac{T_{\infty}}{T}\ . \tag{33}$$

Hence, we find:

$$n(\vec{r})=n_{\infty}\cdot\left[1+C\cdot\frac{\Phi(\vec{r})}{k_{\mathrm{B}}T}\right]^{\nu}\ ,\ T(\vec{r})=T_{\infty}\cdot\left[1+C\cdot\frac{\Phi(\vec{r})}{k_{\mathrm{B}}T}\right]\ ,\ p(\vec{r})=p_{\infty}\cdot\left[1+C\cdot\frac{\Phi(\vec{r})}{k_{\mathrm{B}}T}\right]^{\nu+1}\ . \tag{34}$$

Now let's find the constant *C*. Averaging the temperature profile over the whole positional space is the global temperature, $\langle T(\vec{r})\rangle_{\vec{r}}=T$, and using the definition of the potential degrees of freedom, $\tfrac{1}{2}d_{\Phi}=\langle\Phi(\vec{r})\rangle_{\vec{r}}\,/\,(k_{\mathrm{B}}T)$, we find:

$$C=-(\nu+1)\tfrac{T_{\infty}}{T}=-(\nu+1+\tfrac{1}{2}d_{\Phi})\ . \tag{35}$$

Finally, we observe that the constant *C* coincides with the kappa index κ0 and the formulation of the thermal observables with that of kappa distributions (Figure 2). (For more details, see: [24].)

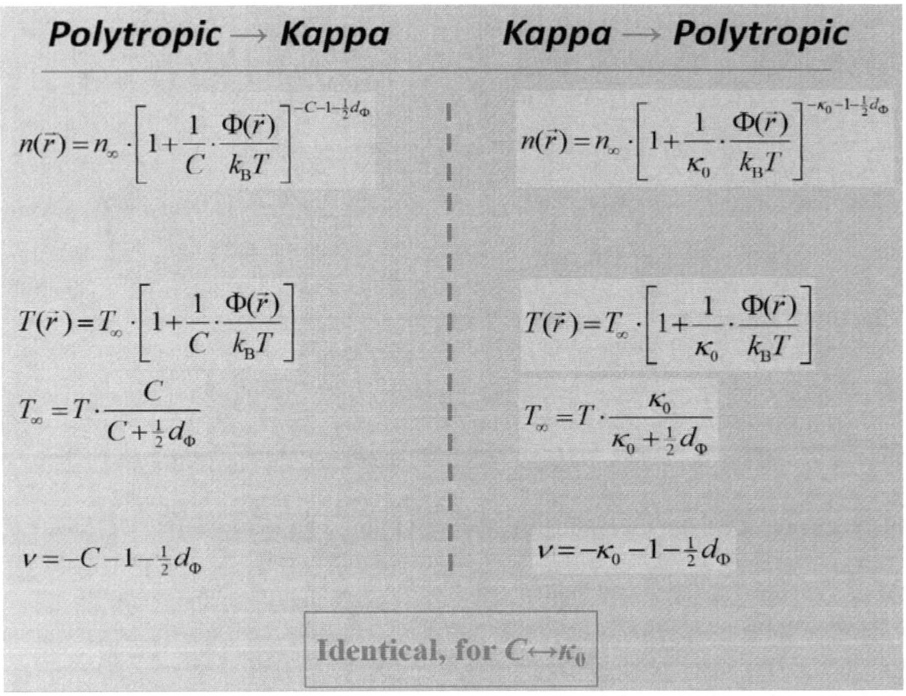

Figure 2. Equivalence between the thermal observables derived from (left) polytropic behavior and (right) kappa distributions. (For more details, see: [24].)

5. Application: The adiabatic process

As it was shown in Section 5, the classical Maxwell-Boltzmann statistical framework does *not* describe the adiabatic or any other polytropic thermodynamic process besides the isothermal one. On the other hand, the adiabatic process is consistent with the framework of kappa distributions specifically for the kappa index given by:

$$\kappa_0=-\tfrac{1}{2}d_{\mathrm{K}}-\tfrac{1}{2}d_{\Phi}-1\Leftrightarrow\text{ adiabatic process }. \tag{36}$$

This relationship can be fulfilled in the cases of (i) positive potential energies and negative kappa indices (e.g., see: negative kappa distributions in [11]), and (ii) negative potential energies (e.g., $\Phi(\vec{r})=\pm\frac{1}{b}k\cdot|\vec{r}|^{\pm b}$ with negative exponent -*b*<0) and either positive or negative kappa indices. In Table, we present several examples of power-law potential energies and the corresponding kappa indices that may lead to the adiabatic process.

Table 1. Connection of kappa and potential energy for adiabatic processes

$\Phi(\vec{r}) \propto \pm\|\vec{r}\|^{\pm b}$	$\frac{1}{2}d_\Phi = \pm d_r / b$	$\nu = -\kappa_0 \mp d_r / b - 1$	**Adiabatic,** $\kappa_0 = \frac{1}{2} \mp d_r / b$
$\Phi(r) \propto -1/r$	-3	$\nu = 2 - \kappa_0$	$\kappa_0 = 3.5$, $\kappa_3 = 5$
$\Phi(r) \propto -1/\sqrt{r}$	-6	$\nu = 5 - \kappa_0$	$\kappa_0 = 6.5$, $\kappa_3 = 8$
$\Phi(x) \propto +x^2$	1/2	$\nu = -1.5 - \kappa_0$	$\kappa_0 = -3$, $\kappa_3 = -1.5$
$\Phi(z) \propto +z$	1	$\nu = -2 - \kappa_0$	$\kappa_0 = -3.5$, $\kappa_3 = -2$

6. Conclusions

Classical collisional particle systems residing in thermal equilibrium have their particle velocity/energy distribution function stabilized into a Maxwell-Boltzmann distribution. On the contrary, space and astrophysical plasmas are exotic collisionless particle systems residing in stationary states characterized by the so-called kappa distribution function. Kappa distributions have become increasingly widespread across the physics of space plasma processes, describing particles in the heliosphere, from the solar wind and planetary magnetospheres to the heliosheath and beyond, the interstellar and intergalactic plasmas. A breakthrough in the field came with the connection of kappa distributions with statistical mechanics and thermodynamics: kappa distributions (i) maximize the entropy of nonextensive statistical mechanics under the constraints of canonical ensemble, (ii) characterize particle systems exchanging heat with each other eventually stabilized at thermal equilibrium, and (iii) constitute the unique description of particle energies consistent with polytropic behavior; while the classical Maxwell-Boltzmann statistical framework cannot describe the adiabatic or any other polytropic thermodynamic process besides the isothermal, the kappa distributions can describe the adiabatic and any other thermodynamic process for specific kappa indices.

References

[1] Livadiotis, G. 2017 *Kappa distribution: Theory & Applications in plasmas* Netherlands, UK, US: Elsevier.
[2] Binsack, J. H. 1966 *Plasma studies with the IMP-2 satellite*, Ph.D. Thesis, MIT.
[3] Olbert, S. 1968 Summary of experimental results from M.I.T. detector on IMP-1, in Physics of the Magnetosphere, R. L. Carovillano, J. F. McClay, and H. R. Radoski (Eds), New York: Springer, 641.
[4] Vasyliūnas, V. M. 1968 A survey of low-energy electrons in the evening sector of the magnetosphere with OGO 1 and OGO 3. *J. Geophys. Res.* **73**, 2839–2884.
[5] Dialynas, K., S. M. Krimigis, D. G. Mitchell, D. C. Hamilton, N. Krupp, and P. C. Brandt 2009 Energetic ion spectral characteristics in the Saturnian magnetosphere using Cassini/MIMI measurements. *J. Geophys. Res.* **114**, A01212.
[6] Livadiotis, G., McComas, D.J, Dayeh, M.A., Funsten, H.O. and Schwadron, N.A. 2011 First sky map of the inner heliosheath temperature using IBEX spectra. *Astrophys. J.* **734**, 1.
[7] Livadiotis, G. 2015 Statistical background and properties of kappa distributions in space plasmas. *J. Geophys. Res.* **120**, 1607–1619.
[8] Wilson III, L. B., Chen, L.-J., Wang, S., Schwartz, S.J., Turner, D.L., Stevens, M.L., Kasper, J.C., Osmane, A., Caprioli, D., Bale, S.D., Pulupa, M.P., Salem, C.S., and Goodrich, K. A. 2019 Electron energy partition across interplanetary shocks. I. Methodology and data product. *Astrophys. J. Suppl. Ser.* **243**, 8.
[9] Lazar, M., V. Pierrard, R. Schlickeiser and S. Poedts 2012 Modeling space plasma dynamics with anisotropic Kappa distributions, *Astrophys. Space Sci.* **33**, 97-107.
[10] Livadiotis, G. and McComas, D.J. 2013 Understanding kappa distributions: A toolbox for space science and astrophysics. *Space Sci. Rev.* **75**, 183–214.
[11] Livadiotis, G. 2015 Kappa distribution in the presence of a potential energy. *J. Geophys. Res.* **120**, 880.
[12] Livadiotis, G. 2015 Kappa and *q* indices: Dependence on the degrees of freedom. *Entropy* **17**, 2062.

[13] Livadiotis, G. 2018 Derivation of the entropic formula for the statistical mechanics of space plasmas. *Nonlin. Processes Geophys.* **25**, 77–88.
[14] Livadiotis, G. 2018 Thermodynamic origin of kappa distributions. *Europhys. Lett.* **122**, 50001.
[15] Beck, C., Cohen. E. G. D. 2003 Superstatistics. *Physica A* **322**, 267.
[16] Schwadron, N., Dayeh, M., Desai, M., Fahr, H., Jokipii, J. R., Lee, M.A. 2010 Superposition of stochastic processes and the resulting particle distributions. *Astrophys. J.* **713**, 1386.
[17] Hanel, R., Thurner, S., Gell-Mann, M. 2011 Generalized entropies and the transformation group of superstatistics. *Proc. Natl. Acad. Sci. U.S.A.* **108**, 6390.
[18] Livadiotis, G., Assas., L., Dennis, B., Elaydi, S., Kwessi, E. 2016 Kappa function as a unifying framework for discrete population modeling. *Nat. Res. Mod.* **29**, 130–144.
[19] Zank, G.P., Li, G., Florinski, V., Hu, Q., Lario, D., Smith, C.W. 2006 Particle acceleration at perpendicular shock waves: Model and observations. *J. Geophys. Res.* **111**, A06108.
[20] Yoon, P.H. 2014 Electron kappa distribution and quasi-thermal noise. *J. Geophys. Res.* **119**, 7074.
[21] Bian, N., Emslie, G.A., Stackhouse, D.J., Kontar, E.P. 2014 The formation of a kappa-distribution accelerated electron populations in solar flares. *Astrophys. J.* **796**, 142.
[22] Livadiotis, G., McComas, D.J. 2011 The influence of pick-up ions on space plasma distributions. *Astrophys. J.* **738**, 64.
[23] Fisk, L.A., Gloeckler, G. 2014 The case for a common spectrum of particles accelerated in the heliosphere: Observations and theory. *J. Geophys. Res.* **119**, 8733.
[24] Livadiotis, G. 2019 On the origin of polytropic behavior in space and astrophysical plasmas. *Astrophys. J.* **874**, 10.
[25] Livadiotis, G. 2016 Superposition of polytropes in the inner heliosheath. *Astrophys. J. Suppl. Ser.* **223**, 13.
[26] Livadiotis, G. 2018 Using kappa distributions to identify the potential energy. *J. Geophys. Res.* **123**, 1050–1060.
[27] Livadiotis, G., Desai, M.I., and Wilson III, L.B. 2018 Generation of kappa distributions in solar wind at 1 AU. *Astrophys. J.* **853**, 142.
[28] Livadiotis, G., and McComas, D.J. 2011 Invariant kappa distribution in space plasmas out of equilibrium, *Astrophys. J.* **741**, 88.
[29] Shizgal, B.D. Kappa and other nonequilibrium distributions from the Fokker-Planck equation and the relationship to Tsallis entropy. *Phys. Rev. E* **2018**, *97*, 052144.
[30] Livadiotis, G. On the simplification of statistical mechanics for space plasmas, *Entropy* **2017**, *19*, 285
[31] Gurnett, D. A.; Bhattacharjee, A. Introduction to Plasma Physics with Space and Laboratory Applications; Cambridge University Press: UK, 2005.
[32] Livadiotis, G.; McComas, D.J. Electrostatic shielding in plasmas and the physical meaning of the Debye length, *J. Plasma Phys.* **2014**, *80*, 341-378.
[33] Livadiotis, G.; McComas, D.J. Evidence of large scale phase space quantization in plasmas. *Entropy* **2013**, *15*, 1118-1132.
[34] Livadiotis, G., and McComas, D.J. 2014 Large-scale quantization from local correlations in space plasmas. *J. Geophys. Res.* **119**, 3247–3258.
[35] Beck, C., Schlögl, F. 1993 *Thermodynamics of Chaotic Systems*; Cambridge University Press: UK.
[36] Tsallis, C., Mendes, R.S., and Plastino, A.R. 1998 The role of constraints within generalized nonextensive statistics. *Physica A* **261**, 534–554.
[37] Tsallis, C. 2009 *Introduction to Nonextensive Statistical Mechanics*; Springer: USA.
[38] Maksimovic, M., Pierrard V., and Lemaire, J. 1997 A kinetic model of the solar wind with Kappa distributions in the corona. *Astron. Astrophys.* **324**, 725-734.
[39] Pierrard, V., Maksimovic, M., Lemaire, J. 1999 Electron velocity distribution function from the solar wind to the corona. *J. Geophys. Res.* **104**, 17021-17032.
[40] Mann, G., Classen, H. T., Keppler, E., Roelof, E.C. 2002 On electron acceleration at CIR related shock waves. *Astron. Astrophys.* **391**, 749–756.
[41] Marsch, E. 2006 Kinetic physics of the solar corona and solar wind. *Living Rev. Sol. Phys.* **3**, 1.
[42] Zouganelis, I. 2008 Measuring suprathermal electron parameters in space plasmas: Implementation of the quasi-thermal noise spectroscopy with kappa distributions using in situ Ulysses/URAP

radio measurements in the solar wind. *J. Geophys. Res.* **113**, A08111.

[43] Štverák, S., Maksimovic, M., Travnicek, P.M., Marsch, E., Fazakerley, A.N., Scime, E.E. 2009 Radial evolution of nonthermal electron populations in the low-latitude solar wind: Helios, Cluster, and Ulysses Observations, *J. Geophys. Res.* **114**, A05104.

[44] Leitner, M., Farrugia, C.J., Vörös, Z. 2011 Change of solar wind quasi-invariant in solar cycle 23—Analysis of PDFs. *J. Atm. Sol.-Terr .Phys.* **73**, 290-293.

[45] Livadiotis, G., and McComas, D.J. 2013 Fitting method based on correlation maximization: Applications in Astrophysics. *J. Geophys. Res.* **118**, 2863-2875.

[46] Pierrard, V., and Pieters, M. 2015 Coronal heating and solar wind acceleration for electrons, protons, and minor ions, obtained from kinetic models based on kappa distributions. *J. Geophys. Res.* **119**, 9441.

[47] Pavlos, G.P., Malandraki, O.E., Pavlos, E.G., Iliopoulos, A.C., Karakatsanis, L.P. 2016 Non-extensive statistical analysis of magnetic field during the March 2012 ICME event using a multi-spacecraft approach. *Physica A* **464**, 149–181.

[48] Nicolaou, G., Livadiotis, G., Owen, C.J., Verscharen, D., Wicks, R.T. 2018 Determining the kappa distributions of space plasmas from observations in a limited energy range. *Astrophys. J.* **864**, 3.

[49] Dzifčáková, E., Dudík, J. 2013 H to Zn ionization equilibrium for the non-Maxwellian electron κ-distributions: Updated calculations. *Astrophys. J. Suppl. Ser.* **206**, 6.

[50] Dzifčáková, E., Dudík, J., Kotrč, P., Fárník, F., Zemanová, A. 2015 KAPPA: A package for synthesis of optically thin spectra for the non-Maxwellian κ-distributions based on the Chianti database. *Astrophys. J. Suppl. Ser.* **217**, 14.

[51] Owocki, S.P., and Scudder, J.D. 1983 The effect of a non-Maxwellian electron distribution on oxygen and iron ionization balances in the solar corona. *Astrophys. J.* **270**, 758-768.

[52] Vocks, C., Mann, G., and Rausche, G. 2008 Formation of suprathermal electron distributions in the quiet solar corona. *Astron. Astrophys.* **480**, 527–536.

[53] Lee, E., Williams, D.R., and Lapenta, G. 2013 Spectroscopic indication of suprathermal ions in the solar corona, arXiv:1305.2939v.

[54] Cranmer, S.R. 2014 Suprathermal electrons in the solar corona: Can nonlocal transport explain heliospheric charge states? *Astrophys. J. Lett.* **791**, L31.

[55] Xiao, F., Shen, C., Wang, Y., Zheng, H., and Whang, S. 2008 Energetic electron distributions fitted with a kappa-type function at geosynchronous orbit. *J. Geophys. Res.* **113**, A05203.

[56] Laming, J.M., Moses, J. D., Ko, Y.-K., Ng, C.K., Rakowski, C.E., and Tylka, A.J. 2013 On the remote detection of suprathermal ions in the solar corona and their role as seeds for solar energetic particle production. *Astrophys. J.* **770**, 73.

[57] Chotoo, K., et al. 2000 The suprathermal seed population for corotaing interaction region ions at 1AU deduced from composition and spectra of H+, He++, and He+ observed by Wind. *J. Geophys. Res.* **105**, 23107–23122.

[58] Mann, G., Warmuth, A., and Aurass, H. 2009 Generation of highly energetic electrons at reconnection outflow shocks during solar flares. *Astron. Astrophys.* **494**, 669-675.

[59] Jeffrey, N.L.S., Fletcher, L., and Labrosse, N. 2016 First evidence of non-Gaussian solar flare EUV spectral line profiles and accelerated non-thermal ion motion. *Astron. Astrophys.* **590**, A99.

[60] Formisano, V., Moreno, G., Palmiotto, F., and Hedgecock, P.C. 1973 Solar Wind Interaction with the Earth's Magnetic Field 1. Magnetosheath, *J. Geophys. Res.* **78**, 3714-3730.

[61] Ogasawara, K., Angelopoulos, V., Dayeh, M.A., Fuselier, S.A., Livadiotis, G., McComas, D.J., and McFadden, J.P. 2013 Characterizing the dayside magnetosheath using ENAs: IBEX and THEMIS observations. *J. Geophys. Res.* **118**, 3126-3137.

[62] Ogasawara, K., Dayeh, M.A., Funsten, H.O., Fuselier, S.A., Livadiotis, G., and McComas, D.J. 2015 Interplanetary magnetic field dependence of the suprathermal energetic neutral atoms originated in subsolar magnetopause. *J. Geophys. Res.* **120**, 964-972.

[63] Grabbe, C. 2000 Generation of broadband electrostatic waves in Earth's magnetotail. *Phys. Rev. Lett.* **84**, 3614.

[64] Pisarenko, N.F., Budnik, E.Yu., Ermolaev, Yu.I., Kirpichev, I.P., Lutsenko, V.N., Morozova, E.I., and Antonova, E.E. 2002 The ion differential spectra in outer boundary of the ring current: November 17, 1995 case study. *J. Atm. Solar-Terr. Phys.* **64**, 573 – 583.

[65] Christon, S.P. 1987 A comparison of the Mercury and earth magnetospheres: electron measurements and substorm time scales. *Icarus* **71**, 448-471.

[66] Wang, C.-P., Lyons, L.R., Chen, M.W., Wolf, R.A., and Toffoletto, F.R. 2003 Modeling the inner plasma sheet protons and magnetic field under enhanced convection. *J. Geophys. Res.* **108**, 1074.

[67] Kletzing, C.A., Scudder, J.D., Dors, E.E., and Curto, C. 2003 Auroral source region: Plasma properties of the high latitude plasma sheet. *J. Geophys. Res.* **108**, 1360.

[68] Hapgood, M., Perry, C., Davies, J., and Denton, M. 2011 The role of suprathermal particle measurements in CrossScale studies of collisionless plasma processes. *Planet. Space Sci.* **59**, 618–629.

[69] Ogasawara, K., Livadiotis, G., Grubbs, G.A., Jahn, J.-M., Michell, R., Samara, M., Sharber, J.R., and Winningham, J.D. 2017 Properties of suprathermal electrons associated with discrete auroral arcs. *Geophys. Res. Lett.* **44**, 3475–3484.

[70] Collier, M.R., and Hamilton, D.C. 1995 The relationship between kappa and temperature in the energetic ion spectra at Jupiter. *Geophys. Res. Lett.* **22**, 303-306.

[71] Mauk, B.H., Mitchell, D.G., McEntire, R.W., Paranicas, C.P., Roelof, E.C., Williams, D.J., Krimigis, S.M., and Lagg, A. 2004 Energetic ion characteristics and neutral gas interactions in Jupiter's magnetosphere. *J. Geophys. Res.* **109**, A09S12.

[72] Nicolaou, G., McComas, D.J., Bagenal, F., Elliott, H.A., and Wilson, R.J. 2015 Plasma properties in the deep Jovian magnetotail. *Plan. Space Sci.* **119**, 222-232.

[73] Kim, T.K., Ebert, R.W., Valek, P.W., Allegrini, F., McComas, D.J., Bagenal, F., Chae, K., Livadiotis, G., Loeffler, C.E., Pollock, C., Ranquist, D.A., Thomsen, M.F., Wilson, R.J., Clark, G., Kollmann, P., Mauk, B. H., Bolton, S., Levin, S., and Nicolaou, G. 2019 Method to derive ion properties from Juno JADE including abundance estimates for O^+ and S^{2+}. *J. Geophys. Res.*, In Press.

[74] Livi, R., Goldstein, J., Burch, J. L., Crary, F., Rymer, A.M., Mitchell, D.G., Persoon, A.M. 2014 Multi-instrument analysis of plasma parameters in Saturn's equatorial, inner magnetosphere using corrections for spacecraft potential and penetrating background radiation. *J. Geophys. Res.* **119**, 3683.

[75] Carbary, J.F., Kane, M., Mauk, B.H., and Krimigis, S.M. 2014 Using the kappa function to investigate hot plasma in the magnetospheres of the giant planets. *J. Geophys. Res.* **119**, 8426–8447.

[76] Dialynas, K., Roussos, E., Regoli, L., Paranicas, C. P., Krimigis, S. M., Kane, M., Mitchell, D.G., Hamilton, D.C., Krupp, N., and Carbary, J.F., 2018 Energetic ion moments and polytropic index in Saturn's magnetosphere using Cassini/MIMI measurements: A simple model based on κ-distribution functions, *J. Geophys. Res.* **123**, 8066-8086.

[77] Mauk, B.H., Krimigis, S.M., Keath, E.P., Cheng, A.F., Armstrong, T.P., Lanzerotti, L.J., Gloeckler, G., and Hamilton, D.C. 1987 The hot plasma and radiation environment of the Uranian magnetosphere. *J. Geophys. Res.* **92**, 15283.

[78] Krimigis, S.M., et al. 1989 Hot plasma and energetic particles in Neptune's magnetosphere. *Science* **246**, 1483.

[79] Moncuquet, M., Bagenal, F., and Meyer-Vernet, N. 2002 Latitudinal structure of the outer Io plasma torus. *J. Geophys. Res.* **108**, 1260.

[80] Jurac, S., McGrath, M. A., Johnson, R. E., Richardson, J. D., Vasyliunas, V. M., and Eviatar, A. 2002 Saturn: Search for a missing water source. *Geophys. Res. Lett.* **29**, 2172.

[81] Broiles, T.W., Livadiotis, G., Burch, J.L., Chae, K., Clark, G., Cravens, T.E., Davidson, R., Eriksson, A., Frahm, R.A., Fuselier, S.A., Goldstein, J., Goldstein, R., Henri, P., Madanian, H., Mandt, K.E., Mokashi, P., Pollock, C., Rahmati, A., Samara, M., Schwartz, S.J. 2016 Characterizing cometary electrons with kappa distributions. *J. Geophys. Res.* **121**, 7407-7422.

[82] Broiles, T.W., Burch, J.L., Chae, K., Clark, G., Cravens, T.E., Eriksson, A., Fuselier, S.A., Frahm, R.A., Gasc, S., Goldstein, R., Henri, P., Koenders, C., Livadiotis, G., Mandt, K.E., Mokashi, P., Nemeth, Z., Rubin, M., and Samara, M. 2016 Statistical analysis of suprathermal electron drivers at 67P/ Churyumov-Gerasimenko, *MNRAS* **462**, S312-S322.

[83] Decker, R.B., and Krimigis, S.M. 2003 Voyager observations of low-energy ions during solar cycle 23. *Adv. Space Res.* **32**, 597–602.

[84] Decker, R.B., et al. 2005 Voyager 1 in the foreshock, termination shock, and heliosheath, *Science* **309**, 2020–2024.

[85] Heerikhuisen, J., Pogorelov, N.V., Florinski, V., Zank, G.P., and le Roux, J.A. 2008 The effects of a k-distribution in the heliosheath on the global heliosphere and ENA flux at 1 AU. *Astrophys. J.* **682**, 679-689.
[86] Heerikhuisen, J., Zirnstein, E., and Pogorelov, N. 2015 κ-distributed protons in the solar wind and their charge-exchange coupling to energetic hydrogen. *J. Geophys. Res.* **120**, 1516–1525.
[87] Zank, G.P., Heerikhuisen, J., Pogorelov, N.V., Burrows, R., and McComas, D.J. 2010 Microstructure of the heliospheric termination shock: implications for energetic neutral atom observations. *Astrophys. J.* **708**, 1092.
[88] Elliott, H.A., McComas, D.J., Zirnstein, E. J., Randol, B. M., Delamere, P. A., Livadiotis, G., Bagenal, F., Barnes, N.P., Stern, S.A., Young, L.A., Olkin, C. B., Spencer, J., Weaver, H. A., Ennico, K., Gladstone, R., and Smith, C.W. 2019 Slowing of the solar wind in the outer heliosphere. Astrophys. J., In Press.
[89] Livadiotis, G., McComas, D.J., Randol, B., Möbius, E., Dayeh, M.A., Frisch, P.C., Funsten, H.O., Schwadron, N.A., Zank, G.P. 2012 Pick-up ion distributions and their influence on ENA spectral curvature. *Astrophys. J.* **751**, 64.
[90] Livadiotis, G., McComas, D.J., Schwadron, N.A., Funsten, H.O., and Fuselier, S.A. 2013 Pressure of the proton plasma in the inner heliosheath. *Astrophys. J.* **762**, 134.
[91] Livadiotis, G., and McComas, D.J. 2010 Exploring transitions of space plasmas out of equilibrium. *Astrophys. J.* **714**, 971.
[92] Livadiotis, G., and McComas, D.J. 2012 Non-equilibrium thermodynamic processes: Space plasmas and the inner heliosheath. *Astrophys. J.* **749**, 11.
[93] Livadiotis, G. 2014 Lagrangian temperature: Derivation and physical meaning for systems described by kappa distributions. *Entropy* **16**, 4290-4308.
[94] Livadiotis, G. 2016 Curie law for systems described by kappa distributions. *Europhys. Lett.* **113**, 10003.
[95] Fuselier, S.A., Allegrini, F., Bzowski, M., Dayeh, M.A., Desai, M., Funsten, H.O., Galli, A., Heirtzler, D., Janzen, P., Kubiak, M.A., Kucharek, H., Lewis, W., Livadiotis, G., McComas, D. J., Möbius, E., Petrinec, S. M., Quinn, M., Schwadron, N., Sokół, J. M., Trattner, K. J., Wood, B. E., and Wurz, P. 2014 Low energy neutral atoms from the heliosheath. *Astrophys. J.* **784**, 89.
[96] Zirnstein, E.J., and McComas, D.J. 2015 Using kappa functions to characterize outer heliosphere proton distributions in the presence of charge-exchange. *Astrophys. J.* **815**, 31.
[97] Zank, G.P. 2015 Faltering steps into the galaxy: The boundary regions of the heliosphere. *Ann. Rev. Astron. Astrophys.* **53**, 449.
[98] Nicholls, D.C., Dopita, M.A., and Sutherland, R.S. 2012 Resolving the Electron Temperature Discrepancies in H II Regions and Planetary Nebulae: κ-distributed Electrons, *Astrophys. J.* **752**, 148.
[99] Nicholls, D.C., Dopita, M.A., Sutherland, R.S., Kewley, L.J., and Palay, E. 2013 Measuring nebular temperatures: the effect of new collision strengths with equilibrium and κ-distributed electron energies. *Astrophys. J. Supp.* **207**, 21.
[100] Zhang, Y., Liu, X.-W., and Zhang B. 2014 H-I free-bound emission of planetary nebulae with large abundance discrepancies: Two-component models vs. κ-distributed electrons. *Astrophys. J.* **780**, 93.
[101] Raymond, J.C., Winkler, P.F., Blair, W.P., Lee, J.-J., and Park, S. 2010 Non-Maxwellian Hα profiles in Tycho's supernova remnant. *Astrophys. J.* **712**, 901.
[102] Hou, S.Q., He, J.J., Parikh, A., Kahl, D., Bertulani, C.A., Kajino, T., Mathews, G.J., and Zhao, G. 2017 Non-extensive statistics to the cosmological lithium problem. *Astrophys. J.* **834**, 165.
[103] Saito, S., Forme, F. R. E., Buchert, S. C., Nozawa, S., and Fujii, R. 2000 Effects of a kappa distribution function of electrons on incoherent scatter spectra. *Ann. Geophys.* **18**, 1216-1223.
[104] Yoon, P. H., Rhee, T., and Ryu, C. M. 2006 Self-consistent formation of electron κ distribution: 1. Theory. *J. Geophys. Res.* **111**, A09106.
[105] Raadu, M. A., and Shafiq, M. 2007 Test charge response for a dusty plasma with both grain size distribution and dynamical charging. *Phys. Plasmas* **14**, 012105.
[106] Hellberg, M. A., Mace, R. L., Baluku, T.K., Kourakis, I., and Saini, N. S. 2009 Comment on

"Mathematical and physical aspects of Kappa velocity distribution" [Phys. Plasmas 14, 110702 (2007)]. *Phys. Plasmas* **16**, 094701.

[107] Livadiotis, G. 2009 Approach on Tsallis statistical interpretation of hydrogen-atom by adopting the generalized radial distribution function. *J. Math. Chem.* **45**, 930-939.

[108] Tribeche, M., Mayout, S., and Amour, R. 2009 Effect of ion suprathermality on arbitrary amplitude dust acoustic waves in a charge varying dusty plasma. *Phys. Plasmas* **16**, 043706.

[109] Baluku, T. K., Hellberg, M.A., Kourakis, I., and Saini, N. S. 2010 Dust ion acoustic solitons in a plasma with kappa-distributed electrons. *Phys. Plasmas* **17**, 053702.

[110] Le Roux, J.A., Webb, G.M., Shalchi, A., and Zank, G.P. 2010 A generalized nonlinear guiding center theory for the collisionless anomalous perpendicular diffusion of cosmic rays. *Astrophys. J.* **716**, 671-692.

[111] Livadiotis, G., and McComas, D.J. 2010 Measure of the departure of the q-metastable stationary states from equilibrium. *Phys. Scr.* **82**, 035003.

[112] Eslami, P., Mottaghizadeh, M., and Pakzad, H.R. 2011 Nonplanar dust acoustic solitary waves in dusty plasmas with ions and electrons following a q-nonextensive distribution. *Phys. Plasmas* **18**, 102303.

[113] Kourakis, I., Sultana, S., and Hellberg, M.A. 2012 Dynamical characteristics of solitary waves, shocks and envelope modes in kappa-distributed non-thermal plasmas: an overview. *Plasma Phys. Control. Fusion* **54**, 124001.

[114] Randol, B.M., and Christian, E.R. 2016 Coupling of charged particles via Coulombic interactions: Numerical simulations and resultant kappa-like velocity space distribution functions. *J. Geophys. Res.* **121**, 1907–1919.

[115] Varotsos, P.A., Sarlis, N.V., and Skordas, E.S. 2014 Study of the temporal correlations in the magnitude time series before major earthquakes in Japan. *J. Geophys. Res.* **119**, 9192–9206.

[116] Viñas, A.F., Moya, P.S., Navarro, R.E., Valdivia, J.A., Araneda, J.A., and Muñoz, V. 2015 Electromagnetic fluctuations of the whistler-cyclotron and firehose instabilities in a Maxwellian and Tsallis-kappa-like plasma. *J. Geophys. Res.* **120**, 3307-3317.

[117] Nicolaou, G., and Livadiotis, G. 2016 Misestimation of temperature when applying Maxwellian distributions to space plasmas described by kappa distributions. *Astrophys. Space Sci.* **361**, 359.

[118] Livadiotis, G. 2016 Non-Euclidean-normed Statistical Mechanics. *Physica A* **445**, 240–255.

[119] Oka, M., Birn, J., Battaglia, M., Chaston, C.C., Hatch, S.M., Livadiotis, G., Imada, S., Miyoshi, Y., Kuhar, M., Effenberger, F., Eriksson, E., Khotyaintsev, Y.V., and Retino, A. 2018 Electron power-law spectra in solar and space plasmas. *Space Sci. Rev.* **214**, 82.

[120] Varotsos, P., Sarlis, N., and Skordas, E.S. 2018 Tsallis entropy index q and the complexity measure of seismicity in natural time under time reversal before the M9 Tohoku Earthquake in 2011. *Entropy* **20**, 757.

[121] Milovanov A.V., and Zelenyi, L. M. 2000 Functional background of the Tsallis entropy: "coarse-grained" systems and "kappa" distribution functions. *Nonlin. Processes Geophys.* **7**, 211–221.

[122] Livadiotis, G., and McComas, D.J. 2009 Beyond kappa distributions: Exploiting Tsallis statistical mechanics in space plasmas. *J. Geophys. Res.* **114**, A11105.

[123] Livadiotis, G. 2019 Theoretical aspects of Hamiltonian kappa distributions. *Phys. Scr.* **94**, 105009.

[124] Livadiotis, G., 2018 Thermal Doppler broadening of spectral emissions by space plasma particles. *Astrophys. J. Suppl. Ser.* **239**, 25.

[125] Nicolaou, G., Livadiotis, G., and Moussas, X. 2014 Long term variability of the polytropic Index of solar wind protons at ~1AU. *Sol. Phys.* **289**, 1371-1378.

[126] Livadiotis, G., and Desai, M.I. 2016 Plasma-field coupling at small length scales in solar wind near 1 au. *Astrophys. J.* **829**, 88.

[127] Livadiotis, G., 2019 On the generalized formulation of Debye shielding in plasmas. *Phys Plasmas* **26**, 050701.

[128] Livadiotis, G. 2018 Long-term independence of solar wind polytropic index to plasma flow speed. *Entropy* **20**, 799.

Acknowledgments
The work was supported in part by the project NNX17AB74G of NASA's HGI Program.

Journal of Physics: Conference Series **1332** (2019) 012011 doi:10.1088/1742-6596/1332/1/012011

Observation of Suprathermal Tails of He^+ Pickup Ions across Solar Wind Compression Regions with STEREO PLASTIC

E Möbius[1], J Bower[1], A Aly[1], L Berger[2], C Farrugia[1], A B Galvin[1], D Keilbach[2] B Klecker[3], M A Lee[1], N A Schwadron[1]

[1]Space Science Center and Department of Physics, University of New Hampshire, Durham, NH, USA
[2]Institut für Angewandte und Experimentelle Physik, Christian-Albrecht Universität, Kiel, Germany
[3]Max-Planck-Institut für Extraterrestrische Physik, Garching, Germany

eberhard.moebius@unh.edu

Abstract. The presence of suprathermal tails of solar wind and pickup ions in interplanetary space has been widely observed, even during quiet times with no simultaneous observation of solar energetic particles. One of the persistent characteristics of these tails have been power law spectra of the velocity distribution function with a v^{-5} dependence in the solar wind reference frame and exponential fall-off at higher energies, but variations in the spectra including those of other species have also been observed. Several attempts to explain the formation of suprathermal tails during quiet times have been made, among them continuing acceleration by compressive fluctuations of the solar wind and the stochastic superposition of exponential, Gaussian, and variable power law spectra from diffusive shock, stochastic, and other acceleration processes. We find here that acceleration is effective within compression regions with and without shocks. In the context of a superposed epoch analysis of the evolution of He^+ pickup ion distributions across compression regions, we report on a related study of He^+ tails, using STEREO PLASTIC data from 2007 through 2014. Quiet times have been selected based on limiting energetic He fluxes above the tail energies and based on the tail fluxes themselves. We find that the suprathermal tail flux is dependent on the compression strength and varies substantially across the compression region. The strongest tails with spectra somewhat steeper than v^{-5} occur in the compressed fast solar wind, and they decrease rapidly with distance from the preceding and following compression into the rarefaction region, when using an unbiased data sample and applying a quiet time criterion based on higher energy ions. This may be consistent with the compressions being a potential source of the tails. When applying a quiet time criterion based on the observed tail fluxes, the temporal evolution disappears, possibly implicating a selection of the lower end of a Poisson distribution of tail count rates, rendering such a selection unusable for temporal evolution studies.

1. Introduction

Initially, the origin of energetic particles in interplanetary space has been associated with either acceleration in solar flares or disturbances in the solar wind (SW). Coronal mass ejections (CME) and SW stream interaction regions were particularly suspected, along with their accompanying shocks [e.g. 1]. While flare related energetic particles were thought to be accelerated directly out of coronal material close to the Sun, CME and SW stream interaction related particles were thought to be accelerated out of

Content from this work may be used under the terms of the Creative Commons Attribution 3.0 licence. Any further distribution of this work must maintain attribution to the author(s) and the title of the work, journal citation and DOI.
Published under licence by IOP Publishing Ltd

the SW. However, in the latter case, the requirement that ions of the source population either need a minimum speed to outrun the shock from downstream or be sufficiently mobile relative to the local bulk speed makes direct acceleration of SW a challenge, which is referred to as the injection problem [2].

Other observational challenges to the original acceleration paradigm, yet potentially mitigating the injection problem, came from composition measurements. He^+ was found overly abundant in interplanetary energetic particles [3], and an unusually high $^3He/^4He$ ratio in many CME related energetic particle events, mimicking the abundance of impulsive flare particles, was reported [4]. Whereas the origin of the former abundance anomaly was unknown at first, the latter immediately implicated impulsive flares as the potential origin of the source population. Indeed, the signature of impulsive flare material was identified in the pre-interplanetary shock population, and it became clear that dependent on the phase of the solar cycle this population and thus the CME particles were more or less peppered with impulsive flare material [5]. The mystery of the huge He^+ abundance resolved itself in favor of interstellar pickup ions (PUIs), when stark overabundances of He^+ were found in the energetic particle populations of co-rotating interaction regions (CIR) [6], with a distinct increase towards the end of a CIR where the observer is connected to it at larger distances from the Sun [7]. Even when the He^+/He^{2+} ratio in the SW was anomalously enhanced in a CME, PUIs are accelerated and not the SW population [8]. In both cases, apparently an already more energetic particle population, or better a much wider velocity distribution in the SW is preferentially accelerated relative to the also contributing SW distribution. This scenario of an already suprathermal source population for further acceleration was briefly summarized [9] and reviewed in great detail [10].

The presence of suprathermal tails in interplanetary space even during quiet times, as selected by energetic and suprathermal particle rates below a pre-defined threshold [11], was pointed out by Gloeckler et al. [12] and led to numerous follow-up work. Gloeckler et al. [13] extended their spectral description with an exponential term that accounts for losses at the high energy end of the distribution due to finite space and time. They demonstrated that the lower portion exhibits a constant v^{-5} slope. Desai et al. [14] and Dayeh et al. [15] investigated systematically the composition of the suprathermal population and the spectra of heavier species. They found a distinct variation with the solar cycle, or the occurrence of impulsive events, and a substantial variability of the spectra of heavy ion species in that population.

Two different types of models have been developed to explain the observed characteristics of the suprathermal tail populations, either through continuous local acceleration in undisturbed SW or as remnants from acceleration in CME and stream interaction related compressions or shocks. The widely promoted constancy of the power law portion of the suprathermal tail spectra and their presence during quiet times has inspired models of continuous local acceleration. Compressional turbulence in the SW [11, 16, 17] is suggested to provide naturally a v^{-5} spectrum. But Jokipii & Lee [18] pointed to several weaknesses in the derivation of this model. Wave particle interaction [19] and stochastic acceleration by waves [20] leads to a spectrum close to v^{-5}, when adiabatic cooling is included. Also, reconnection in stochastically distributed magnetic islands in the SW leads to acceleration into a power law, whose index is still Mach number dependent, but tends toward v^{-5} in the outer heliosphere [21, 22] and is strongest downstream of interplanetary shocks [23]. The model by Schwadron et al. [24] is a hybrid between the two model types, as it addresses acceleration in CIRs through transit time damping, which appears to produce spectra close to a power law, with indices between -5 and -6, but it could also provide acceleration in quiet SW, just at lower fluctuation levels. Similar to this model, the ones by Fisk & Lee [25], Giacalone et al. [26], and Richardson [27] address acceleration into the suprathermal regime at CIR shocks and compressions. Acceleration at CME related shocks also produce power law spectra in the lower energy portion [28, 29], but mostly at v^{-4} in the observer frame, which amounts to about v^{-6} in the SW frame for ions in the suprathermal speed range, with variations according to the shock strength. While these sources can explain the variability of the heavy ion spectra and, together with contributions from impulsive flares the varying composition, Fisk & Gloeckler [17] point out that a separate continuous acceleration is needed to explain the often-observed v^{-5} spectra during quiet times. However, Schwadron et al. [30] have shown that a stochastic combination of different power law, exponential, and

Gaussian spectra typically leads to a spectrum that is close to v^{-5}, which could equally explain the constant behavior during quiet times.

Popecki et al. [31] showed that He^+ tails steepen from the slow compressed wind to the fast CME related wind in CME compressions, with large intrinsic variability, which could provide a variety of source spectra for a superposition as discussed by Schwadron et al. [30]. Tessein et al. [32] have demonstrated that the fluxes of the suprathermal tails at compressions correlate with the strength of the turbulence in its immediate vicinity. These findings raise the question, whether, perhaps, the suprathermal tail fluxes decrease and their spectra vary systematically as a function of distance (and/or time) from CIR or CME related sources. In other words, could, in principle, the reservoir of suprathermal tail populations be replenished just from several, stochastically distributed, localized sources in the heliosphere?

In this contribution, we will take advantage of a previous systematic study of the evolution of the PUI velocity distribution across SW compression regions and its dependence on the strength of the compression [33] to perform a similar systematic study of suprathermal He^+ tail fluxes and spectra. This study should provide further information on the question whether the compressions and shocks, which are known to energize ions could be sources of these suprathermal tails or whether an additional process may be needed that operates anywhere in the SW. In Section 2 we will briefly summarize the data analysis method, followed by the presentation of the results in Section 3, and a discussion of them along with their implications in Section 4. We will conclude with a summary of the results, conclusions, and an outlook toward future investigations.

This paper is in tribute of Dr. Ed Stone on his 80th birthday, whose leadership of the ACE [34] and Voyager Interstellar Mission [35] has contributed substantially to the studies of suprathermal tails in the heliosphere.

2. Superposed Epoch Analysis Method to Study Suprathermal He^+

In recent work, Bower et al. [33] have performed a superposed epoch (SPE) analysis of the behavior of the PUI cutoff across SW compression regions, from the slow compressed via the fast compressed and peak speed region all the way into the rarefaction region. They showed that the PUIs experience a maximum positive cutoff shift at the end of the fast compressed wind, with sustained positive shifts into the first half of the rarefaction region. The magnitude of this shift correlates with the mean SW speed gradient across the compression. The unexpected positive shift in the rarefaction region appears to be connected with the continuing presence of compressional fluctuations from the fast compression region well into the rarefaction region, which thus are a likely contributor to the energization of the PUIs observed through the cutoff shift. In this study, also substantial suprathermal tails were observed in the compression region, which we will follow up on here.

For the study, we use the He^+ count rate distributions for 2007 – 2014 from the PLASTIC instrument on STEREO A [36] binned as a function of speed and angle, transformed into the SW frame [37, 38], which facilitates the direct comparison with previous suprathermal particle spectra. In order to provide spectra of the velocity distribution function of these ions, we make use of the fact that the differential energy flux density $E\Delta J/\Delta E$ as measured by an instrument that uses an electrostatic analyzer for energy determination is directly proportional to the observed count rate in the spacecraft frame, assuming a detection efficiency that is constant over energy. Given that the species, here He^+, is determined by the instrument the observed energy E is directly translated into the particle speed in the spacecraft frame v'. Because we will compare the suprathermal tail fluxes directly with the concurrent PUI fluxes (the expected source population) there is no need to obtain absolute fluxes, and thus only a correction for the energy dependence of the PLASTIC He^+ detection efficiency is applied here to the observed count rates.

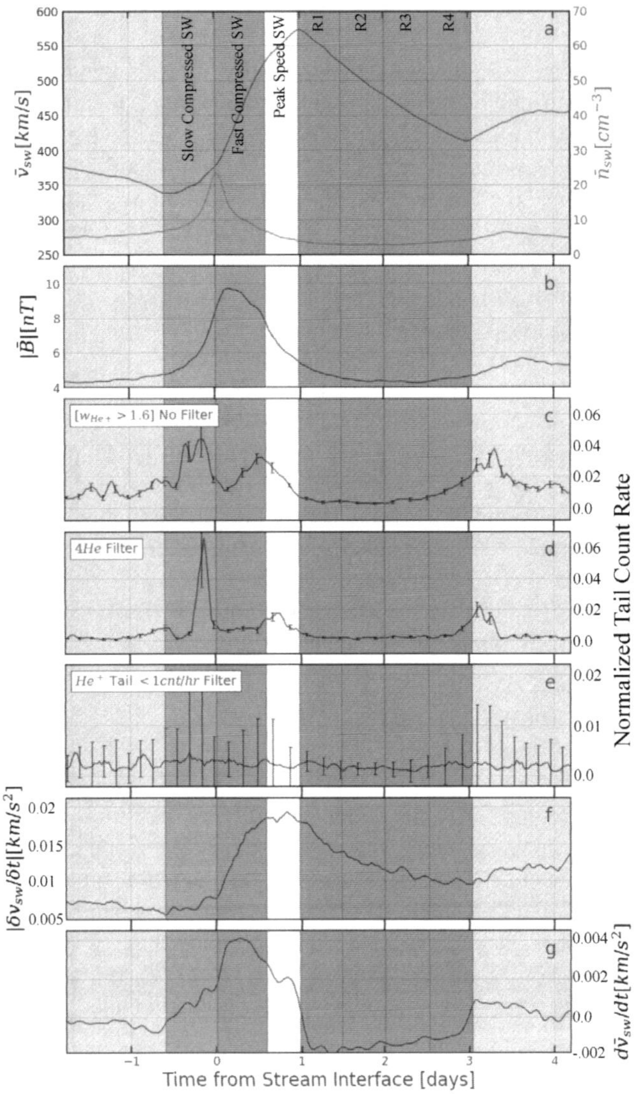

Figure 1: Evolution of SW V_{sw} and density n_{sw} (a), magnetic field strength B (b), counts in the suprathermal tail, normalized to the PUI maximum (c, d, e), along with statistical error bars, strength of SW speed fluctuations $|\delta V_{sw}/\delta t|$ (f), and local SW speed gradient dV_{sw}/dt (g) as function of time across a superposed SW compression as described in the text [33].

To obtain this efficiency, the ratios of single and coincidence rates in the SW main channel of the PLASTIC time-of-flight system are used, at an energy just below the He^+ cut-off energy. Thus, the variation of the He^+ cut-off with solar wind speed provides the energy dependence of the He^+ efficiency. For the PLASTIC calibration to He^+ and a direct comparison with ACE SWICS see also Möbius et al. [39]. Furthermore, $E\Delta J/\Delta E = v'^4 f/2$, where f is the velocity distribution function and m the mass of the observed species. Because we are using solely observations in the SW section of PLASTIC, we take $v' = v + V_{sw}$, where v is the He^+ speed in the SW frame and compute f from the observed efficiency corrected count rate.

Let us briefly review the method used in the SPE analysis here, which is described in more detail elsewhere [33]. After applying a sliding average, the compression regions are identified by maxima in the SW density that coincide with SW speed increases. The density peak is taken as the stream interface, or the boundary between slow and fast compressed SW, and thus used for the epoch of the time scale. From here, the start of the slow compressed wind is identified by the preceding minimum of the SW speed, the end of the fast compressed wind/start of the peak region by the maximum of the speed gradient, and the end of the peak region/start of the rarefaction by the maximum of the SW speed. The rarefaction region ends when another speed minimum is encountered. To synchronize the individual compressions for the superposition, the average length of each of these regions is computed and then used as the time scale for the combined data. For each of the individual compressions, the length of each region is compressed or stretched accordingly.

Figure 1 shows the superposition of the strongest 25% of the compression regions during the observation period, with each region highlighted by a specific color, whose scheme will be maintained in the Figures presented below: slow compressed wind (red), fast compressed wind (purple), peak region (white), and rarefaction region (blue), which is subdivided into four equal consecutive time intervals to follow the evolution of the tail spectra. Shown are from top to bottom: SW speed V_{sw} (blue) and density n_{sw} (red) (panel a), magnetic field strength B (panel b), tail count rate for $w \geq 1.6$, normalized to the maximum PUI count rate, with no quiet time criterion (panel c), with a threshold of the STEREO SIT

^{4}He fluxes (panel d), with a threshold of the PLASTIC tail count rate (panel e), absolute strength of the SW speed fluctuations $|\delta V_{sw}/\delta t|$ (panel f), and average local SW speed gradient dV_{sw}/dt (panel g).

In all cases the start of the suprathermal tail is defined at $v/V_{sw} \geq 1.6$ in the SW frame. We use a somewhat larger value than Gloeckler et al. [12], who used 2.39 in the spacecraft frame, because the cutoff is varying substantially with ecliptic longitude due to the interstellar gas flow velocity pattern at 1 AU which produces a modulation by about 15% between downwind and upwind [40]. To obtain the unbiased evolution of the tail fluxes across the compression regions we first show the normalized tail distributions as observed, without quiet time criterion, in panel c. Then we apply two different criteria to select quiet times, whose importance may become clear when discussing the results. In panel d, we select from the original dataset, only those 5-minute time intervals, for which the concurrent ^{4}He flux at 0.16-0.24 MeV/nuc, measured with STEREO SIT is <0.5 (cm^2 sr s MeV/nuc)$^{-1}$. In panel e, we apply a criterion in analogy to the one used by Fisk & Gloeckler [11], who required $\leq$1 H$^+$ count/4h for $w' >$ 2.39 (in the spacecraft frame) in the Ulysses SWICS data. Because of the larger PLASTIC geometric factor, we select only those 1-hour PLASTIC data time intervals, which contain $\leq$ 1 He$^+$ counts for $w \geq$ 1.6. It should be noted that Dayeh et al. [15] applied a similar criterion for their quiet time selection, although defining their absolute threshold based on the fraction of time intervals with the lowest fluxes in the complete flux distribution over the observation time. We keep a fixed count rate selection for now. After introducing these data selection schemes for the total tail count rates relative to the maximum PUI count as a function of time across the superposed compression in Figure 1, we use them also to construct spectra of tail distributions, separately for each regional subdivision.

3. Analysis Results

Figure 1 shows the evolution of the superposed compression in the same way as used in [33] for the bulk PUI evolution. In fact, panels a, b, and f show the same quantities, yet for a slightly different selection of compressions. For context, panel g is added to show the average local SW speed gradient. As can be seen from panels c and d, with the unbiased tail counts and selected quiet times from the SIT ^{4}He fluxes, the normalized tail count rate shows two distinct maxima in the slow and fast compressed wind, followed by another maximum after the end of the rarefaction region, probably indicating the arrival of the next compression. Based on this temporal evolution of the normalized tail count rate, it is obvious that the rate systematically decreases with distance from the compression, but it increases again toward the end of the rarefaction region, when apparently the next compression arrives. This behavior is very similar in panel c and d. In contrast, the normalized tail count rate is much reduced through the entire time interval, after the quiet time count rate threshold criterion has been applied using the observed PLASTIC He$^+$ tail count rate. Also, except for statistical fluctuations, the normalized count rate remains almost constant throughout.

Next, we evaluate the normalized tail distribution spectra as a function of the normalized speed $w = v/V_{sw}$ in the SW frame across the different regions, starting with the compression itself. Figure 2 shows the spectra in a log-log representation for the slow compressed (red), fast compressed (purple), and peak speed (black) regions for three ranges of increasing average SW speed gradients across the combined slow and fast compressed SW from the left to the right (at the bottom of each panel). Error bars represent the statistical uncertainties and are mostly smaller than the symbols, except for the highest energies. Also shown are Chi-squared fit lines, assuming a power law, with resulting indices and fit errors shown in the upper right corner of each panel. Spectral indices between about -5.2 and -8.2 are found, a range that is consistent with the findings by Popecki et al. [31] for two sample compressions and shocks. Generally, the spectra are harder in the slow compressed SW than in the other two regions. They appear to soften systematically from the smallest to the largest average speed gradients. That the spectra appear to extend further in the slow compressed SW is due to the upper PLASTIC energy limit of 80 keV/q. For higher SW speed the observations extend to lower values in w.

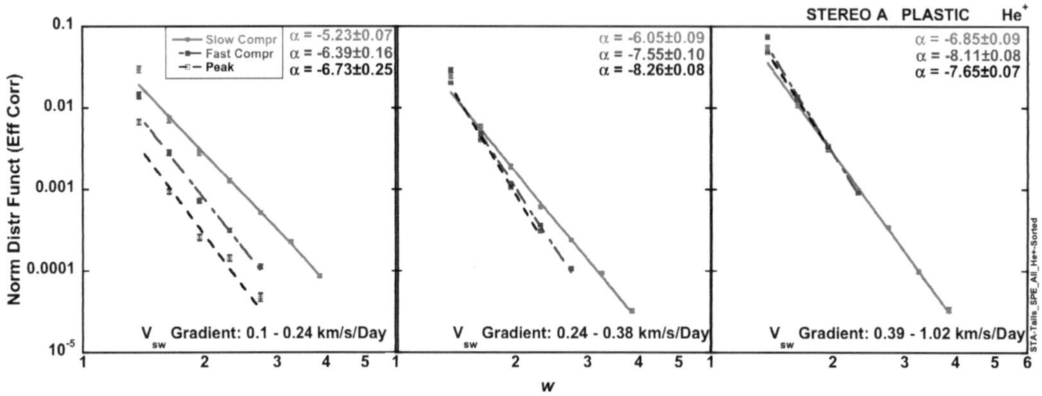

Figure 2: Spectra of suprathermal tails, without any quiet time condition imposed, in the slow compressed SW (red), fast compressed SW (blue), and peak SW (black) for three different ranges in the strength of the SW speed gradient across superposed compression regions, both slow and fast compressed SW combined.

In Figure 3 we turn to the normalized tail distribution spectra in each of the four consecutive sections of the rarefaction region for the largest speed gradient range from Figure 2. The four sections are marked with different symbols and shades of blue. Here, we compare the results for different quiet time selections, unbiased samples (left), only time intervals with ≤0.5 $(cm^2\ sr\ s\ MeV/nuc)^{-1}$ of ^4He at 0.16-0.24 MeV/nuc in SIT selected (center), and only time intervals with ≤1 count/h He^+ for $w \geq 1.6$ in PLASTIC selected (right), for the range with the largest solar SW gradients from Figure 2. For all cases, the power law indices range from -6.2 to -9.0. For the unbiased selection and the quiet time selection that uses STEREO SIT data, the normalized count rates start highest (with ≈0.01 at the lowest speed) in section

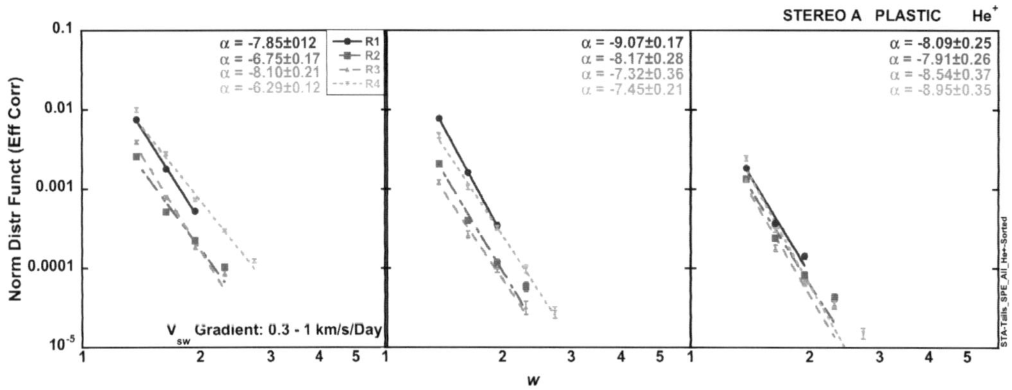

Figure 3. Spectra of suprathermal tails in four consecutive sections (R1, 2, 3, and 4) of the rarefaction region based a superposed epoch sample of compression regions for the strongest SW speed gradients.
Left: Unbiased sample, with no quiet time criterion applied
Center: Only time intervals with ≤0.5 $(cm^2\ sr\ s\ MeV/nuc)^{-1}$ of ^4He at 0.16-0.24 MeV/nuc in SIT selected
Right: Only time intervals with ≤1 count/h He^+ for $w \geq 1.6$ in PLASTIC selected

R1, then decrease by about a factor of 3 to R2 and R3, and increase to about the same level again in R4, thus mimicking the behavior of the total normalized tail fluxes in Figure 1. There is no visible difference between these two selections in the power law indices. In contrast, the spectra in all four sections appear to be almost identical at approximately the level of R2 and R3 from the previous selections when using PLASTIC tail count data for the quiet time criterion. The spectral indices are similar to those in the left two panels, perhaps, with slightly softer spectra though.

4. Discussion

Let us start the discussion with the regions that appear to feature most prominently suprathermal tails, i.e. the slow and fast compression as well as the peak speed region. Here, the total tail fluxes peak separately in both regions, and the spectra appear harder in slow wind than in the fast wind, similar to findings in [31]. Also, the spectra appear to harden for smaller SW speed gradients, which seems counterintuitive at first glance. On the other hand, the tail count rate at the low speed end of the spectra increases as a function of the speed gradient. Apparently, a steeper gradient, indicative of a stronger compression, may produce a higher tail flux, but it appears as if the acceleration to higher energies is less effective. The situation looks similar for the steeper (and thus softer) spectra in the fast-compressed and peak wind compared to the slow-compressed wind. At this point this is not understood and may only be addressed quantitatively in a combined simulation and data analysis. A marked dip between the two maxima is also visible, when inspecting the time evolution of the total tail count rate (not shown here), but it is probably enhanced when choosing the normalized tail count rate. As shown in [33], the total PUI count rate peaks near the boundary between slow- and fast-compressed wind, thus suppressing the normalized tail count rate here. Overall, the normalized tail count rates do not follow closely the speed gradient, nor the fluctuation level, although they are generally high where these two indicators are high.

There are small, but visible differences between the time evolution of the unbiased normalized tail count rates and those selected for quiet times using STEREO SIT ^{4}He. The latter probably indicate where times with higher energy particles have been eliminated from the slow- and fast-compressed wind, as expected for such a selection. However, the evolution in the rarefaction region is largely unchanged. For the He$^+$ tail count based quiet time selection analogous to the one used by Fisk & Gloeckler [11] and similar to Dayeh et al. [15], the temporal evolution is starkly different. Here, any visible temporal evolution over the superposed compression is removed, and the count rates reflect tail fluxes close to the lowest levels in the center of the rarefaction region, but with large, most likely purely statistical, variations. When integrated over the time intervals into spectra, this behavior is reinforced. All spectra in the rarefaction region taken with all three criteria show power law indices that vary visibly around values steeper than -5, values and variations similar to those found by Popecki et al. [31]. While the unbiased and SIT related selections reflect the systematic tail flux decrease toward the center of the rarefaction region (note, the tails increase toward the preceding and the following compression), all spectra contain similarly low fluxes for the selection with a threshold for the PLASTIC tail counts. It is also very surprising that apparently even during the compression itself quiet time intervals seem to be selected, with normalized tail count rates similar to those in the rarefaction region.

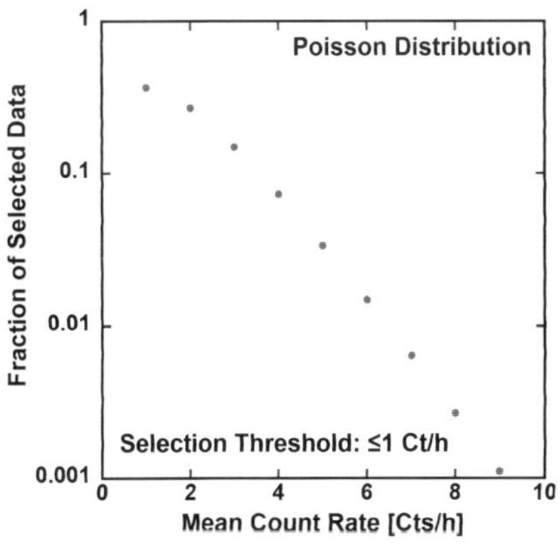

Figure 4: Fraction of selected data time periods as a function of mean count rate for a threshold of ≤ 1 Ct/h, if the events are Poisson distributed.

How can such a surprising result, at first sight, be understood? To make progress with this question, let us emphasize that with the latter quiet time criterion data time intervals are selected and accumulated that either contain one or no count, individually. Of course, accumulated over a large portion of a multi-year data set collectively they still generate meaningful count rates and spectra. Barring any systematic effects, individual counts, as selected here, are distributed between the time intervals that are accumulated according to a Poisson distribution. With a threshold of ≤ 1 count/h, therefore, all those time intervals are selected for accumulation that contain 0 or 1 count for any mean count rate that prevails during the observation period of interest. Figure 4 shows the fraction of time intervals (or data) that are selected with this criterion from Poisson distributions for mean count rates ranging from

0 to 10 counts/h. Even for 9 counts/h, 0.001 of the time intervals fulfill the selection criterion.

It becomes immediately clear that fluxes over quite a range in magnitude contribute to the accumulated population, thus even allowing tails in the compressions to contribute, if no further condition is applied. In other words, the underlying fluxes of the observed population must have a substantial uncertainty. Secondly, due to the required threshold only a constant count rate of ≈ 1 count/h is returned.

With this explanation in mind, the unbiased tail count selection and the one screened for energetic ^{4}He fluxes with STEREO SIT appear to provide a realistic temporal evolution across the superposed compression region. Both indicate in the total normalized tail fluxes and the spectra a well visible gradient from the preceding and following compression into the rarefaction region. This observation appears consistent with the compressions being a source for the tails here. The spectra in the rarefaction region appear to be closer to the ones from the fast-compressed wind and peak regions. Why the harder spectra of the slow compressed wind don't seem to show up more prominently here is not understood and should be studied further in the future. Furthermore, the quiet time criteria used previously appear to produce a stochastic superposition of particles from regions with a variety of mean fluxes, which may have different power law indices from -5 as found here. A stochastic superposition would be consistent with the prediction by Schwadron et al. [30] that a composite spectrum with about v^{-5} results in this scenario.

While quiet time selections based on a threshold in the observed total tail count rate have intrinsic value, when analyzing average spectral [12, 13] and compositional [14, 15] characteristics of suprathermal tails, one needs to be aware of potential shortcomings of this method when attempting to extract spatial and temporal evolutions, even with an SPE method as used here. Firstly, such a threshold selects particles from a rather wide range of mean fluxes, as an intrinsic property of the Poissonian distribution of the particles. Secondly, any temporal evolution that may point to potential localized sources is eliminated.

5. Conclusions and Outlook

Building on a related SPE analysis of the properties of the bulk PUIs across SW compressions [33], we have performed the first systematic study of the evolution of the total normalized suprathermal tail count rates as a function of average temporal separation from the stream interface. We found the strongest tails in the slow-compressed, fast-compressed, and peak speed wind regions, largely consistent with the findings by Tessein et al. [32], with a harder spectrum in the slow wind than in the other two regions, consistent with the findings by Popecki et al. [31]. The spectra in the compression region tend to soften for steeper average SW speed gradients. The total normalized tail count rates and the levels of the spectra decrease in the rarefaction region as a function of distance from the preceding and following compression, with a minimum in the center of the rarefaction. However, a constant low tail flux is returned when applying a quiet time selection with a threshold in the observed PLASTIC He^+ tail count rates, which is explained with the intrinsic behavior of Poisson distributed tail counts for such low count rates. The distinct temporal variation of the tail count rates across the compression region is found for the unbiased data and for a selection of quiet times using SIT ^{4}He fluxes, which appears consistent with a source of the tails in the compressions, and per extension near shocks of the interaction region. A v^{-5} tail spectrum may then emerge as a stochastic superposition of these sources, as predicted by Schwadron et al. [30].

To assess this emergent hypothesis quantitatively, a combination of simulations of the acceleration of ions into the suprathermal tails and their transport to the observer together with a detailed data analysis for the parameter dependencies found in the simulations will have to be undertaken. This extension of the current pilot study is planned for the near future.

Acknowledgments

The work in this paper was supported at UNH under NASA Grants NNX16AF79G, 80NSSC17K000, NNX13AI75G, NNX14AG13A, and NNX14B965 as well as by the DFG under Wi2139/7-1 and DLR under 50 0C 1103 at the Universität Kiel. E. Möbius is thankful for the support and hospitality at ISSI during the finalization of the results and preparation of the manuscript.

References

[1] Reames D V 1990 *Astrophys J. Suppl.* **73** 235
[2] Scholer M, et al. 1999 *Space Sci. Rev.* **89** 369
[3] Hovestadt D, Gloeckler G, Klecker B and Scholer M 1984 *Astrophys. J.* **281** 463
[4] Mason G M, Mazur J E and Dwyer J R 1999 *Astrophys. J.* **525** L133
[5] Desai M I, Mason G M, Dwyer J R, et al. 2003 *Astrophys. J.* **588** 1149
[6] Gloeckler G, Jokipii J R, Giacalone J and Geiss J 1994 *J. Geophys. Res.* **99** 17637
[7] Morris D, Möbius E, Lee M A, Popecki M A, B. Klecker, Kistler L M and Galvin A B 2001 *AIP Conf. Proc.* **598** 201
[8] Kucharek H, Möbius E, Li W, et al. 2003 *J. Geophys. Res.* **108** 8030
[9] Möbius E and Kallenbach R 2005 In: *The Heliosphere and Beyond/10 Years of ISSI*, J. Geiss and B. Hultqvist eds., **ISSI-SR003** 165
[10] Desai M I and Giacalone J 2016 *Living Rev. Sol. Phys.* **13** 3
[11] Fisk L A and Gloeckler G 2006 *Astrophys. J.* **640** L79
[12] Gloeckler G, Fisk L A Zurbuchen T H and Schwadron N A 2000 *AIP Conf. Proc.* **528**, 221
[13] Gloeckler G, Fisk L A, Mason G M and Hill M E 2008 *AIP Conf. Proc.* **1039** 367
[14] Desai M I, Mason G M, Mazur J E and Dwyer J R 2006 *Astrophys. J. Lett.* **645** L81
[15] Dayeh M A, Desai M I, Mason G M, Ebert R W and Farahat A 2017 *Astrophys. J.* **835** 155
[16] Fisk L A and Gloeckler G 2008 *Astrophys. J.* **686** 1466
[17] Fisk L A and Gloeckler G 2014 *J. Geophys. Res.* **119** 8773
[18] Jokipii J R and Lee M A 2010 *Astrophys. J.* **713** 475
[19] Zhang M 2010 *J. Geophys. Res.* **115** A12102
[20] Zhang M and Lee M A 2013 *Space Sci. Rev.* **176** 133
[21] Drake J F, Swisdak M and Fermo R 2012 *Astrophys. J.* **763** L5
[22] Zank G P, le Roux J A, Webb G M, Dosch A and Khabarova O 2014 *Astrophys. J.* **797** 28
[23] Zhao L-L, Zank G P, Kharbarova O, et al. 2018 *Astrophys. J. Lett* **864** L34
[24] Schwadron N A, Fisk L A and Gloeckler G 1996 *Geophys. Res. Lett.* **23** 2871
[25] Fisk L A and Lee M A 1980 *Astrophys. J.* **237** 620
[26] Giacalone J, Jokipii J R and Kóta J 2002 *Astrophys. J.* **573** 845
[27] Richardson I G 2004 *Space Sci. Rev.* **111** 267
[28] Zank G P, Rice W K M and Wu C C 2000 *J. Geophys. Res.* **105** 25079
[29] Lee M A 2005 *Astrophys. J. Suppl.* **158** 38
[30] Schwadron N A, Dayeh M A, Desai M, Fahr H-J, Jokipii J R and Lee M A 2010 *Astrophys. J.* **713** 1386
[31] Popecki M A, Klecker B, Simunac K D C, Galvin A B and Kucharek H 2013 *AIP Conf. Proc.* **1539** 255
[32] Tessein J A, Matthaeus W H, Wan M, Osman K T, Ruffolo D and Giacalone J 2013 *Astrophys. J. Lett.* **776** L8
[33] Bower J, Möbius E, Berger L, et al. 2019 *J. Geophys. Res.*, in press
[34] Stone E C, Frandsen A M, Mewaldt R A, et al. 1998 *Space Sci. Rev.*, **86** 1
[35] Stone E C, Cummings A C, McDonald F B, Heikkila B C, Lal N and Webber W R 2005 *Science* **309** 2017
[36] Galvin, A B, et al. 2008 *Space Sci. Rev.* **136** 437
[37] Drews C, Berger L, Taut A, Peleikis T and Wimmer-Schweingruber R F 2015 *Astron. Astrophys.* **575** A97
[38] Taut A, Berger L, Drews C, et al. 2018 *Astron. Astrophys.* **611** A61
[39] Möbius E, Klecker B, Bochsler P, et al. 2010 *AIP Conf. Proc.* **1302** 37
[40] Möbius E, Lee M A and Drews C 2015 *Astrophys. J.* **815** 20

Journal of Physics: Conference Series **1332** (2019) 012012 doi:10.1088/1742-6596/1332/1/012012

Energetic Particle Acceleration in Shearing Flows

Frank M. Rieger[1,2]

[1] Zentrum für Astronomie (ZAH), Institut für Theoretische Astrophysik, Universität Heidelberg, Philosophenweg 12, 69120 Heidelberg, Germany
[2] Max-Planck-Institut für Kernphysik, P.O. Box 103980, 69029 Heidelberg, Germany

E-mail: f.rieger@uni-heidelberg.de

Abstract. Shear flows are ubiquitously present in space and astrophysical plasmas, and are known to be conducive to the acceleration of energetic charged particles. In particular, for relativistic flow speeds, efficient Fermi-type particle acceleration can be achieved, capable of producing momentum power-law particle distributions and sustaining energetic particles all along the shear. This paper reviews some of the key results concerning the stochastic acceleration of energetic particles in gradual shear flows and reports on some recent developments in the field.

1. Introduction

Shear flows are naturally expected in astrophysical environments; prominent examples include the rotating accretion flows around compact objects and the relativistic outflows (jets) in gamma-ray bursts (GRBs) or Active Galactic Nuclei (AGN) [1]. From a phenomenological point of view the latter have turned out to be particularly interesting. Spatially, the fast jets of powerful AGN can extend over several hundred kilo-parsecs (kpc). Bright hot spots are formed and significant back flows induced when these jets eventually interact with the ambient medium. Often these large-scale jets appear laminar despite large fluid Reynolds numbers ($Re = \rho u L/\mu > 10^{20}$), see Ref. [2] for orientation. The Chandra detection of extended non-thermal X-ray emission in several of these AGN jets points to the presence of energetic charged particles that have experienced continuous acceleration within them [3–5]. The favoured electron synchrotron origin of the X-ray emission in fact implies the presence of relativistic electrons with Lorentz factors up to $\gamma \sim 10^8$, e.g. [6]. Since their synchrotron cooling timescale $t_{\rm cool} \propto 1/\gamma$ and length, $c\,t_{\rm cool}$, are very short ($\ll 1$ kpc), some efficient and distributed (re-)acceleration mechanism is required to keep electrons energized throughout these jets [7]. It seems in principle conceivable that the same mechanism could also facilitate the acceleration of cosmic rays (CRs) to extreme energies, making large-scale AGN jets to possible ultra-high-energy (UHE)CR acceleration sites [8; 9].

These and other observational findings have given fresh impetus to a variety of shear particle acceleration and emission scenarios. In general, the acceleration and transport of charged particles in shearing flows can be explored on multiple scales, from the plasma skin depth (electron inertial-scale) [10] via the relativistic gyro-scale [4; 11; 12] to large turbulent length scales [13]. The present paper focuses on the second one which exhibits a history dating back to the 1980s [1; 4; 7; 11; 12; 14–26]. We will review some of the key results concerning the stochastic acceleration of energetic particles in gradual shear flows, highlight expected spectral characteristics, and report on some recent developments.

Content from this work may be used under the terms of the Creative Commons Attribution 3.0 licence. Any further distribution of this work must maintain attribution to the author(s) and the title of the work, journal citation and DOI.
Published under licence by IOP Publishing Ltd

2. Basic Concept

Shear particle acceleration can be understood as a stochastic, Fermi-type [27] acceleration mechanism in which particle energization occurs as a results of elastically scattering off moving magnetic inhomogeneities [4; 28]. The energy change in an elastic scattering event is simply given by $\Delta\epsilon = \epsilon_2 - \epsilon_1 = 2\gamma_u^2(\epsilon_1 u^2/c^2 - \vec{p}_1 \cdot \vec{u})$, where $\vec{u}$ is the characteristic scattering center speed (u its magnitude, γ_u the Lorentz factor), and ϵ_1 and $\vec{p}_1$ denote the initial particle energy and momentum, respectively. Accordingly, a particle will gain energy in head-on ($\vec{p}_1 \cdot \vec{u} < 1$) collisions, and lose energy in following ($\vec{p}_1 \cdot \vec{u} < 1$) collisions. Stochastic acceleration (i.e., an average energy gain) results from the fact the interaction probability for head-on collisions is higher than the one for following collisions, retaining a second-order dependence in u, i.e., $< \Delta\epsilon > \propto (u/c)^2\epsilon$. Similarly, shear acceleration can be understood as a stochastic acceleration process, yet with the conventional scattering center speed replaced by an effective velocity u_e, given by the gradient in the flow speed multiplied by the particle free path, $\lambda = c\,\tau$. Hence, for a non-relativistic gradual shear flow with, e.g., $\vec{u} = u_z(x)\vec{e}_z$, $u_e = (\partial u_z/\partial x)\,\lambda$, and the fractional energy changes becomes

$$\frac{< \Delta\epsilon >}{\epsilon} \propto \left(\frac{u_e}{c}\right)^2 \propto \left(\frac{\partial u_z}{\partial x}\right)^2 \lambda^2. \tag{1}$$

This suggests a characteristic acceleration timescale

$$t_{\rm acc} = \frac{\epsilon}{(d\epsilon/dt)} \sim \frac{\epsilon}{< \Delta\epsilon >} \times \frac{\lambda}{c} \propto \frac{1}{\lambda}, \tag{2}$$

which, in contrast to classical 1st and 2nd order Fermi acceleration, is inversely depending on the particle mean free path. In the case of e.g. a gyro-dependence with $\lambda \sim r_g \propto \gamma$, $t_{\rm acc}$ would reveal the same scaling as the synchrotron loss timescale $t_{\rm cool}$, with interesting implications [22]. The principal reason for this unusual behaviour is related to the fact that as a particle increases its energy ($\epsilon \simeq \gamma mc^2$) and thereby its mean free path ($\lambda \propto \gamma^\alpha, \alpha > 0$), a higher effective velocity u_e is experienced. For shear to start operating efficiently within a system, one thus usually requires injection of some pre-accelerated seed particles, either from acceleration at shocks or via classical 2nd order Fermi processes [7]. The situation is typically more favourable, though, for the case of protons or ions.

3. Microscopic Picture - Fokker-Planck type diffusion for non-relativistic flows

A more detailed picture of the particle evolution becomes apparent in a microscopic approach via an explicit calculation of the average rate of momentum change and dispersion (i.e., the corresponding Fokker Planck coefficients). Consider for simplicity a non-relativistic shear flow with $\vec{u} = u_z(x)\vec{e}_z$. While travelling across the flow, the momentum of a particle relative to the flow will change by $\vec{p}_2 = \vec{p}_1 + m\,\delta\vec{u}$, where $\delta\vec{u} = (\partial u_z/\partial x)\delta x\,\vec{e}_z$ with $\delta x = v_x\,\tau$, and where the timescale for collisions (scattering time) might be a function of momentum, i.e., $\tau \equiv \tau(p) = \tau_0\,p^\alpha$. Expanding $(p_2 - p_1)$ to second order in δu and averaging over an isotropic particle distribution, the Fokker Planck coefficients then become [16; 22]

$$\begin{aligned} \left\langle \frac{\Delta p}{\Delta t} \right\rangle &\propto p \left(\frac{\partial u_z}{\partial x}\right)^2 \tau\,, \\ \left\langle \frac{(\Delta p)^2}{\Delta t} \right\rangle &\propto p^2 \left(\frac{\partial u_z}{\partial x}\right)^2 \tau\,. \end{aligned} \tag{3}$$

The Fokker Planck coefficients can be shown to be related by the equation

$$\left\langle \frac{\Delta p}{\Delta t} \right\rangle = \frac{1}{2p^2}\frac{\partial}{\partial p}\left[p^2 \left\langle \frac{(\Delta p)^2}{\Delta t} \right\rangle\right] = \frac{\Gamma}{p^2}\frac{\partial}{\partial p}\left(p^4\tau\right)\,, \tag{4}$$

i.e., to satisfy the principle of detailed balance (scattering being reversible). Here Γ in eq. (4) denotes the shear flow coefficient, and for the employed flow profile is given by $\Gamma = (1/15)(\partial u_z/\partial x)^2$. Under the condition of detailed balance, the associated Fokker-Planck equation reduces to a diffusion equation in momentum space. Hence in the absence of radiative losses, the particle phase-space distribution function $f(p,t)$ undergoing shear acceleration might be cast in a simple Fokker-Planck type diffusion equation

$$\frac{\partial f(p,t)}{\partial t} = \frac{1}{p^2}\frac{\partial}{\partial p}\left(p^2 D \frac{\partial f}{\partial p}\right), \tag{5}$$

where D denotes the momentum space diffusion coefficient, and for the considered application is given by

$$D = \Gamma p^2 \tau = \frac{1}{15}\left(\frac{\partial u_z}{\partial x}\right)^2 p^2 \tau\,, \tag{6}$$

where $\tau = \tau_0\, p^\alpha$. Figure 1 shows an exemplary solution of eq. (5) for the case $\tau \propto p$ [22]. At

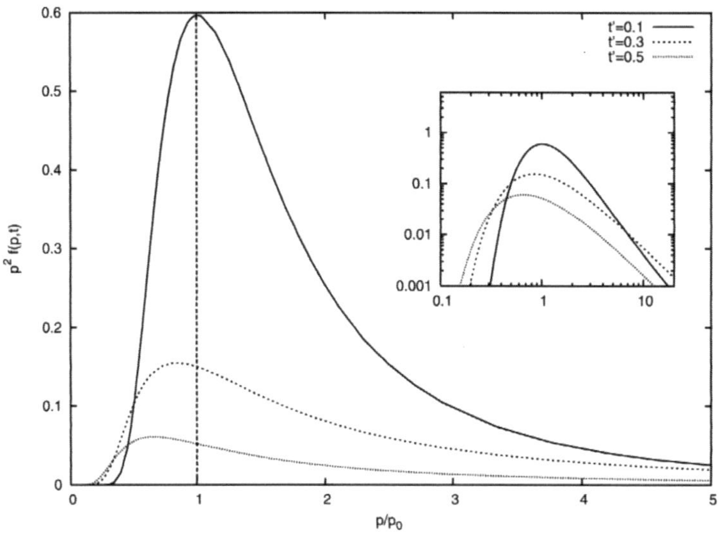

Figure 1. Time-dependent solution $f(p,t)$ of the Fokker-Planck diffusion equation for non-relativistic shear assuming (impulsive) mono-energetic injection with p_0 at $t_0 = 0$. A momentum-dependence $\alpha = 1$ has been used. The distribution broadens with time due to momentum dispersion. The inlet illustrates the formation of a power law like distribution $n(p) \propto p^2 f(p) \propto p^{-2}$ above p_0 for $t' \geq 0.3$.

sufficiently large times, the particle distribution above the injection momentum p_0 approaches a power law shape

$$n(p) \propto p^2 f(p) \propto p^{-(1+\alpha)}\,, \tag{7}$$

for $\alpha > 0$, with power law index depending on the scaling of the particle mean free path ($\lambda \simeq c\tau \propto p^\alpha$) [11; 14; 22]. In particular, for a gyro-dependence, $\alpha = 1$, one has $n(p) \propto p^{-2}$, similar to first-order Fermi acceleration at non-relativistic shocks. The quasi steady-state (time-integrated) distribution $f(p)$ for continuous injection then becomes constant below p_0, and takes on the noted power-law shape ($\propto p^{-(3+\alpha)}$) above it.

4. Non-relativistic particle transport equation

In order to take spatial transport into account, an extended particle transport equation needs to be derived. In the non-relativistic limit this has been done by Earl et al. [12], starting from the non-relativistic Boltzmann equation for the phase-space distribution $f(\vec{x}, \vec{p}, t)$ and assuming a simple BKG-type collision term, $(\partial f/\partial t)_s = (f - \langle f \rangle)/\tau$. The approach relies on a mixed system of phase-space coordinates such that quantities which are operated upon the scattering operator, i.e. the momentum, are evaluated in the comoving flow frame (allowing to treat the scattering physics in simple ways), while time and space coordinates are still measured in the laboratory frame. Given non-relativistic flow speeds the particle momentum components are then related by a Galilean transformation $p_i = p'_i + m\, u_i$ (background flow speed u_i, comoving p'_i). Scattering is assumed to be strong enough to guarantee the diffusion approximation, i.e., to ensure that the departure from isotropy is small, $f = f_0 + f_1$, with $< f_1 >= 0$ and $f_1 \ll f_0$. The full non-relativistic particle transport equation for $f_0(\vec{x}, p', t)$ then becomes [12]

$$\begin{aligned} \frac{\partial f_0}{\partial t} + u_i \frac{\partial f_0}{\partial x_i} & - \frac{p'}{3} \frac{\partial u_i}{\partial x_i} \frac{\partial f_0}{\partial p'} - \frac{\partial}{\partial x_i} \left(\kappa \frac{\partial f_0}{\partial x_i} \right) \\ & + \frac{2\,\tau\, p'}{3} A_i \frac{\partial^2 f_0}{\partial x_i \partial p'} + \frac{1}{3\, p'^2} \frac{\partial\, (\tau\, p'^3)}{\partial p'} A_i \frac{\partial f_0}{\partial x_i} \\ & - \frac{\Gamma}{p'^2} \frac{\partial}{\partial p'} \left(\tau\, p'^4 \frac{\partial f_0}{\partial p'} \right) + \frac{p'}{3} \frac{\partial (\tau A_i)}{\partial x_i} \frac{\partial f_0}{\partial p'} = 0\,, \end{aligned} \tag{8}$$

$(i = 1, 2, 3)$, where κ denotes the isotropic spatial diffusion coefficient, $\kappa(p) = \tau(p) c^2/3$, and

$$A_i = \frac{\partial u_i}{\partial t} + u_l \frac{\partial u_i}{\partial x_l} \tag{9}$$

is the viscous shear flow coefficient given by

$$\Gamma = \frac{1}{30} \left(\frac{\partial u_i}{\partial x_k} + \frac{\partial u_k}{\partial x_i} \right)^2 - \frac{2}{45} \frac{\partial u_i}{\partial x_i} \frac{\partial u_k}{\partial x_k}\,. \tag{10}$$

The second, third and fourth term in eq. (8) relate to the well-known effects of convection, adiabatic energy change and spatial diffusion, while the terms involving A_i describe the effects of inertial drifts (cf. also [29] for incorporation of a mean magnetic field). The additional term involving Γ characterizes the energy changes due to flow shear and divergence. For a steady (non-relativistic) shear flow of the form $\vec{u} = u_z(x)\vec{e}_z$, the adiabatic and inertial terms vanishes ($\partial u_i/\partial x_i = 0$ and $A_i = 0$), Γ reduces to $\Gamma = (1/15)(\partial u_z/\partial x)^2$, and the space-independent part of eq. (8) becomes

$$\frac{\partial f_0}{\partial t} = \frac{\Gamma}{p'^2} \frac{\partial}{\partial p'} \left(\tau\, p'^4 \frac{\partial f_0}{\partial p'} \right)\,, \tag{11}$$

thus coinciding with eq. (5).

5. Relativistic Generalization

To compete with diffusive particle escape, efficient shear particle acceleration typically requires relativistic flow speeds (see also below). This then demands a generalization of the particle transport to the relativistic regime as has been obtained by e.g. Webb et al. [15; 18]. Assuming isotropic diffusion and denoting the (covariant) metric tensor by $g_{\alpha\beta}$, the zero component of the (comoving) particle momentum four vector by $p'^0 = E'/c$, the fluid four velocity by u_α and the fluid four acceleration by $\dot{u}_\alpha := u^\beta \nabla_\beta u_\alpha$, with $\nabla_\beta u_\alpha$ indicating covariant derivation, the full particle transport equation for the isotopic distribution function $f_0(x^\alpha, p')$ with $x^\alpha = (ct, x, y, z,)$ becomes [15]

$$\nabla_\alpha \left[c u^\alpha f_0 - \kappa \left(g^{\alpha\beta} + u^\alpha u^\beta \right) \left(\frac{\partial f_0}{\partial x^\beta} - \dot{u}_\beta \frac{(p^0)^2}{p'} \frac{\partial f_0}{\partial p'} \right) \right]$$
$$+ \frac{1}{p'^2} \frac{\partial}{\partial p'} \left[-\frac{p'^3}{3} c \nabla_\beta u^\beta f_0 + p'^3 \left(\frac{p'^0}{p'} \right)^2 \kappa \, \dot{u}^\beta \left(\frac{\partial f_0}{\partial x^\beta} - \dot{u}_\beta \frac{(p'^0)^2}{p'} \frac{\partial f_0}{\partial p'} \right) - \Gamma \tau p'^4 \frac{\partial f_0}{\partial p'} \right] = Q \,, \tag{12}$$

where Q denotes the source term and Greek indices (α, β) run from 0 to 3. Γ denotes the (generalized) relativistic shear coefficient, which in the strong scattering limit is given by

$$\Gamma = \frac{c^2}{30} \sigma_{\alpha\beta} \sigma^{\alpha\beta} \,, \tag{13}$$

where $\sigma_{\alpha\beta}$ is the (covariant) fluid shear tensor. For a cylindrical jet with a steady (relativistic) shear flow profile $\vec{u} = u(r)\vec{e}_z$, the fluid four acceleration ($\dot{u}_\alpha = 0$) and divergence ($\nabla_\beta u^\beta = 0$) vanish, retaining only the shear term with $\Gamma = (\gamma_u^4/15)\,(du/dr)^2$ in the second line of eq. (12). z-independent, steady-state solutions of eq. (12) for such a flow profile and a specific form of the radial dependence of $\kappa(r, p)$ have been recently presented by Webb et al. [26].

6. Application

In the following some recent applications and results in the context of shear particle acceleration are briefly reported.

6.1. Particle acceleration in expanding flows

Both, the jets in AGN and GRBs exhibit relativistic speeds, and are likely to be (quasi-conically) expanding, motivating an analysis of particle acceleration in expanding, relativistic shear flows [21; 24]. For the simplest case of a radial shear flow profile $u^\alpha = \gamma_v\,(\theta)(1, v_r(\theta)/c, 0, 0)$, where θ denotes the polar angle and r the radial coordinate, the impact of different functional dependencies for $\gamma_v(\theta)$, i.e., a power-law-, Gaussian- or Fermi-Dirac-type can be explored, see Fig. 2. The associated characteristic (comoving) acceleration timescales for a particle with mean free path λ' [24],

$$t_{\rm acc}(r, \theta)' \sim \frac{2\,c\,r^2}{\gamma_v^2\,(v_r^2 + 0.75\,\gamma_v^2\,[\partial v_r/\partial\theta]^2)\,\lambda'} \,, \tag{14}$$

can then become a strong function of θ, allowing for the generation of some prominent non-axis emission features (e.g., 'ridge line'). To overcome adiabatic losses ($\propto \gamma_v v_r/r$) and experience efficient acceleration, relativistic flow speeds and sufficient energetic seed particles ($\lambda'/r > 10^{-3}$ for the application presented in Fig. 2) are needed. It seems possible that such a process could lead to a weak and long-duration leptonic emission component in GRBs, and facilitate cosmic-ray acceleration to ultra high energies [21].

6.2. Generating multi-component particle distributions

As a two-stage process, stochastic shear acceleration could lead to the formation of multi-component particle distributions. A basic example considering radiative-loss-limited acceleration in a mildly relativistic shearing flow is shown in Fig. 3 [7]. The figure is based on a solution of the Fokker-Planck equation for $f(p, t)$ (or equivalently, $f(\gamma, t)$) including classical 2nd order Fermi and shear particle acceleration as well as radiative synchrotron losses. Employing a mean

Journal of Physics: Conference Series **1332** (2019) 012012 doi:10.1088/1742-6596/1332/1/012012

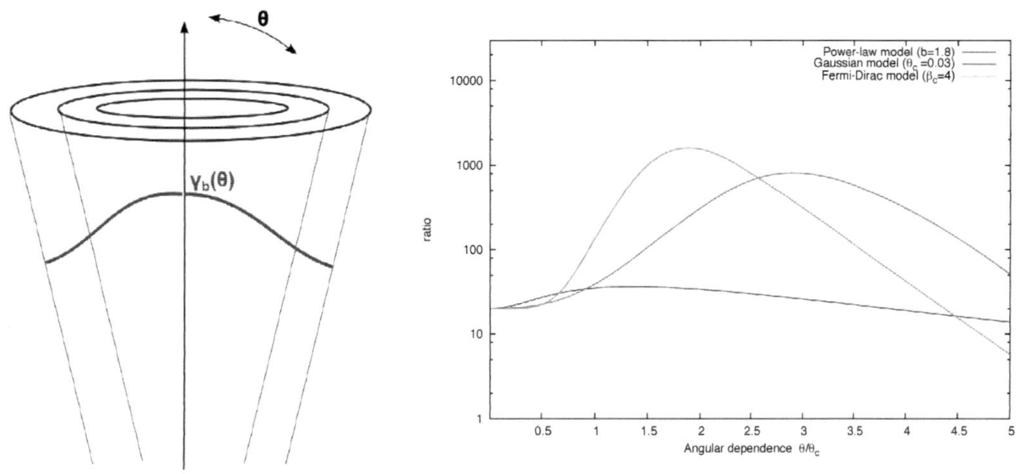

Figure 2. Left: Illustration of a simple radially expanding flow. Right: Ratio of viscous shear gain versus adiabatic losses multiplied by (r/λ'), illustrated for $\theta_c = 0.03$ rad and a flow Lorentz factor $\gamma_v = 30$ on the axis [24]. A non-axis preference becomes particularly evident for a Gaussian or Fermi-Dirac shaped flow profile.

free path scaling $\lambda = \xi^{-1} r_g\, (r_g/\Lambda_{\max})^{1-q} \propto \gamma^{2-q}$, with $q = 5/3$ (Kolmogorov) and $\xi = 0.1$, and choosing parameters applicable to the large-scale jets in AGN, electron acceleration up to Lorentz factors $\gamma \sim 10^9$ seems feasible. Stochastic 2nd order Fermi acceleration dominates particle energization up to $\gamma \sim 10^4$ above which efficient shear acceleration becomes operative leading to a somewhat flatter spectral slope (changing by 2/3 for the noted application). Synchrotron radiation eventually introduces a spectral cut-off at high energies. As shearing conditions are likely to prevail along astrophysical jets, these findings illustrate that the considered processes could be of relevance for understanding the extended X-ray emission in the large-scale jet of AGN (cf. Sec 1). Moreover, by the same means relativistic large-scale AGN jets could also facilitate the acceleration of cosmic rays to ultra-high energies [7; 26].

6.3. Spatial transport and diffusive escape

The above considerations did not account for details of the spatial transport, and possible modifications introduced by the diffusive escape of particles from the system. Implications of these could in principle be analyzed with resort to the relativistic particle transport equation (12). Deriving analytical, steady-state solutions $f_0(r, p')$ for a cylindrical jet, assuming z-independence and a special form for the radial dependence of the scattering time, $\tau(r,p) = \tau_0\, (p/p_0)^\alpha/[r\, d\xi(r)/dr]$ where $\xi(r) \equiv 0.5 \ln(1 + \beta(r)/[1 - \beta(r)])$, with $\beta(r) = u(r)/c$ the flow speed, Webb et al. [26] showed that diffusive escape could counter-act efficient acceleration. While the local particle distribution still follows a power law $f(p') \propto p'^{-\mu}$, its momentum index μ becomes dependent on the maximum flow speed β_0 on axis, and significantly steepens with decreasing β_0 (with $\mu \to \infty$ for $\beta_0 \to 0$). Though possible limitations due to the chosen τ-dependence may deserve some further studies, these results imply that efficient shear particle acceleration requires relativistic flow speeds. This becomes also apparent when comparing the corresponding timescale for acceleration, $t'_{\rm acc} \sim 1/[\gamma_u^4 (du/dr)^2 \tau']$, with the one for cross-field escape $t'_e \sim r^2/[c^2\tau']$.

The analytical solutions [26] can be used to gain insights into the radial evolution of the particle transport. Below an example is given for the particle transport in the presence of a

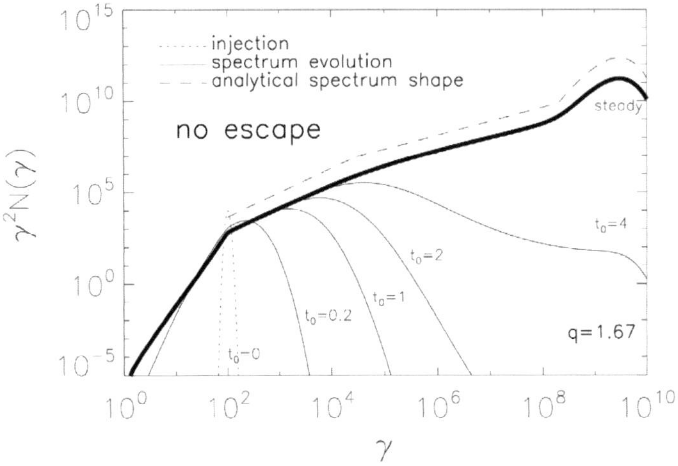

Figure 3. Time-evolution of the electron spectrum $\gamma^2 n(\gamma)$ in the presence of stochastic-shear particle acceleration, where $n(\gamma) \propto \gamma^2 f(\gamma)$ represents a solution of the corresponding Fokker Planck equation given a linearly decreasing shear flow profile of width $\Delta r \sim r_j/10$ and an Alfven speed $\beta_A \sim 0.007$ [7]. Above particle Lorentz factors of a few times 10^4 the spectrum is shaped by shear acceleration, with a high-energy spectral cut-off around $\gamma \sim 10^9$ being introduced by synchrotron losses.

hyperbolic, relativistic shear flow as shown in Fig. 4.

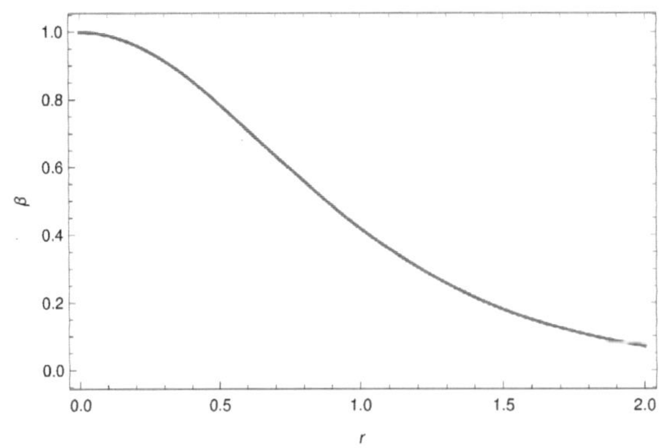

Figure 4. Relativistic flow profile $\beta(r)\,\vec{e}_z$ chosen for evaluating the particle distribution function in Fig. 5. A maximum flow Lorentz factor $\gamma_\beta = 20$ on the axis has been assumed.

Figure 5 provides an illustration that away from injection the 'classical' power-law momentum dependence, eq. (7), is recovered at high flow speeds ($\beta_0 \to 1$). Clearly, advancing our understanding of the (radial) diffusion properties in astrophysical jets will be important to better characterize the efficiency of shear particle acceleration.

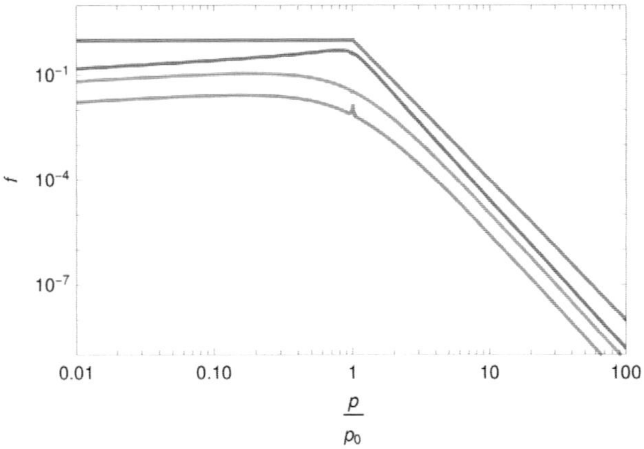

Figure 5. Steady-state solution $f(r, p')$ as a function of momentum p'/p'_0 for the flow profile in Fig. 4 and three different spatial locations, $r = 0.06$ (blue), $r = 0.50$ (orange), $r = 1.20$ (green). Mono-energetic particle injection with p'_0 at $r_1 = 0.02$ and an outer (escape) boundary at $r_2 = 2$ have been assumed. A momentum-dependence of $\alpha = 1$ has been employed for the scattering time. The red curve (top) designates the classical power-law dependence, with $f(p') \propto p'^{-4}$ above p'_0, as inferred from the Fokker-Planck approach in Sec. 3.

7. Conclusions

Particle acceleration in fast shearing flows could represent a promising mechanism for the continued acceleration of charged particles, capable of producing (local) power-laws in momentum. As such it can provide a plausible explanation for, e.g., the origin of the extended high-energy emission seen in the large-scale jets of AGN, and contribute to the energization of extreme cosmic rays.

In many circumstances, injection of energetic seed particles (in the case of electrons) and relativistic flow speeds will be required for the mechanism to operate efficiently. In the case of AGN and GRBs the former condition are likely to be met by first-order shock and/or stochastic second-order Fermi processes. In this regard, shear acceleration thus resembles a two-stage process for facilitating particle acceleration beyond the common limit.

To further improve our understanding, high-resolution studies of astrophysical jets and a quantitative analysis of the stability and radial transport properties of fast shearing flows are of particular interest.

Acknowledgments

I would like to thank Gary Zank and the organizers for the invitation to an inspiring and pleasant 18th AIAC conference. I am also grateful to Peter Duffy, Felix Aharonian and Ruoyu Liu for collaboration and discussion. Financial support by a DFG Heisenberg Fellowship under RI 1187/6-1 is acknowledged.

References

[1] Rieger F M and Duffy P 2004 *ApJ* **617** 155–161 (*Preprint* astro-ph/0410269)

[2] Worrall D M and Birkinshaw M 2006 *Physics of Active Galactic Nuclei at all Scales* (*Lecture Notes in Physics, Berlin Springer Verlag* vol 693) ed Alloin D p 39 (*Preprint* astro-ph/0410297)

[3] Harris D E and Krawczynski H 2006 *ARA&A* **44** 463–506 (*Preprint* astro-ph/0607228)
[4] Rieger F M, Bosch-Ramon V and Duffy P 2007 *Ap&SS* **309** 119–125 (*Preprint* astro-ph/0610141)
[5] Georganopoulos M, Meyer E and Perlman E 2016 *Galaxies* **4** 65
[6] Sun X N, Yang R Z, Rieger F M, Liu R Y and Aharonian F 2018 *A&A* **612** A106 (*Preprint* 1712.06390)
[7] Liu R Y, Rieger F M and Aharonian F A 2017 *ApJ* **842** 39 (*Preprint* 1706.01054)
[8] Aharonian F A, Belyanin A A, Derishev E V, Kocharovsky V V and Kocharovsky V V 2002 *Phys. Rev. D* **66** 023005 (*Preprint* astro-ph/0202229)
[9] Kotera K and Olinto A V 2011 *ARA&A* **49** 119–153 (*Preprint* 1101.4256)
[10] Alves E P, Grismayer T, Fonseca R A and Silva L O 2014 *New Journal of Physics* **16** 035007 (*Preprint* 1404.0555)
[11] Berezhko E G and Krymskii G F 1981 *Soviet Astronomy Letters* **7** 352
[12] Earl J A, Jokipii J R and Morfill G 1988 *ApJL* **331** L91–L94
[13] Ohira Y 2013 *ApJL* **767** L16 (*Preprint* 1301.6573)
[14] Berezhko E G 1982 *Soviet Astronomy Letters* **8** 403–405
[15] Webb G M 1989 *ApJ* **340** 1112–1123
[16] Jokipii J R and Morfill G E 1990 *ApJ* **356** 255–258
[17] Ostrowski M 1990 *A&A* **238** 435–438
[18] Webb G M, Jokipii J R and Morfill G E 1994 *ApJ* **424** 158–180
[19] Rieger F M and Mannheim K 2002 *A&A* **396** 833–846 (*Preprint* astro-ph/0210286)
[20] Ostrowski M 1998 *A&A* **335** 134–144 (*Preprint* astro-ph/9803299)
[21] Rieger F M and Duffy P 2005 *ApJL* **632** L21–L24 (*Preprint* astro-ph/0509495)
[22] Rieger F M and Duffy P 2006 *ApJ* **652** 1044–1049 (*Preprint* astro-ph/0610187)
[23] Dempsey P and Rieger F M 2009 *International Journal of Modern Physics D* **18** 1651–1654
[24] Rieger F M and Duffy P 2016 *ApJ* **833** 34 (*Preprint* 1611.04342)
[25] Kimura S S, Murase K and Zhang B T 2018 *Phys. Rev. D* **97** 023026 (*Preprint* 1705.05027)
[26] Webb G M, Barghouty A F, Hu Q and le Roux J A 2018 *ApJ* **855** 31
[27] Fermi E 1949 *Physical Review* **75** 1169–1174
[28] Lemoine M 2019 *arXiv e-prints* (*Preprint* 1903.05917)
[29] Williams L L and Jokipii J R 1991 *ApJ* **371** 639–647

Journal of Physics: Conference Series **1332** (2019) 012013 doi:10.1088/1742-6596/1332/1/012013

Structure of the IBEX Ribbon from Distributed Sources

N. A. Schwadron
University of New Hampshire, Durham, NH 03824
Princeton University, Princeton, NJ, 08544

E-mail: nschwadron@unh.edu

D. J. McComas
Princeton University, Princeton, NJ, 08544

E-mail: dmccomas@princeton.edu

Abstract. The Interstellar Boundary Explorer (IBEX) mission discovered the presence of a global structure in energetic neutral atom emissions from the outer heliosphere not predicted by any model – the IBEX "ribbon". In the search for possible explanations, observations have pointed to a likely source from neutral atoms produced through charge-exchange with the outflowing solar wind. The secondary (neutral) solar wind then undergoes charge-exchange beyond the heliopause to produce the proton population that forms the ribbon. Here we study the plasma structure beyond the heliopause created from distributed secondary neutral atom sources. We provide a framework to analyze the IBEX ribbon taking into account distributed neutral atom sources within the local interstellar medium.

1. Introduction

The Interstellar Boundary Explorer (IBEX) was launched in 2008 to study the global heliosphere [1]. The subsequent discovery was made by IBEX of a narrow, almost circular structure of energetic neutral atoms from ~300 eV to 6 keV [1, 2], and extending in position from beneath the nose of the heliosphere toward the north polar region, then slightly above the tail and back down beneath the nose [2–5]. This structure is now referred to as the "IBEX ribbon", and is believed to be centered approximately on the direction of the interstellar magnetic field [2, 5, 6].

There are a broad array of proposed sources for the IBEX ribbon [2, 7]. IBEX observations [2] increasingly point to a source mechanism for the IBEX ribbon from the secondary (neutral) solar wind, which travels out beyond the heliopause, charge-exchanges with protons within the very local interstellar medium (VLISM), and then through subsequent charge-exchange collisions becomes energetic neutral atoms that travel back toward the Sun where they can be detected by IBEX. This mechanism is fundamentally tied to the secondary solar wind and has been modeled in numerous studies [8–13].

Models of the IBEX ribbon involving the secondary solar wind are roughly organized into two variants: (1) ring-beam models in which secondary solar wind atoms populate a ring-beam about the interstellar magnetic field after ionization beyond the heliopause. Energetic Neutral Atoms (ENAs) are then created from the portion of the ring beam that projects neutral atoms

Content from this work may be used under the terms of the Creative Commons Attribution 3.0 licence. Any further distribution of this work must maintain attribution to the author(s) and the title of the work, journal citation and DOI.
Published under licence by IOP Publishing Ltd

back toward the Sun, and therefore toward IBEX where they are detected [8–10, 12, 13]; (2) models involving the spatial retention of protons beyond the heliopause on magnetic field lines [11, 14, 15]. In the first of these variants of secondary solar wind models, very low scattering rates are required so that ring beams remain stable over multiple years [8, 11]. Conversely, for spatial retention models, we require ion scattering from magnetic field variations and Alfvén waves to retrain ions within the ribbon [11, 14, 15].

Here, we discuss the concept that the ribbon is composed of a superposition of spatial distributions integrated along the line-of-sight. In §2, we derive the kappa function that results from a superposition of randomly distributed sources. In §3, we fit observations using our derived kappa function and provide conclusions in §4.

2. Structure of the Ribbon from Distributed Sources

In the case of the retention region or ring beam distributions, the line-of-sight integrated observations of ENAs includes an array of plasma structures beyond the heliopause. The spatial distributions beyond the heliopause can be modeled either using a spatial retention region or in the limit of weak scattering. In these cases, we expect a peaked distribution close to the source with a spatial tail that decays exponentially, $f \propto \exp(-\beta x)$. Here, x is distance along a magnetic field line from the center of the source region. The scale-length of the exponential tail is $1/\beta$. In the specific case of spatial retention [11], $1/\beta = 2rv_A/v$, where r is the distance to the source. In the case of weak scattering, the scale-length is related to the scattering coefficient $\kappa_{\|}$ along the magnetic field: $1/\beta = 2\sqrt{\kappa_{\|}\tau_x}$ where τ_x is the charge-exchange rate, $\tau_x = 1/(n_{\mathrm{H-OSH}}\sigma v)$, $n_{\mathrm{H-OSH}}$ is the neutral hydrogen density in the outer heliosheath, v is the particle speed and σ is the charge-exchange cross-section.

The superposition of sources along the line-of-sight leads to the formation of kappa distributions [16]. Statistics involving a collection of variable states has been considered in many different areas of study [17, 18]. For example, in the case of solar flares, waiting times between subsequent solar flares are distributed according to a power-law that emerges from randomly distributed processes [19]. A given random (or Poisson) process with an average occurrence rate λ yields an exponential probability distribution of waiting time t between subsequent events, and there is an equal probability for an event occurring after time t for $t < 1/\lambda$. This uniform probability distribution of waiting times is required for any Poisson process. The superposition of exponential waiting time distributions leads to a kappa distribution, which has the observed power-law behavior for waiting times.

In analogy to the waiting time distribution of flaring processes, we consider the superposition of individual exponential spatial distributions. Each spatial distribution is defined by an exponential $f \propto \exp(-\beta x)$ with inverse scale-length β centered on the secondary ENA source beyond the heliopause. In this analog, the parameter β is the counterpart of an occurrence rate for a Poisson process.

The remainder of the derivation to describe the superposition of states follows previous work [16]. We identify each spatial exponential distribution of ENAs as a state, and consider the superposition of these states along the line-of-sight. The number of states for inverse scale-lengths in the range β and $\beta + d\beta$ is defined $F(\beta)d\beta$. The entropy of states, defined as $S = -\int d\beta F(\beta)\ln[F(\beta)]$, is maximized for the functional form $F(\beta) = \exp(-\beta/\beta_0)/\beta_0$. In this case, the superposed distribution $\tilde{f}$ of source protons within the ribbon takes the following form:

$$\tilde{f} = A(|\beta_0 x| + 1)^{-3} \tag{1}$$

where A is the amplitude of the distribution.

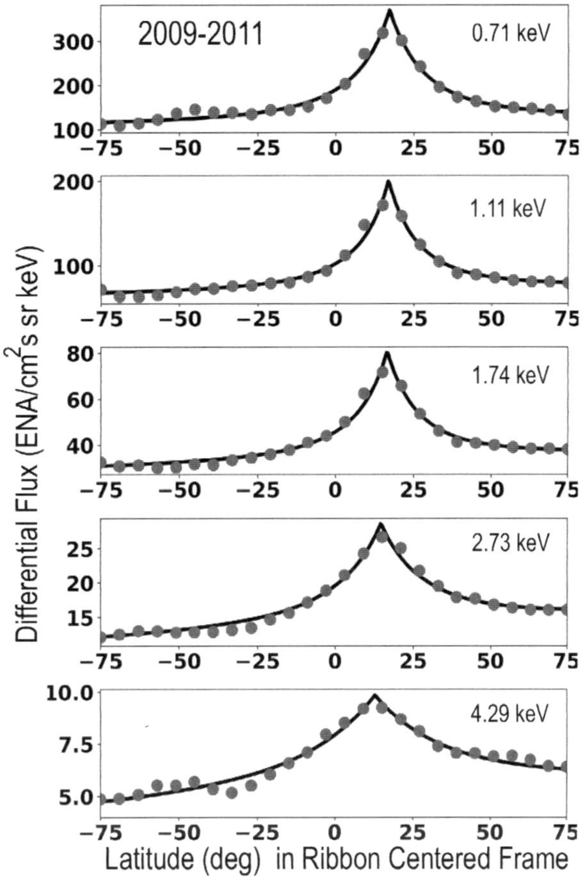

Figure 1. Fits to the ribbon distribution using equation (2) and a sloped line for the globally distributed flux in the reference frame centered on the IBEX ribbon.

The observations of IBEX are mapped by angle over the plane of the sky. The angular distribution corresponding to equation (1) is:

$$\tilde{f}(\theta) = A(|\gamma_0(\theta - \theta_0)| + 1)^{-3}. \quad (2)$$

Here γ_0 is the average inverse angular width of the ribbon, θ_0 is the central latitude in the reference frame centered on the ribbon, and θ represents a given latitude in the reference frame center on the ribbon. In the following section, we use the derived kappa distribution to provide fits to the observed ENA distributions.

3. Spatial Fitting of the IBEX Ribbon

We demonstrate the fitting methodology using the first three years of IBEX observations from 2009 through 2011, and order observations as a function of latitude in the reference frame centered on the ribbon. In this reference reference frame, the pole at 90° latitude corresponds to the ribbon center (taken as ecliptic longitude 221° and ecliptic latitude 39°. In this coordinate system, we average over longitude and display the data as a function of latitude. A uniform circular ribbon would appear at a fixed latitude, and if the ribbon formed a great circle, it would appear at 0° latitude.

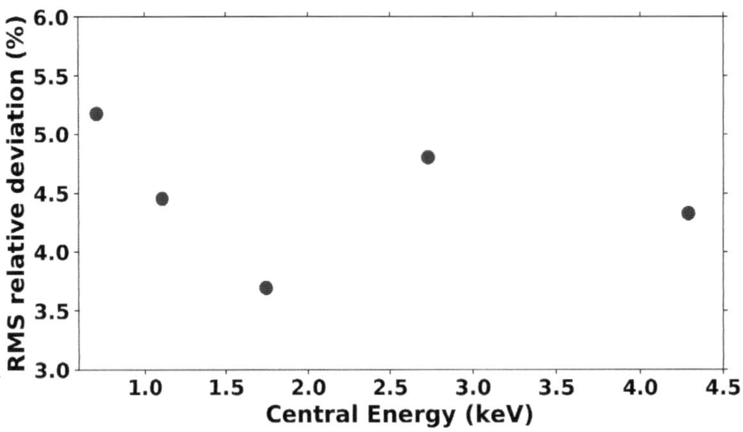

Figure 2. RMS deviations of the fits to the ribbon distribution shown in Figure 1.

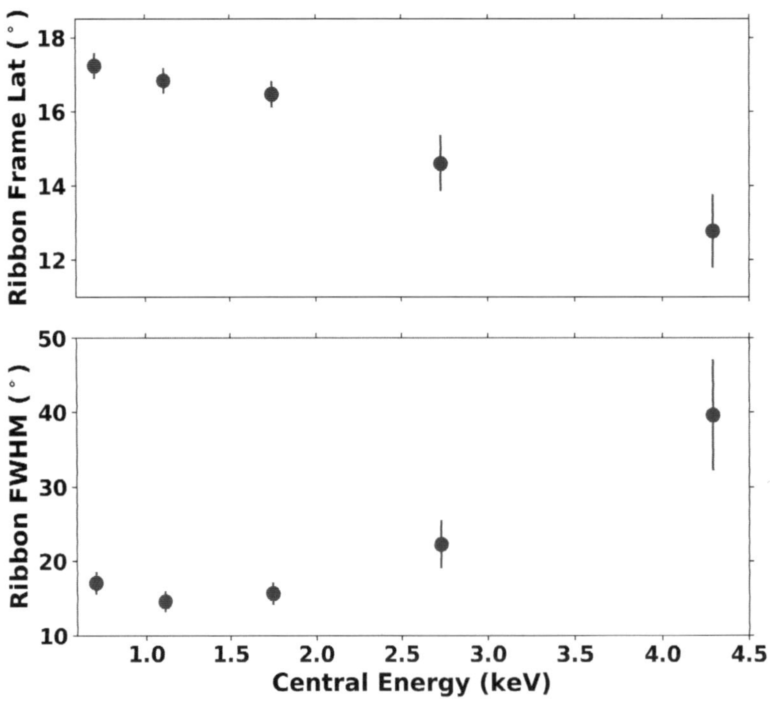

Figure 3. Fit parameters derived from the superposition of states that yields a kappa distribution for the ribbon (2): (top) the latitude in the frame centered on the ribbon, (bottom) the ribbon's FWHM.

Figure 1 shows the kappa function fits using the functional form in equation (2) for the ribbon and a sloped line for the globally distributed flux (GDF). There are five parameters for each fit: the height and slope of the GDF linear fit, and the center, width, and height of the kappa distribution for the ribbon (from equation 2). The root-mean-square (RMS) deviation between the fits and the observations is very low, typically less than 5% (Figure 2).

The fitting parameters from equation (2) include the average latitude of the ribbon in the ribbon-centered reference frame, and the full-width-half-max of the ribbon as a function of energy. These fit parameters and their uncertainties are shown in Figure 3.

The quality of the fits in Figure 1 and the relatively low RMS deviations in Figure 2 suggest that the functional form, equation (2), associated with the superposition of multiple sources along the line-of-sight provides reasonable agreement with IBEX observations in the 2009–2011 timeframe. In the next section we utilize the results of these fits to aid in our interpretation of the properties of the IBEX ribbon.

4. Conclusions

The fits (Figure 1) and the fitting parameters (Figure 3) show a number of important features that elucidate the properties of the ribbon. We observe the center of the ribbon at latitudes (Figure 3, top panel) in the ribbon frame that are relatively close to the equator ($< 18°$). This suggests that the ribbon is close to being a great-circle. Note that the ribbon center shifts steadily from $17.2° \pm 0.3°$ latitude at ~ 0.71 keV down to $12.8° \pm 1.0°$ latitude at ~ 4.29 keV. This shift suggests that we are observing increasingly large line-of-sight distances within the ribbon at increasing energies [20,], which suppresses the effect of magnetic field draping near the heliopause.

We find that the ribbon FWHM minimizes at the energy step centered on ~ 1.1 keV (with energy range 0.84–1.55 keV), with a large increase of the ribbon width at increasing energy. This behavior suggests that the we are observing a source with an intrinsic width at low energies, and that the width of the ribbon increases at higher energies due to increasing ion mobility.

The ribbon FWHM from the 0.84–1.55 keV energy step is found to be $\Delta\alpha_A = 14.6° \pm 1.4°$. We consider the speed of ENAs at the low end of this IBEX ENA energy range, near 0.84 keV, since the energy spectrum preferentially weights the lower energy ENAs within the observed energy range. With a speed of $v = 400$ km/s at 0.84 keV, we can use the relationship between the width of the ribbon and Alfvén wave production in the retention region to derive an Alfvén speed, v_A. Specifically, with $\Delta\alpha_A \approx 2v_A/v$ [11], we find an Alfvén speed $v_A \sim 50$ km s^{-1}.

Given a proton density of 0.07 cm^3 [21] in the plasma beyond the heliopause where the ribbon is generated, we find a magnetic field strength of $B \sim 6$ μG, which is similar to the 4.5 – 6 μG found in Voyager 1 observations beyond the heliopause [22, 23]. The ribbon width observed by IBEX is likely broadened due to variations along the line-of-sight and due to the angular response of IBEX-Hi. Therefore, our calculation slightly overestimates both the Alfvén speed and the magnetic field strength within the ribbon.

The broadening of the ribbon at higher energies from IBEX may be associated in part with increased ion mobility along the magnetic field. In the case of weak scattering, the width of the ribbon is related to the scattering mean free path, $\Delta\alpha \sim 2\sqrt{\kappa_{\|}\tau_x}/r$. Given $\tau_x \sim 2$ years at 4.29 keV and a ribbon width $\Delta\alpha = 39.6° \pm 7.5°$, we find a scattering mean free path parallel to the magnetic field given by $\lambda_{\|} = \lambda_0 r^2/R_{130}^2$ where $\lambda_0 = 15$ au and R_{130} is 130 au. Note that the width of the ribbon at this energy corresponds to a physical distance of 90 au for a ribbon that is produced near 130 au. The derived scattering mean free path is therefore smaller, but of the rough order-of-magnitude of the ribbon width. Therefore, the effect of scattering is significant within the ribbon, and the increased mobility at higher energies causes broadening.

We have analyzed the properties of the ribbon over the first three years of IBEX observations by considering the superposition of spatially distributed sources along the line-of-sight. We find that this superposition yields a specific kappa function that provides excellent fits to the IBEX observations. These fits are used to determine properties associated with ribbon source. The width of the ribbon provides for an estimate of the Alfvén speed, $v_A \sim 50$ km s^{-1}, beyond the heliopause near the ribbon source, and the increased mobility of ions at the higher energies observed by IBEX leads to ribbon broadening. Thus, we provide a framework to analyze the IBEX ribbon taking into account distributed neutral atom sources within the very local interstellar medium.

4.1. Acknowledgments
We are very grateful to the many individuals who have made the IBEX project possible. This work is supported by the Interstellar Boundary Explorer mission as a part of NASA's Explorer Program and partially by NASA SR&T Grant NNG06GD55G.

References
[1] McComas D J, Allegrini F, Bochsler P, Bzowski M, Collier M, Fahr H, Fichtner H, Frisch P, Funsten H O, Fuselier S A, Gloeckler G, Gruntman M, Izmodenov V, Knappenberger P, Lee M, Livi S, Mitchell D, Möbius E, Moore T, Pope S, Reisenfeld D, Roelof E, Scherrer J, Schwadron N, Tyler R, Wieser M, Witte M, Wurz P and Zank G 2009 *Space Sci. Rev.* **146** 11–33
[2] McComas D J, Allegrini F, Bochsler P, Bzowski M, Christian E R, Crew G B, DeMajistre R, Fahr H, Fichtner H, Frisch P C, Funsten H O, Fuselier S A, Gloeckler G, Gruntman M, Heerikhuisen J, Izmodenov V, Janzen P, Knappenberger P, Krimigis S, Kucharek H, Lee M, Livadiotis G, Livi S, MacDowall R J, Mitchell D, Möbius E, Moore T, Pogorelov N V, Reisenfeld D, Roelof E, Saul L, Schwadron N A, Valek P W, Vanderspek R, Wurz P and Zank G P 2009 *Science* **326** 959–
[3] Fuselier S A, Bochsler P, Chornay D, Clark G, Crew G B, Dunn G, Ellis S, Friedmann T, Funsten H O, Ghielmetti A G, Googins J, Granoff M S, Hamilton J W, Hanley J, Heirtzler D, Hertzberg E, Isaac D, King B, Knauss U, Kucharek H, Kudirka F, Livi S, Lobell J, Longworth S, Mashburn K, McComas D J, Möbius E, Moore A S, Moore T E, Nemanich R J, Nolin J, O'Neal M, Piazza D, Peterson L, Pope S E, Rosmarynowski P, Saul L A, Scherrer J R, Scheer J A, Schlemm C, Schwadron N A, Tillier C, Turco S, Tyler J, Vosbury M, Wieser M, Wurz P and Zaffke S 2009 *Space Sci. Rev.* **146** 117–147
[4] Funsten H O, Allegrini F, Bochsler P, Dunn G, Ellis S, Everett D, Fagan M J, Fuselier S A, Granoff M, Gruntman M, Guthrie A A, Hanley J, Harper R W, Heirtzler D, Janzen P, Kihara K H, King B, Kucharek H, Manzo M P, Maple M, Mashburn K, McComas D J, Moebius E, Nolin J, Piazza D, Pope S, Reisenfeld D B, Rodriguez B, Roelof E C, Saul L, Turco S, Valek P, Weidner S, Wurz P and Zaffke S 2009 *Space Sci. Rev.* **146** 75–103
[5] Funsten H O, DeMajistre R, Frisch P C, Heerikhuisen J, Higdon D M, Janzen P, Larsen B A, Livadiotis G, McComas D J, Möbius E, Reese C S, Reisenfeld D B, Schwadron N A and Zirnstein E J 2013 *Astrophys. J.* **776** 30
[6] Schwadron N A, Bzowski M, Crew G B, Gruntman M, Fahr H, Fichtner H, Frisch P C, Funsten H O, Fuselier S, Heerikhuisen J, Izmodenov V, Kucharek H, Lee M, Livadiotis G, McComas D J, Moebius E, Moore T, Mukherjee J, Pogorelov N V, Prested C, Reisenfeld D, Roelof E and Zank G P 2009 *Science* **326** 966–
[7] McComas D J, Lewis W S and Schwadron N A 2014 *Reviews of Geophysics* **52** 2013RG000438 ISSN 1944-9208 URL `http://onlinelibrary.wiley.com/doi/10.1002/2013RG000438/abstract`
[8] Heerikhuisen J, Pogorelov N V, Zank G P, Crew G B, Frisch P C, Funsten H O, Janzen P H, McComas D J, Reisenfeld D B and Schwadron N A 2010 *Astrophys. J. Lett.* **708** L126–L130
[9] Chalov S V, Alexashov D B, McComas D, Izmodenov V V, Malama Y G and Schwadron N 2010 *Astrophys. J. Lett.* **716** L99–L102 (*Preprint* `1003.4826`)
[10] Möbius E, Liu K, Funsten H, Gary S P and Winske D 2013 *Astrophys. J.* **766** 129
[11] Schwadron N A and McComas D J 2013 *Astrophys. J.* **764** 92
[12] Zirnstein E J, Heerikhuisen J, Funsten H O, Livadiotis G, McComas D J and Pogorelov N V 2016 *Astrophys. J. Lett.* **818** L18
[13] Zirnstein E J, Heerikhuisen J and Dayeh M A 2018 *Astrophys. J.* **855** 30
[14] Isenberg P A 2014 *Astrophys. J.* **787** 76 (*Preprint* `1404.2170`)
[15] Giacalone J and Jokipii J R 2015 *Astrophys. J. Lett.* **812** L9
[16] Schwadron N A, Dayeh M A, Desai M, Fahr H, Jokipii J R and Lee M A 2010 *Astrophys. J.* **713** 1386–1392
[17] Dragulescu A and Yakovenko V M 2000 "*Eur. Phys. J. B* **17** 723
[18] Yakovenko V M and Rosser, Jr J B 2009 "*Reviews of Modern Physics*" **81** 1703

Journal of Physics: Conference Series **1332** (2019) 012013 doi:10.1088/1742-6596/1332/1/012013

[19] Wheatland M S 2000 *Astrophys. J. Lett.* **536** L109–L112 (*Preprint* arXiv:astro-ph/0005121)
[20] Schwadron N A, Moebius E, Fuselier S A, McComas D J, Funsten H O, Janzen P, Reisenfeld D, Kucharek H, Lee M A, Fairchild K, Allegrini F, Dayeh M, Livadiotis G, Reno M, Bzowski M, Sokół J M, Kubiak M A, Christian E R, DeMajistre R, Frisch P, Galli A, Wurz P and Gruntman M 2014 *Astrophys. J. Suppl.* **215** 13
[21] Frisch P C, Bzowski M, Grün E, Izmodenov V, Krüger H, Linsky J L, McComas D J, Möbius E, Redfield S, Schwadron N, Shelton R, Slavin J D and Wood B E 2009 *Space Sci. Rev.* **146** 235–273
[22] Burlaga L F and Ness N F 2016 *Astrophys. J.* **829** 134
[23] Schwadron N A and McComas D J 2017 *Astrophys. J.* **849** 135

Time-Dependent Perpendicular Transport of Energetic Particles

A. Shalchi

Department of Physics and Astronomy, University of Manitoba, Winnipeg, Manitoba R3T 2N2, Canada

E-mail: `andreasm4@yahoo.com`

Abstract. The motion of energetic particles in magnetic turbulence across a mean magnetic field can be explored analytically. The approach discussed in this paper allows for a full time-dependent description of the transport, including compound sub-diffusion. For the first time it is shown systematically that as soon as there is transverse structure of the turbulence, diffusion is restored even if no Coulomb collisions are invoked. Compared to other non-linear theories the new approach has the advantage that a diffusion approximation is no longer part of that theory. Criteria for sub-diffusion and normal Markovian diffusion are provided as well. A comparison with test-particle simulations is also discussed.

1. Introduction

An important problem in space science is the understanding of energetic particles moving through a magnetized plasma. In the interplanetary space one finds a magnetic field configuration of the form

$$\vec{B}(\vec{x},t) = B_0\hat{z} + \delta\vec{B}(\vec{x},t) \tag{1}$$

where we assumed that the mean magnetic field B_0 points into the z-direction and is constant. This mean field deflects energetic and electrically charged particles forcing them to follow a helical path. Furthermore, there is a complicated turbulent component $\delta\vec{B}$ of the magnetic field. This component is, in general, time-dependent and strongly depends on the different directions of space. Some simplification can be achieved by considering the incompressible case $\delta B_z = 0$ and/or the static case where we neglect the time-dependence of the turbulent field.

The turbulent magnetic field scatters the energetic particles and leads to a diffusive motion. In this case the transport of particles can be described by Parker's famous transport equation (see Parker (1965))

$$\frac{\partial f}{\partial t} + \sum_i u_i \frac{\partial f}{\partial x_i} = \sum_{i,j} \frac{\partial}{\partial x_i}\left(\kappa_{ij}\frac{\partial f}{\partial x_j}\right) + \frac{1}{3}\sum_i \left(\frac{\partial u_i}{\partial x_i}\right)\frac{\partial f}{\partial \ln p} + S. \tag{2}$$

Here we have used the plasma bulk velocity $\vec{u}$ and the components of the spatial diffusion tensor κ_{ij}. It should be noted, that more general versions of this equation can be found in the literature (see, e.g., Schlickeiser (2002)) involving further terms such as momentum diffusion. Furthermore, non-diffusive transport, also known as anomalous diffusion, was discussed in several papers such as Perri & Zimbardo (2009a) and Perri & Zimbardo (2009b). Assuming normal or Markovian

Content from this work may be used under the terms of the Creative Commons Attribution 3.0 licence. Any further distribution of this work must maintain attribution to the author(s) and the title of the work, journal citation and DOI.
Published under licence by IOP Publishing Ltd

diffusion, and axi-symmetry with respect to the mean magnetic field allows us to use a diffusion tensor which has, in this case, the following form

$$\kappa = \begin{pmatrix} \kappa_\perp & \kappa_A & 0 \\ -\kappa_A & \kappa_\perp & 0 \\ 0 & 0 & \kappa_\parallel \end{pmatrix} \tag{3}$$

where we have used the parallel diffusion coefficient $\kappa_\parallel$, the perpendicular diffusion coefficient $\kappa_\perp$, and the drift coefficient κ_A. In the current paper we focus on the perpendicular diffusion coefficient and treat the parallel diffusion coefficient as a variable entering non-linear theories for perpendicular transport. The drift coefficient and its importance in solar modulation studies were explored before in the literature (see, e.g., Engelbrecht & Burger (2015) and Engelbrecht et al. (2017)) but is not further discussed in the current paper.

The theoretical exploration of perpendicular transport has a long history starting with the pioneering paper of Jokipii (1966). In the latter article the first time a perpendicular diffusion coefficient was calculated based on quasi-linear theory (QLT). In this case perpendicular diffusion is entirely caused by particles following magnetic field lines which themselves behave diffusively. Further important steps leading to an improved understanding of perpendicular transport which are reviewed in the current paper are:

- The compound-diffusion model describes particles which are tied to magnetic field lines while they move diffusively in the parallel direction (see, e.g., Kóta & Jokipii (2000) and Webb et al. (2006)). As a consequence perpendicular transport behaves sub-diffusively.
- The non-linear guiding center (NLGC) theory of Matthaeus et al. (2003) was the first systematic theory for perpendicular transport. Furthermore, the theory showed agreement with several simulations and solar wind observations if combined with a two-component turbulence model (see Bieber et al. (2004)).
- The extended non-linear guiding center (ENLGC) theory was developed in Shalchi (2006) for describing perpendicular diffusion in two-component turbulence. While the NLGC theory does not provide a sub-diffusive result for slab turbulence, the ENLGC theory describes perpendicular transport due to slab modes sub-diffusively while the two-dimensional contribution is as in NLGC theory.
- The unified non-linear transport (UNLT) theory of Shalchi (2010) was developed with the aim to obtain a diffusion theory which works for all cases ranging from slab to full three-dimensional turbulence. Furthermore, UNLT theory contains the Matthaeus et al. (1995) theory of field line random walk as well as other theories and limits.
- The time-dependent UNLT theory was developed in Shalchi (2017) and Lasuik & Shalchi (2017). The theory does not longer rely on a diffusion approximation and can describe the initial ballistic regime, sub-diffusion, as well as the turnover to the diffusive regime. Furthermore, this theory provides a condition which needs to be satisfied in order to obtain normal diffusion.

The analytical description of perpendicular transport in not just important in order to improve our fundamental understanding of processes in nature. There are several applications of analytical forms of particle diffusion coefficients. Examples are the acceleration of particles at shock waves (see, e.g., Zank et al. (2004), Dosch & Shalchi (2010), Li et al. (2012), Ferrand et al. (2014), and Hu et al. (2017)) and solar modulation studies (see, e.g. Hitge & Burger (2010), Engelbrecht & Burger (2010), Wawrzynczak & Alania (2010), Alania et al. (2011), Potgieter & Nndanganeni (2013), Engelbrecht & Burger (2014), Manuel et al. (2014), Potgieter et al. (2014), Ahluwalia & Ygbuhay (2015), and Shen & Qin (2018)).

It is the purpose of the current article to review the analytical description of the perpendicular transport of energetic particles. The organization is as follows. In Sect. 2 we discuss simple models and previous theories for perpendicular diffusion. In Sect. 3 and 4, diffusive and time-dependent UNLT theories are discussed, respectively. In Sect. 5 a comparison between analytical results and test-particle simulations are shown as an example. In Sect. 6 we summarize and conclude.

2. Simple Models and Previous Theories for Perpendicular Diffusion

Before more advanced theories for perpendicular diffusion are discussed, we briefly review three previous approaches, namely the field line random walk (FLRW) limit, compound sub-diffusion, and the non-linear guiding center (NLGC) theory.

2.1. The FLRW Limit

One of the first descriptions of perpendicular transport was entirely based on the effect of FLRW and was presented in Jokipii (1966). This approach is simple but does not work in the general case as shown later. As an approximation we assume that field line random walk is diffusive for all distances meaning that we neglect the initial free-streaming regime entirely. Thus, we can employ

$$\langle (\Delta x)^2 \rangle_{FL} = 2\kappa_{FL} |z| \quad \forall \quad z \tag{4}$$

with the diffusion coefficient of FLRW κ_{FL}. As a simple model for the perpendicular transport of energetic particles we assume that the particle follows a single magnetic field line while it moves with constant velocity in the parallel direction meaning that

$$z_p = v\mu t \tag{5}$$

where we have used the pitch-angle cosine μ and particle speed v. Furthermore, if particles follow field lines, the perpendicular mean square displacements of particle trajectories and field lines are the same. Therefore, we can combine Eqs. (4) and (5) to find

$$\langle (\Delta x)^2 \rangle_p = 2vt\kappa_{FL} |\mu| . \tag{6}$$

After averaging over all $-1 \leq \mu \leq +1$, this becomes

$$\langle (\Delta x)^2 \rangle_p = vt\kappa_{FL} \tag{7}$$

corresponding to normal diffusion with the perpendicular diffusion coefficient of the energetic particle

$$\kappa_\perp = \frac{v}{2}\kappa_{FL}. \tag{8}$$

We can clearly see that the perpendicular diffusion coefficient of the energetic particles is directly proportional to the diffusion coefficient of random walking magnetic field lines. The only property of the particle entering this simple equation is the particle speed v. Alternatively, one can compute the perpendicular mean free path

$$\lambda_\perp = \frac{3}{v}\kappa_\perp = \frac{3}{2}\kappa_{FL}. \tag{9}$$

Obviously the latter quantity does not depend on any particle properties. This means that the perpendicular mean free path does even not depend on particle energy, momentum, or rigidity. Later in this review paper we will argue that this behavior can typically be found for high particle energies.

2.2. Compound Sub-diffusion

Real particles do not perform an unperturbed motion in the parallel direction as assumed in the previous paragraph. Therefore, we now assume a diffusive parallel motion. However, we still assume that the particle is tied to a single magnetic field line as before. A very comprehensive description of compound sub-diffusion was presented in Webb et al. (2006). The latter authors employed an approach based on the so-called Chapman-Kolmogorov equation (see, e.g., Gardiner (1985))

$$f_{\perp}(x,y;t) = \int_{-\infty}^{+\infty} dz \; f_{FL}(x,y;z) f_{\parallel}(z;t) \tag{10}$$

with the particle distributions in the parallel direction $f_{\parallel}(z;t)$, in the perpendicular direction $f_{\perp}(x,y;t)$, and the field line distribution function $f_{FL}(x,y;z)$. Whereas Webb et al. (2006) computed the function $f_{\perp}(x,y;t)$ by solving Eq. (10), we follow Shalchi & Kourakis (2007) and only consider the second moment of $f_{\perp}(x,y;t)$ leading to

$$\langle (\Delta x(t))^2 \rangle_P = \int_{-\infty}^{+\infty} dz \, \langle (\Delta x(z))^2 \rangle_{FL} f_{\parallel}(z,t) \tag{11}$$

giving us a relation between particle and field line mean square displacements. In the following we employ a Gaussian particle distribution and assume that parallel transport is diffusive for all times. Thus, the parallel distribution function has the form

$$f_{\parallel}(z,t) = \frac{1}{\sqrt{4\pi\kappa_{\parallel} t}} e^{-\frac{z^2}{4\kappa_{\parallel} t}}. \tag{12}$$

Using this and Eq. (4) in Eq. (11) leads to the following integral

$$\langle (\Delta x)^2 \rangle_P = \frac{\kappa_{FL}}{\sqrt{\pi\kappa_{\parallel} t}} \int_{-\infty}^{+\infty} dz \; |z| \, e^{-\frac{z^2}{4\kappa_{\parallel} t}}. \tag{13}$$

This integral can easily be solved yielding

$$\langle (\Delta x)^2 \rangle_P = 4\kappa_{FL} \sqrt{\frac{\kappa_{\parallel} t}{\pi}} \tag{14}$$

corresponding to sub-diffusive transport. Webb et al. (2006) computed not just the second moment but also the perpendicular distribution function $f_{\perp}(x,y;t)$. They found a so-called *Fox function* which is more sharply peaked at the center compared to the usual Gaussian found in diffusive cases. Arendt & Shalchi (2018) found this type of distribution via test-particle simulations confirming the work of Webb et al. (2006). A velocity correlation function based approach to describe compound sub-diffusion was presented in Kóta & Jokipii (2000).

Compound sub-diffusion is highly relevant for particle transport in slab turbulence. According to the so-called theorem on reduced dimensionality presented in Jokipii et al. (1993) and Jones et al. (1998), particles are tied to magnetic field lines if the turbulence has reduced dimensionality. Undoubtedly this is the case for slab turbulence. If parallel transport is diffusive, this means that for slab turbulence we find compound sub-diffusion. Numerically it was confirmed via simulations by Qin & Shalchi (2015) that for slab turbulence particles follow indeed field lines.

2.3. The Non-linear Guiding Center Theory

Matthaeus et al. (2003) developed a more systematic approach in order to describe perpendicular diffusion called the non-linear guiding center (NLGC) theory. To derive a simple equation of motion for perpendicular transport, we combine the Newton-Lorentz equation for a purely magnetic system

$$\frac{d\vec{p}}{dt} = \frac{q}{c}\vec{v} \times \vec{B}(\vec{x}) \tag{15}$$

with the guiding center transformation (see, e.g., Schlickeiser (2002))

$$\vec{X} = \vec{x} + \frac{1}{\Omega}\left(\vec{v} \times \hat{z}\right). \tag{16}$$

The gyrofrequency is given by

$$\Omega = \frac{qB_0}{mc\gamma} \tag{17}$$

where we have used the Lorentz factor γ, the particle's electric charge q and rest mass m, and the speed of light c. After some straightforward algebra, we find the following equation of motion

$$V_x = v_z \frac{\delta B_x\left(\vec{x}\right)}{B_0} \tag{18}$$

for the guiding center velocity component $V_x = \dot{X}$. A similar equation can be obtained for V_y but this is not needed due to the assumption of axi-symmetry. If Eq. (18) is combined with the TGK formula (see Taylor (1922), Green (1951), and Kubo (1957))

$$\kappa_\perp = \int_0^\infty dt\ \langle v_x(t) v_x(0)\rangle\,, \tag{19}$$

one obtains

$$\kappa_\perp = \frac{a^2}{B_0^2}\int_0^\infty dt\ \langle v_z(t)v_z(0)\delta B_x(t)\delta B_x^*(0)\rangle\,. \tag{20}$$

Here we have used an additional parameter a^2. In the original paper by Matthaeus et al. (2003), the parameter a was introduced directly in Eq. (18) but it was shown in Qin & Shalchi (2016) that Eq. (18) is correct as it is without containing any type of additional parameter. It is still unclear what the exact value of this parameter is, nor is it known what the physics of a^2 is (see, however, the work of Qin & Zhang (2014)). In the following we just keep this parameter in our equations and assume that this parameter is there in order to balance out some inaccuracies of the theory related to some of the used approximations (e.g., the parameter a^2 can balance out the inaccuracy of the Corrsin approximation used below).

To proceed, Matthaeus et al. (2003) assumed that the fourth-order correlation in Eq. (20) can be replaced by a product of two second-order correlations

$$\langle v_z(t)v_z(0)\delta B_x(t)\delta B_x^*(0)\rangle \approx \langle v_z(t)v_z(0)\rangle\,\langle \delta B_x(t)\delta B_x^*(0)\rangle \tag{21}$$

leading to

$$\kappa_\perp = \frac{a^2}{B_0^2}\int_0^\infty dt\ \langle v_z(t)v_z(0)\rangle\,\langle \delta B_x(t)\delta B_x^*(0)\rangle\,. \tag{22}$$

Approximation (21) is problematic and does not work in the general case as shown in the numerical work of Qin & Shalchi (2016). Therefore, this approximation is dropped in the next section.

Furthermore, Matthaeus et al. (2003) modeled the parallel velocity correlation function by the isotropic exponential model

$$\langle v_z(t)v_z(0)\rangle = \frac{v^2}{3}e^{-vt/\lambda_\parallel} \tag{23}$$

where we used the parallel mean free path $\lambda_\parallel$. It was shown in Shalchi (2011a) that Eq. (23) is only exact for an isotropic pitch-angle Fokker-Planck coefficient. However, for other forms of $D_{\mu\mu}$, Eq. (23) should still provide an accurate approximation. To proceed, the magnetic correlation function $\langle \delta B_x(t) \delta B_x^*(0) \rangle$ has to be replaced by a Fourier representation leading to

$$\kappa_\perp = \frac{a^2 v^2}{3B_0^2} \int_0^\infty dt \int d^3k \int d^3k' \left\langle \delta B_x(\vec{k}) \delta B_x^*(\vec{k}') e^{i\vec{k}\cdot\vec{x}} \right\rangle e^{-vt/\lambda_\parallel} \tag{24}$$

where we used $\vec{x}(t=0)=0$. Now we employ Corrsin's independence hypothesis (see Corrsin (1959))

$$\left\langle \delta B_x(\vec{k}) \delta B_x^*(\vec{k}') e^{i\vec{k}\cdot\vec{x}} \right\rangle \approx \left\langle \delta B_x(\vec{k}) \delta B_x^*(\vec{k}') \right\rangle \left\langle e^{i\vec{k}\cdot\vec{x}} \right\rangle \tag{25}$$

and assume homogeneous turbulence

$$\langle \delta B_n(\vec{k}) \delta B_m^*(\vec{k}') \rangle = P_{nm}(\vec{k}) \delta(\vec{k} - \vec{k}') \tag{26}$$

to derive

$$\kappa_\perp = \frac{a^2 v^2}{3B_0^2} \int d^3k \, P_{xx}(\vec{k}) \int_0^\infty dt \left\langle e^{i\vec{k}\cdot\vec{x}} \right\rangle e^{-vt/\lambda_\parallel}. \tag{27}$$

Next, the characteristic function $\langle e^{i\vec{k}\cdot\vec{x}} \rangle$ must be approximated. Matthaeus et al. assumed a Gaussian distribution of the particles and that the particle motion is diffusive for all times. Therefore, one can use

$$\left\langle e^{i\vec{k}\cdot\vec{x}} \right\rangle = e^{-\kappa_\perp k_\perp^2 t - \kappa_\parallel k_\parallel^2 t} \tag{28}$$

corresponding to the characteristic function of a usual diffusion equation for the axi-symmetric case. Combining Eq. (27) with Eq. (28) and evaluating the time-integral, yields

$$\kappa_\perp = \frac{a^2 v^2}{3B_0^2} \int d^3k \, \frac{P_{xx}(\vec{k})}{v/\lambda_\parallel + \kappa_\perp k_\perp^2 + \kappa_\parallel k_\parallel^2}. \tag{29}$$

This is the non-linear integral equation of the NLGC theory as originally derived by Matthaeus et al. (2003). This integral equation can be used for turbulence models with purely magnetic fluctuations, where $\delta B_z = 0$ corresponding to the incompressible case. Furthermore, we assumed static turbulence here but the theory can easily be generalized to allow for dynamical turbulence (see Matthaeus et al. (2003) and Shalchi et al. (2004)).

As shown in Matthaeus et al. (2003) and Bieber et al. (2004), Eq. (29) agrees with some simulations as well as solar wind observations as long as a two-component turbulence model is used and one sets $a^2 = 1/3$. This was undoubtedly a breakthrough in the theory of perpendicular diffusion. However, Eq. (29) cannot be seen as the final solution to this problem. One of the major problems is that NLGC theory does not work for slab turbulence where perpendicular transport should be sub-diffusive.

3. Diffusive UNLT Theory

After NLGC theory was developed it was pointed out quickly that the theory has some weaknesses. In Shalchi (2006) a modification was presented called the extended non-linear guiding center (ENLGC) theory. The latter theory was developed specifically for slab/2D turbulence and describes the slab contribution as sub-diffusive whereas the 2D contribution is as in NLGC theory. Historically the next step was the formulation of the unified non-linear transport (UNLT) theory (see Shalchi (2010)) which is now called diffusive UNLT theory.

In order to avoid using Eq. (21) one has to be able to compute higher order correlations analytically. In Shalchi (2010) this was done based on the pitch-angle dependent Fokker-Planck equation (see, e.g., Schlickeiser (2002) for a more complete version of this equation)

$$\frac{\partial f}{\partial t}+v\mu\frac{\partial f}{\partial z}=\frac{\partial}{\partial \mu}\left(D_{\mu\mu}\frac{\partial f}{\partial \mu}\right)+D_{\perp}\left(\frac{\partial^2 f}{\partial x^2}+\frac{\partial^2 f}{\partial y^2}\right). \tag{30}$$

In principle the solution of this equation $f = f(\vec{x}, \mu, t)$ allows one to determine higher order correlations. After using a Fourier representation in Eq. (20) and after employing Corrsin's approximation, one ends up with a correlation of the form

$$\langle v_z(t)v_z(0)e^{i\vec{k}\cdot\vec{x}}\rangle=\frac{v^2}{4}\int d^3x\int_{-1}^{+1}d\mu_0\int_{-1}^{+1}d\mu\;\mu_0\mu e^{i\vec{k}\cdot\vec{x}}f(\vec{x},\mu,t) \tag{31}$$

where f is the solution of Eq. (30) and μ_0 is the initial pitch-angle cosine. This idea was used in Shalchi (2010) in order to find an improved theory for perpendicular diffusion. However, the derivation is lengthy and not straightforward. Therefore, we only provide the result, namely

$$\kappa_{\perp}=\frac{a^2v^2}{3B_0^2}\int d^3k\,\frac{P_{xx}(\vec{k})}{F(k_{\parallel},k_{\perp})+(4/3)\kappa_{\perp}k_{\perp}^2+v/\lambda_{\parallel}} \tag{32}$$

with the function

$$F(k_{\parallel},k_{\perp})=(vk_{\parallel})^2/(3\kappa_{\perp}k_{\perp}^2). \tag{33}$$

As mentioned above, the original derivation of this result is not easy. However, a simpler derivation from time-dependent UNLT theory developed by Shalchi (2017) and Lasuik & Shalchi (2017) can be found in Sect. 4.4 of this paper.

Although Eq. (32) has some similarity with Eq. (29), it is not equal nor does it provide the same solution in the general case. For pure slab turbulence, for instance, Eq. (32) provides the solution $\kappa_{\perp} = 0$, corresponding to sub-diffusion, whereas NLGC theory provides a finite result, corresponding to normal diffusion. One can easily generalize Eq. (32) to allow for dynamical turbulence (see, e.g., Shalchi (2011b)).

In the following we discuss asymptotic limits which are contained in this theory. This will link Eq. (32) to previously derived and discussed theories such as the FLRW limit. It will be shown that the form of the solution of Eq. (32) depends on two parameters, namely the Kubo number (Kubo (1963))

$$K=\frac{\ell_{\parallel}}{\ell_{\perp}}\frac{\delta B_x}{B_0} \tag{34}$$

as well as the ratio $\lambda_{\parallel}/\ell_{\parallel}$. Here we have used the correlation lengths[1] in the different directions of space $\ell_{\parallel}$ and $\ell_{\perp}$, respectively.

3.1. The FLRW-Limit and the Matthaeus et al. (1995) Theory

First we consider the limit $\lambda_{\parallel}/\ell_{\parallel} \to \infty$ corresponding to the case that pitch-angle scattering and therewith parallel diffusion are suppressed. In this case we can set $v/\lambda_{\parallel} = 0$ in Eq. (32) and we find the solution

$$\kappa_{\perp}=\frac{v}{2}\kappa_{FL} \tag{35}$$

[1] It has to be noted that magnetic turbulence depends on several characteristic lengths scales. Besides the bendover scales denoting the turnover from the energy to the inertial range in the spectrum, there are also integral scales, the ultra-scale, and the Kolmogorov scale. Therefore, special care is required if those scales are used and how the Kubo number is defined.

where the field line diffusion coefficient κ_{FL} is given by

$$\kappa_{FL} = \frac{1}{\tilde{B}_0^2} \int d^3k \; P_{xx}(\vec{k}) \frac{\kappa_{FL} k_\perp^2}{k_\parallel^2 + \left(\kappa_{FL} k_\perp^2\right)^2} \tag{36}$$

where we have used the effective mean field $\tilde{B}_0 = B_0/a$. Eq. (35) agrees with Eq. (8) corresponding to the FLRW limit. Eq. (36), on the other hand, is in perfect agreement with the Matthaeus et al. (1995) theory for field line wandering. Obviously the latter theory is contained in Eq. (32). Therefore, UNLT theory can be seen as a unified transport theory for magnetic field lines and energetic particles, hence the name unified non-linear transport theory.

3.2. The Quasi-Linear Regime

The quasi-linear limit can be obtained if we additionally consider $\kappa_{FL} \to 0$ at the right hand side of Eq. (36). Using the relation (Zwillinger (2012))

$$\lim_{\xi \to 0} \frac{\xi}{\xi^2 + x^2} = \pi \delta(x) \tag{37}$$

in Eq. (36) yields

$$\kappa_{FL} = \frac{\pi}{\tilde{B}_0^2} \int d^3k \; P_{xx}(\vec{k}) \delta(k_\parallel) \tag{38}$$

corresponding to the well-known quasi-linear result. As shown here, QLT for perpendicular diffusion can be obtained in the limit of long parallel mean free paths and small field line diffusion coefficients corresponding to small Kubo numbers. It can be shown (see Shalchi (2015)) that for most turbulence models Eq. (35) with (38) leads to the scaling law

$$\lambda_\perp \propto \ell_\parallel \frac{\delta B_x^2}{B_0^2} \tag{39}$$

where we used the parallel correlation scale of the turbulence $\ell_\parallel$. As discussed above, this result should be valid for small Kubo numbers but also for long parallel mean free paths. It is usually assumed that the parallel mean free path increases with increasing magnetic rigidity. Therefore, the quasi-linear scaling should become relevant if high energy particles propagating through small Kubo number turbulence is considered.

3.3. The Kadomtsev & Pogutse Scaling

Here we still consider the case of $\lambda_\parallel/\ell_\parallel \to \infty$ and, thus, Eqs. (35) and (36) are still valid. However, we now assume a large field line diffusion coefficient corresponding to large values of the Kubo number. Considering $\kappa_{FL} \to \infty$ at the right hand side of Eq. (36), we derive (see again Shalchi (2015) for the mathematical details)

$$\kappa_{FL} = \frac{1}{\tilde{B}_0} \left[\int d^3k \; P_{xx}\left(\vec{k}\right) k_\perp^{-2} \right]^{1/2}. \tag{40}$$

The latter limit is sometimes called the non-linear regime or the Bohm limit of field line diffusion and was originally obtained by Kadomtsev & Pogutse (1978). In this case particles still follow magnetic field lines but the field lines are highly non-linear. The more experienced reader can easily see the occurrence of the so-called ultra-scale in Eq. (40). For most spectral tensors Eq. (40) leads to the scaling law

$$\lambda_\perp \propto \ell_\perp \frac{\delta B_x}{B_0} \tag{41}$$

where we used the perpendicular correlation scale of the turbulence $\ell_\perp$. This is very different compared to the quasi-linear scaling given by Eq. (39). This result should be valid for high particle energies and large Kubo number turbulence.

Journal of Physics: Conference Series **1332** (2019) 012014 doi:10.1088/1742-6596/1332/1/012014

3.4. The Owens-Zybin-Istomin Scaling

We now consider the limit $\lambda_\parallel/\ell_\parallel \to 0$ corresponding to strong pitch-angle scattering. In this case Eq. (32) becomes

$$\kappa_\perp = \frac{a^2 v^2}{3B_0^2} \int d^3k \, \frac{P_{xx}(\vec{k})}{F(k_\parallel, k_\perp) + v/\lambda_\parallel}. \tag{42}$$

Here we kept the function $F(k_\parallel, k_\perp)$ in the equation because for a small parallel diffusion coefficient, we also expect that the perpendicular diffusion coefficient is small. This means that we can neglect the term directly proportional to $\kappa_\perp k_\perp^2$ in Eq. (32) but not the term $F(k_\parallel, k_\perp)$ given by Eq. (33). Replacing the parallel mean free path via $\lambda_\parallel = 3\kappa_\parallel/v$ and using Eq. (33) allows us to write

$$\kappa_\perp = \frac{a^2}{B_0^2} \int d^3k \, P_{xx}(\vec{k}) \frac{\kappa_\perp k_\perp^2}{k_\parallel^2 + \frac{\kappa_\perp}{\kappa_\parallel} k_\perp^2}. \tag{43}$$

In order to simplify this further, we consider two cases. First we assume that the ratio $\kappa_\perp/\kappa_\parallel$ is large leading to

$$\kappa_\perp = \frac{a^2}{B_0^2} \kappa_\parallel \int d^3k \, P_{xx}(\vec{k}) = a^2 \kappa_\parallel \frac{\delta B_x^2}{B_0^2}. \tag{44}$$

Important here is to note that this result does not depend on the details of turbulence. Only the ratio of the turbulent field with respect to the mean field enters the latter equation. The result obtained here was derived before by Owens (1974) as well as Zybin & Istomin (1985).

3.5. The Collisionless Rechester & Rosenbluth Scaling

Here we still consider the case $\lambda_\parallel/\ell_\parallel \to 0$ and, thus, Eq. (43) is still valid. First we write Eq. (43) as

$$\frac{\kappa_\perp}{\kappa_\parallel} = \frac{a^2}{B_0^2} \int d^3k \, P_{xx}(\vec{k}) \frac{\sqrt{\frac{\kappa_\perp}{\kappa_\parallel} k_\perp^2}}{k_\parallel^2 + \frac{\kappa_\perp}{\kappa_\parallel} k_\perp^2} \sqrt{\frac{\kappa_\perp}{\kappa_\parallel} k_\perp^2}. \tag{45}$$

Now we assume that the ratio $\kappa_\perp/\kappa_\parallel$ is small. Using Eq. (37) therein yields

$$\sqrt{\frac{\kappa_\perp}{\kappa_\parallel}} = \pi \frac{a^2}{B_0^2} \int d^3k \, P_{xx}\left(\vec{k}\right) k_\perp \delta(k_\parallel). \tag{46}$$

The latter formula can easily be combined with any turbulence model as long as the occurring integrals are convergent. For most spectral tensors we now find the scaling law

$$\lambda_\perp \propto \lambda_\parallel \frac{\ell_\parallel^2}{\ell_\perp^2} \frac{\delta B_x^4}{B_0^4} \tag{47}$$

which sensitively depends on the scale ratio $\ell_\parallel/\ell_\perp$ as well as the magnetic field ratio $\delta B_x/B_0$. This result should be valid for short parallel mean free paths corresponding to strong pitch-angle scattering and small Kubo numbers. Characteristic here is that the ratio $\kappa_\perp/\kappa_\parallel$ does not depend on magnetic rigidity. Furthermore, we expect to find small values of that ratio in this limit (see next subsection for an example).

Spatschek (2008) discussed different transport regimes as well. Besides quasi-linear and Kadomtsev & Pogutse scalings, the famous work of Rechester & Rosenbluth (1978) was also discussed. Rechester & Rosenbluth derived a formula for the perpendicular diffusion coefficient for the case that deviations from the magnetic field lines occur. Assuming that the so-called *Kolmogorov length* L_K is the characteristic scale in the parallel direction for which this deviation occurs, one can estimate the corresponding mean square displacement of the field lines via

$\langle(\Delta x)^2\rangle \propto \kappa_{FL} L_K$. The parallel motion is assumed to be diffusive so that $L_K^2 \propto \kappa_\| t_c$ where t_c is the time the particles need to travel the distance L_K in the parallel direction. The corresponding perpendicular diffusion coefficient of the particle is then given by

$$\kappa_\perp = \frac{\langle(\Delta x)^2\rangle}{2t_c} \propto \frac{\kappa_{FL} L_K}{t_c} \propto \frac{\kappa_{FL}\kappa_\|}{L_K}. \tag{48}$$

This formula was also derived in Shalchi (2019) by using arguments based on field line separation theory. According to Neuer & Spatschek (2006, equation (46)) the Kolmogorov scale is given by

$$L_K \propto \frac{\ell_\perp^2}{\ell_\|}\frac{B_0^2}{\delta B_x^2}. \tag{49}$$

Using this in Eq. (48) yields

$$\kappa_\perp \propto \frac{\ell_\|^2}{\ell_\perp^2}\frac{\delta B_x^4}{B_0^4}\kappa_\| \tag{50}$$

in perfect agreement with the scaling obtained in Eq. (47) from diffusive UNLT theory. Therefore, we refer to the limit (47) as the collisionless Rechester & Rosenbluth (CLRR) scaling. As shown here and originally in Shalchi (2015) this limit can be derived in a very systematic way from diffusive UNLT theory.

It also has to be emphasized that the case considered here (strong pitch-angle scattering and small Kubo number turbulence) includes the case of pure slab turbulence. However, for slab turbulence, we have by definition $\ell_\perp = \infty$. Using this in Eq. (47) yields $\kappa_\perp = 0$ indicating that perpendicular transport in slab turbulence is sub-diffusive. Therefore, diffusive UNLT theory correctly describes transport in slab turbulence in the sense that $\kappa_\perp = 0$. Later we shall see that time-dependent UNLT theory agrees perfectly with compound sub-diffusion for this case.

3.6. An Illustrative Example

Three out of the four limits derived above from diffusive UNLT theory can easily be understood in terms of FLRW and parallel diffusion. Eq. (46), however, is an exception and one can assume that this type of diffusion requires the particles to get scattered away from the original magnetic field line (see Shalchi (2019) for some ideas supporting this assumption). Therefore, we focus more on the CLRR limit and consider an example. This example is the noisy slab model defined via

$$P_{nm}^{ns}(\vec{k}) = g^{slab}(k_\|)\frac{2\ell_\perp}{k_\perp}\Theta(1-k_\perp\ell_\perp)\Big(\delta_{nm} - \frac{k_n k_m}{k_\perp^2}\Big). \tag{51}$$

This model was used in Shalchi (2015) and is useful to study field line and particle transport in turbulence with non-vanishing but weak transverse complexity. Because of the Heaviside step function in Eq. (51) the usual slab model is broadened so that the wave vectors are no longer aligned perfectly parallel with respect to the mean field. In this case Eq. (46) becomes

$$\sqrt{\frac{\kappa_\perp}{\kappa_\|}} = \frac{\pi^2 a^2}{\ell_\perp B_0^2} g^{slab}\left(k_\| = 0\right). \tag{52}$$

For $g^{slab}(k_\|)$ we employ the model spectrum given by Bieber et al. (1994)

$$g^{slab}(k_\|) = \frac{1}{2\pi} C(s)\delta B_{slab}^2 \ell_\| \left[1 + \left(k_\|\ell_\|\right)^2\right]^{-s/2} \tag{53}$$

with the normalization function

$$C(s) = \frac{\Gamma\left(\frac{s}{2}\right)}{2\sqrt{\pi}\Gamma\left(\frac{s-1}{2}\right)}. \tag{54}$$

In the model spectrum we have used the turbulent magnetic field δB_{slab}, the bendover scale $\ell_{\parallel}$ denoting the turnover from the energy to the inertial range, and the inertial range spectral index s. Using this spectrum in Eq. (52) leads to

$$\frac{\kappa_{\perp}}{\kappa_{\parallel}} = \left[\frac{\pi}{2} C(s) a^2 \frac{\ell_{\parallel}}{\ell_{\perp}} \frac{\delta B^2}{B_0^2}\right]^2 \tag{55}$$

which is clearly a special case of the scaling law given by Eq. (47). In order to estimate a number for the ratio $\kappa_{\perp}/\kappa_{\parallel}$ in the limit considered here, we use $\delta B^2/B_0^2 \approx 0.5$, $\ell_{\parallel}/\ell_{\perp} \approx 0.75$, and $a^2 \approx 1$. For $s = 5/3$ (Kolmogorov (1941)) we find $C(s = 5/3) \approx 0.12$ and, thus, Eq. (55) provides $\kappa_{\perp}/\kappa_{\parallel} \approx 0.005$. Perpendicular transport in noisy slab turbulence is discussed in more detail in the next section. However, we can already see that we found a finite result for $\kappa_{\perp}$ corresponding to normal diffusion. Furthermore, the ratio $\kappa_{\perp}/\kappa_{\parallel}$ is small and energy independent.

4. Time-Dependent Perpendicular Transport

For certain applications, a time-dependent description of perpendicular transport could be needed. Perpendicular transport is described by the auto-correlation function $\langle V_x(t)V_x(0)\rangle$ where the guiding center velocity is given by Eq. (18). It is useful to write the latter equation as

$$V_x(t) = \frac{1}{B_0} \int d^3k \; \delta B_x(\vec{k}, t) v_z(t) e^{izk_{\parallel}} e^{i\vec{x}_{\perp}\cdot\vec{k}_{\perp}} \tag{56}$$

where we have used the notation $\vec{x}\cdot\vec{k} = zk_{\parallel} + \vec{x}_{\perp}\cdot\vec{k}_{\perp}$ with the two-dimensional vectors $\vec{x}_{\perp} = (x, y)$ and $\vec{k}_{\perp} = (k_x, k_y)$. As before, we employ a Fourier representation to find for the velocity correlation function

$$\langle V_x(t)V_x(0)\rangle = \frac{1}{B_0^2} \int d^3k \int d^3k' \; \langle \delta B_x(\vec{k}, t)\delta B_x(\vec{k}', 0) v_z(t) v_z(0) e^{iz(t)k_{\parallel}} e^{i\vec{x}_{\perp}(t)\cdot\vec{k}_{\perp}}\rangle \tag{57}$$

where we have set $\vec{x}(t = 0) = 0$. The central idea of time-dependent UNLT theory is to use the following approximation

$$\langle v_z(t)v_z(0)\delta B_x(\vec{k})\delta B_x^*(\vec{k}')e^{i\vec{x}(t)\cdot\vec{k}}\rangle \approx \langle v_z(t)v_z(0)e^{ik_z z(t)}\rangle\langle\delta B_x(\vec{k})\delta B_x^*(\vec{k}')\rangle\langle e^{i\vec{x}_{\perp}(t)\cdot\vec{k}_{\perp}}\rangle \tag{58}$$

meaning that we group together magnetic fields, all particle properties associated with their parallel motion, and their perpendicular motion. This can be understood as an extension of the Corrsin approximation used before in this review. Clearly this is different compared to approximation (21) used during the derivation of NLGC theory. Using this in Eq. (57) and employing Eq. (26) allows us to write the velocity auto-correlation function as

$$\langle V_x(t)V_x(0)\rangle = \frac{1}{B_0^2} \int d^3k \; P_{xx}(\vec{k}, t)\xi(k_{\parallel}, t)\langle e^{i\vec{x}_{\perp}\cdot\vec{k}_{\perp}}\rangle \tag{59}$$

where we have used the parallel correlation function

$$\xi\left(k_{\parallel}, t\right) = \left\langle v_z(t)\, v_z(0)\, e^{iz(t)k_{\parallel}}\right\rangle. \tag{60}$$

For the perpendicular characteristic function in Eq. (59) we set

$$\langle e^{i\vec{x}_\perp \cdot \vec{k}_\perp} \rangle = e^{-\langle (\Delta x)^2 \rangle k_\perp^2 /2} \tag{61}$$

corresponding to a Gaussian distribution with vanishing mean. In Lasuik & Shalchi (2018) other forms of distribution functions, such as kappa distributions, were employed but it was shown that the assumed statistics has only a minor influence on the resulting perpendicular diffusion parameter.

After combining Eqs. (59) and (61), we can write the velocity correlation function as

$$\langle V_x(t) V_x(0) \rangle = \frac{1}{B_0^2} \int d^3k \; P_{xx}(\vec{k}, t) \xi \left(k_\parallel, t \right) e^{-\frac{1}{2} \langle (\Delta x)^2 \rangle k_\perp^2} \tag{62}$$

and after employing the relation

$$\frac{d^2}{dt^2} \langle (\Delta x)^2 \rangle = 2 \langle v_x(t) v_x(0) \rangle \tag{63}$$

we find

$$\frac{d^2}{dt^2} \langle (\Delta x)^2 \rangle = \frac{2}{B_0^2} \int d^3k \; P_{xx} \left(\vec{k}, t \right) \xi \left(k_\parallel, t \right) e^{-\frac{1}{2} \langle (\Delta x)^2 \rangle k_\perp^2}. \tag{64}$$

This ordinary differential equation can be evaluated for any given turbulence model described by the spectral tensor P_{nm} as long as the parallel correlation function $\xi(k_\parallel, t)$ is specified as well. It has to be emphasized that at no point have we assumed that perpendicular transport is diffusive. Within a two-dimensional sub-space approximation, it was derived in Shalchi et al. (2011) as well as Lasuik & Shalchi (2019) that $\xi(k_\parallel, t)$ is given by[2]

$$\xi \left(k_\parallel, t \right) = \frac{v^2}{3} \frac{1}{\omega_+ - \omega_-} \left[\omega_+ e^{\omega_+ t} - \omega_- e^{\omega_- t} \right] \tag{65}$$

with the parameters

$$\omega_\pm = -\frac{v}{2\lambda_\parallel} \pm \sqrt{\left(\frac{v}{2\lambda_\parallel} \right)^2 - \frac{1}{3} \left(v k_\parallel \right)^2}. \tag{66}$$

Since the function $\xi(k_\parallel, t)$ is known, the ordinary differential equation (64) can be evaluated numerically for any given turbulence model including dynamical turbulence. In the following we consider some limits and special cases.

4.1. The Initial Free-Streaming Regime

First we focus on early times. From Eq. (65) it follows for $t \to 0$ that

$$\xi \left(k_\parallel, t \to 0 \right) = \frac{v^2}{3}. \tag{67}$$

Also assuming $\langle (\Delta x)^2 \rangle \to 0$ allows us to approximate Eq. (64) by

$$\frac{d^2}{dt^2} \langle (\Delta x)^2 \rangle = \frac{2v^2}{3B_0^2} \int d^3k \; P_{xx} \left(\vec{k}, t \right) = \frac{2}{3} v^2 \frac{\delta B_x^2}{B_0^2}. \tag{68}$$

Integrating this over time twice yields

$$\langle (\Delta x)^2 \rangle = \frac{v^2}{3} \frac{\delta B_x^2}{B_0^2} t^2 \tag{69}$$

corresponding to ballistic perpendicular transport at early times.

[2] Please note that Eq. (65) is the correct solution of the problem considered here. Eq. (98) of Lasuik & Shalchi (2019), where this result was also derived, has the wrong sign.

4.2. Pure Slab Turbulence

For pure slab turbulence we expect that particles follow magnetic field lines and, thus, compound sub-diffusion should occur. This has to come out of time-dependent UNLT theory. For slab turbulence, defined via

$$P_{nm}^{slab}(\vec{k}) = g^{slab}(k_\parallel)\frac{\delta(k_\perp)}{k_\perp}\delta_{nm}, \tag{70}$$

Eq. (62) becomes

$$\langle V_x(t)V_x(0)\rangle = \frac{4\pi}{B_0^2}\int_0^\infty dk_\parallel\; g^{slab}(k_\parallel)\xi(k_\parallel, t). \tag{71}$$

Using the time-dependent version of the TGK formula (19)

$$d_\perp(t) = \int_0^t d\tau\; \langle v_x(\tau)v_x(0)\rangle \tag{72}$$

and employing Eq. (65) yields for the running perpendicular diffusion coefficient

$$d_\perp(t) = \frac{4\pi v^2}{3B_0^2}\int_0^\infty dk_\parallel\; g^{slab}(k_\parallel)\frac{1}{\omega_+ - \omega_-}\left(e^{\omega_+ t} - e^{\omega_- t}\right). \tag{73}$$

In Fig. 1 we show a comparison of the latter formula with test-particle simulations showing an almost perfect agreement between time-dependent UNLT theory and the simulations. In order to understand Eq. (73) better, we consider the late time limit $t \to \infty$. One can easily see by considering the exponentials in Eq. (73) together with Eq. (66), that the main contribution to the $k_\parallel$-integral comes from very small wave numbers. Therefore, we find

$$d_\perp(t) \approx \frac{4\pi\kappa_\parallel}{B_0^2}g^{slab}(k_\parallel = 0)\int_0^\infty dk_\parallel\; e^{-\kappa_\parallel k_\parallel^2 t}. \tag{74}$$

For the turbulence spectrum we employ Eq. (53) and this becomes

$$d_\perp(t) = C(s)\ell_\parallel\frac{\delta B_{slab}^2}{B_0^2}\sqrt{\frac{\pi\kappa_\parallel}{t}} \tag{75}$$

corresponding to compound sub-diffusion discussed above. Clearly we can see how compound sub-diffusion is correctly described by time-dependent UNLT theory.

4.3. Diffusive Perpendicular Transport in Noisy Slab Turbulence

As demonstrated above, time-dependent UNLT theory provides compound sub-diffusion for slab turbulence. In the following we try to explore what ingredients are needed in order to restore normal diffusion. Therefore, we employ a turbulence model which is very close to the slab model but contains some transverse structure. This model is the noisy slab model defined via Eq. (51). Combining this with Eq. (64) yields

$$\frac{d^2}{dt^2}\langle(\Delta x)^2\rangle = \frac{8\pi\ell_\perp}{B_0^2}\int_0^{1/\ell_\perp} dk_\perp\; e^{-\frac{1}{2}\langle(\Delta x)^2\rangle k_\perp^2}\int_0^\infty dk_\parallel\; g^{slab}(k_\parallel)\xi(k_\parallel, t). \tag{76}$$

The $k_\perp$-integral can be expressed by an error function

$$\int_0^{1/\ell_\perp} dk_\perp\; e^{-\frac{1}{2}\langle(\Delta x)^2\rangle k_\perp^2} = \sqrt{\frac{\pi}{2\langle(\Delta x)^2\rangle}}\mathrm{Erf}\left(\sqrt{\frac{\langle(\Delta x)^2\rangle}{2\ell_\perp^2}}\right) \tag{77}$$

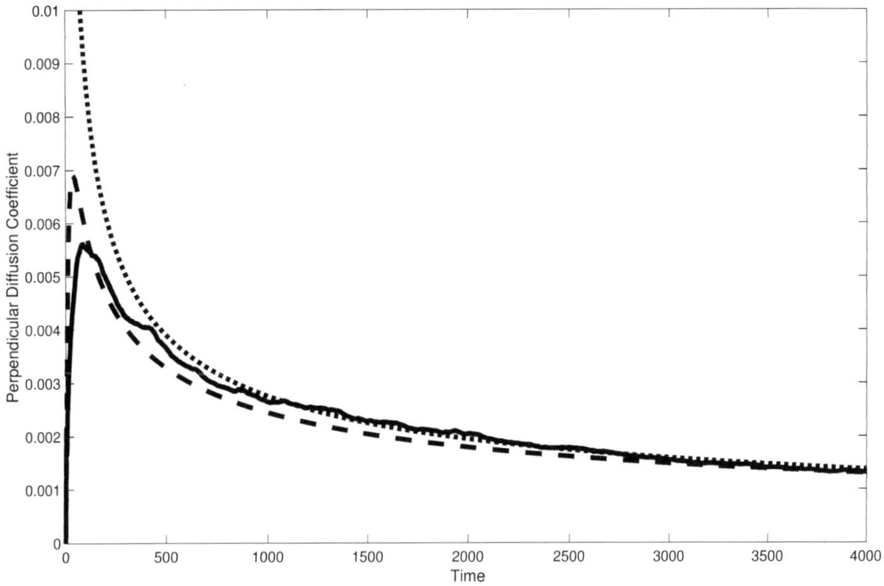

Figure 1. The perpendicular diffusion coefficient for slab turbulence. Shown is the numerical solution of Eq. (73) for $s = 5/3$ and $q = 3$ (dashed line), compound sub-diffusion as given by Eq. (75) (dotted line), as well as the test-particle simulations (solid line). The simulations are from Arendt & Shalchi (2018).

so that

$$\frac{d^2}{dt^2}\langle(\Delta x)^2\rangle = \frac{8\pi\ell_\perp}{B_0^2}\sqrt{\frac{\pi}{2\langle(\Delta x)^2\rangle}}\mathrm{Erf}\left(\sqrt{\frac{\langle(\Delta x)^2\rangle}{2\ell_\perp^2}}\right)\int_0^\infty dk_\parallel \, g^{slab}\left(k_\parallel\right)\xi\left(k_\parallel, t\right). \tag{78}$$

The error function has the following asymptotic properties (see, e.g., Abramowitz & Stegun (1974))

$$\mathrm{Erf} \approx \begin{cases} \frac{2}{\sqrt{\pi}}x & \text{for} \quad x \ll 1 \\ 1 & \text{for} \quad x \gg 1. \end{cases} \tag{79}$$

Therewith, it follows from Eq. (78) that we find compound sub-diffusion as long as the condition $\langle(\Delta x)^2\rangle \ll 2\ell_\perp^2$ is satisfied. A numerical solution of Eq. (78) is visualized in Shalchi (2017) where one can see that diffusion is restored for later times. From Eq. (78) we can learn that in order to recover normal diffusion, we need to satisfy the condition $\langle(\Delta x)^2\rangle \gg 2\ell_\perp^2$. For slab turbulence we have $\ell_\perp = \infty$ and we never satisfy this. As soon as there is transverse complexity, corresponding to a finite $\ell_\perp$, diffusion is restored as soon as the perpendicular particle mean square displacement exceeds the perpendicular bendover scale squared. Everything which is needed in order to get Markovian diffusion in the perpendicular direction is transverse structure in magnetic turbulence and the condition $\langle(\Delta x)^2\rangle \gg 2\ell_\perp^2$ needs to be satisfied.

4.4. Diffusive UNLT Theory
Of course the question arises how diffusive and time-dependent UNLT theories are related to each other. In terms of equations, this means one has to explore how Eqs. (32) and (64) are related. Diffusive UNLT theory is based on the diffusion approximation meaning that $\langle(\Delta x)^2\rangle = 2\kappa_\perp t$

for all t. Using this at the right hand side of Eq. (64) and integrating over all times yields for the perpendicular diffusion coefficient

$$\begin{aligned} \kappa_\perp &= \frac{a^2 v^2}{3B_0^2} \int d^3k \, P_{xx}(\vec{k}) \frac{1}{\omega_+ - \omega_-} \int_0^\infty dt \, \left[\omega_+ e^{\left(\omega_+ - \kappa_\perp k_\perp^2\right)t} - \omega_- e^{\left(\omega_- - \kappa_\perp k_\perp^2\right)t} \right] \\ &= \frac{v^2}{3B_0^2} \int d^3k \, P_{xx}(\vec{k}) \frac{\kappa_\perp k_\perp^2}{\omega_+ \omega_- - (\omega_+ + \omega_-) \kappa_\perp k_\perp^2 + \kappa_\perp^2 k_\perp^4} \end{aligned} \tag{80}$$

in the case of magnetostatic turbulence. Using Eq. (66) in order to replace the parameters ω_+ and ω_- therein, and after performing some straightforward algebra we can derive

$$\kappa_\perp = \frac{a^2 v^2}{3B_0^2} \int d^3k \, \frac{P_{xx}(\vec{k})}{F(k_\parallel, k_\perp) + \kappa_\perp k_\perp^2 + v/\lambda_\parallel} \tag{81}$$

where $F(k_\parallel, k_\perp)$ is still given by Eq. (33). This result agrees with Eq. (32) apart from a factor 4/3. The derivation presented here has the advantage that one can see it as a special case of the time-dependent theory. The only difference between Eqs. (81) and (32) is the factor 4/3 in the denominator. Although this factor is a minor difference, strictly speaking Eq. (32) should be more accurate than Eq. (81).

5. An Example for Three-dimensional Turbulence: The Goldreich-Sridhar Model

In the recent years spectral tensors based on the Goldreich & Sridhar (1995) model became more popular and it was used in test-particle simulations in the paper by Sun & Jokipii (2011). For our analytical considerations, we use the following spectral tensor

$$P_{nm}^{GS}(\vec{k}) = g^{GS}(k_\parallel, k_\perp) \left(\delta_{nm} - \frac{k_n k_m}{k^2} \right). \tag{82}$$

It needs to be emphasized that this corresponds to the compressible case $\delta B_z \neq 0$ but the original work of Goldreich & Sridhar is based on incompressible MHD. For the spectrum Shalchi (2013) suggested the following form

$$g^{GS}(k_\parallel, k_\perp) = \frac{D(s,q)}{2\pi} \ell^3 \delta B^2 \frac{(k_\perp \ell)^{q-s}}{\left[1 + (k_\perp \ell)^2\right]^{(s+q)/2}} e^{-\ell^{2-s} |k_\parallel| k_\perp^{1-s}} \tag{83}$$

which is only valid for $s = 5/3$. The exponential function therein ensures that the critical balance condition of Goldreich & Sridhar $|k_\parallel| \propto k_\perp^{2/3} \ell^{-1/3}$ is satisfied. Furthermore, the spectrum used here contains an energy range where the spectral index is given by q.

Sun & Jokipii (2011) used a model spectrum which corresponds to the one given by Eq. (83) for the case $q = 0$. Diffusive UNLT theory was combined with this spectral tensor and spectrum to test the theory via a comparison with the aforementioned simulations in Shalchi & Hussein (2015). The comparison is visualized in Fig. 2. As shown, we find good agreement between analytical theory and the simulations. Interesting here is that we find good agreement for $a^2 = 1$ meaning that the correction factor a^2 is not needed at all. This was also found for slab as well as noisy slab turbulence but not for two-component turbulence where we need to set $a^2 = 1/3$ (see, e.g., Arendt & Shalchi (2018)). This suggests that UNLT theory works well for small and intermediate Kubo numbers and only for large Kubo numbers one needs the correction factor a^2.

Furthermore, we can clearly see by considering Fig. 2 that $\lambda_\perp \propto \lambda_\parallel$ for small $\lambda_\parallel$ whereas the perpendicular mean free path becomes independent of the parallel mean free path for large $\lambda_\parallel$.

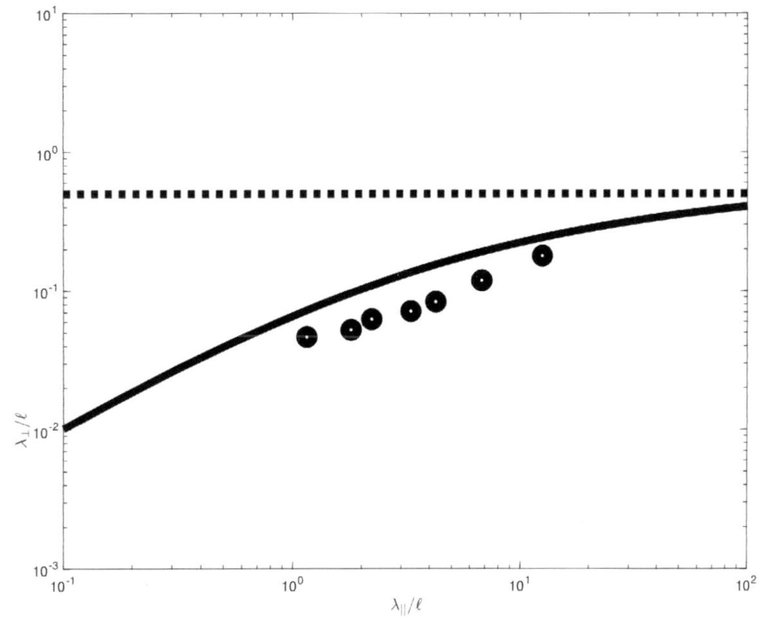

Figure 2. The perpendicular mean free path $\lambda_\perp$ versus the parallel mean free path $\lambda_{||}$ for Goldreich-Sridhar turbulence. Both parameters are normalized with respect to the bendover scale ℓ. Shown are the mean free paths obtained from the simulations (dots) performed by Sun & Jokipii (2011) and the results derived from diffusive UNLT theory (solid line). Also the quasi-linear perpendicular mean free path (see Eq. (3) from Sun & Jokipii (2011)) is shown (dotted line).

This behavior can also be seen for the noisy slab model and two-component turbulence (see, e.g., Hussein et al. (2015)). Thus there is some universality in the transport of particles across the mean magnetic field (see, e.g., Shalchi (2014) and Hussein et al. (2015)). However, this universal behavior of the perpendicular mean free path requires that all fundamental turbulence scales such as the integral scales and the ultra-scale are finite (see Matthaeus et al. (2007) for scales in two-dimensional turbulence and Shalchi (2014) for scales in different three-dimensional models).

6. Summary and Conclusion

The problem of perpendicular transport has a long history going back to the famous paper of Jokipii (1966) where a quasi-linear theory has been developed. A more advanced non-linear tool is provided by the diffusive unified non-linear transport theory and its time-dependent generalization. One important feature of this theory is that it contains several known limits as special cases. Those are listed in the following:

- In the limit of long parallel mean free paths, the unified non-linear transport equation (32) turns into the field line random walk limit and one obtains automatically the non-linear field line diffusion theory of Matthaeus et al. (1995). Therefore, the latter theory is contained in the non-linear integral equation provided by diffusive UNLT theory.
- For magnetostatic slab turbulence time-dependent UNLT theory provides compound sub-

diffusion.

- For long parallel mean free path and small Kubo numbers, UNLT theory is identical compared to quasi-linear theory.
- For long parallel mean free path and large Kubo numbers, on the other hand, UNLT theory provides the limit known as Kadomtsev & Pogutse (1978) limit.
- For short parallel mean free paths and small Kubo numbers, one can derive a Rechester & Rosenbluth (1978) type of scaling from UNLT theory. Since this process does not require to incorporate collisions, this limit can be called the collisionless Rechester & Rosenbluth scaling. This result is interesting because it provides a ratio $\kappa_\perp/\kappa_\parallel$ which is small and does not depend on particle energy.
- For short parallel mean free paths and large Kubo numbers one finds $\kappa_\perp/\kappa_\parallel \approx \delta B_x^2/B_0^2$ which could be explained by assuming that particles diffusive along ballistic magnetic field lines.

Furthermore, UNLT theory agrees well with most test-particle simulations. This is in particular the case for full three-dimensional turbulence and slab turbulence. For instance, UNLT theory agree almost perfectly with simulations performed for Goldreich-Sridhar turbulence.

References

[1] Abramowitz M and Stegun I A 1974 *Handbook of Mathematical Functions* (New York: Dover Publications)
[2] Ahluwalia H S and Ygbuhay R C 2015 *Solar Physics* **290** 635
[3] Alania M V, Modzelewska R and Wawrzynczak A 2011 *Solar Physics* **270** 629
[4] Arendt V and Shalchi A 2018 *Astrophysics and Space Science* **363** 116
[5] Bieber J W, Matthaeus W H, Smith C W, Wanner W, Kallenrode M-B and Wibberenz G 1994 *Astrophys. J.* **420** 294
[6] Bieber J W, Matthaeus W H, Shalchi A and Qin G 2004 *Geophys. Res. Lett.* **31** L10805
[7] Corrsin S 1959, *in Atmospheric Diffusion and Air Pollution, Advances in Geophysics*, **Vol. 6**, ed. F. Frenkiel & P. Sheppard (New York: Academic), 161
[8] Dosch A and Shalchi A 2010 *Advances in Space Research* **46** 1208
[9] Engelbrecht N E and Burger R A 2010 *Advances in Space Research* **45** 1015
[10] Engelbrecht N E and Burger R A 2014 *Brazilian Journal of Physics* **44** 512
[11] Engelbrecht N E and Burger R A 2015 *Advances in Space Research* **55** 390
[12] Engelbrecht N E, Strauss R D, le Roux J A and Burger R A 2017 *Astrophys. J.* **841** 107
[13] Ferrand G, Danos R, Shalchi A, Safi-Harb S, Edmon P and Mendygral P 2014 *Astrophys. J.* **792** 133
[14] Gardiner C W 1985 *Handbook of Stochastic Methods for Physics, Chemistry and the Natural Sciences* 2nd edn (Berlin: Springer)
[15] Goldreich P and Sridhar S 1995 *Astrophys. J.* **438** 763
[16] Green M S 1951 *The Journal of Chemical Physics* **19** 1036
[17] Hitge M and Burger R A 2010 *Advances in Space Research* **45** 18
[18] Hu J, Li G, Ao X, Zank GP and Verkhoglyadova O 2017 *Journal of Geophysical Research* **122** 938
[19] Hussein M, Tautz R C and Shalchi A 2015 *Journal of Geophysical Research* **120** 4095
[20] Jokipii J R 1966 *Astrophys. J.* **146** 480
[21] Jokipii J R, Kóta J and Giacalone J 1993 *Geophysical Research Letters* **20** 1759
[22] Jones F C, Jokipii J R and Baring M G 1998 *Astrophys. J.* **509** 238
[23] Kadomtsev B B and Pogutse O P 1978 *Plasma Phys. Controlled Nucl. Fusion Res.* **40** 38
[24] Kolmogorov A N 1941 *Dokl. Akad. Nauk. SSSR* **30** 301
[25] Kóta J and Jokipii J R 2000 *Astrophys. J.* **531** 1067
[26] Kubo R 1957 *Journal of the Physical Society of Japan* **12** 570
[27] Kubo R 1963 *J. Math. Phys.* **4** 174
[28] Lasuik J and Shalchi A 2017 *Astrophys. J.* **847** 9
[29] Lasuik J and Shalchi A 2018 *Advances in Space Research* **61** 2827
[30] Lasuik J and Shalchi A 2019 *Monthly Notices of the Royal Astronomical Society* **485** 1635
[31] Li G, Shalchi A, Ao X, Zank G P and Verkhoglyadova O P 2012 *Advances in Space Research* **49** 1067
[32] Manuel R, Ferreira S E S and Potgieter M S 2014 *Solar Physics* **289** 2207
[33] Matthaeus M W, Gray P C, Pontius Jr D H, and Bieber J W 1995 *Phys. Rev. Lett.* **75** 2136

[34] Matthaeus W H, Qin G, Bieber J W, and Zank G P 2003 *Astrophys. J.* **590** L53
[35] Matthaeus W H, Bieber J W, Ruffolo D, Chuychai P and Minnie J 2007 *Astrophys. J.* **667** 956
[36] Neuer M and Spatschek K H 2006 *Phys. Rev. E* **73** 26404
[37] Owens A J 1974 *Astrophys. J.* **191** 235
[38] Parker E N 1965 *Planet. Space Sci* **13** 9
[39] Perri S and Zimbardo G 2009a *Astrophys. J. Lett.* **693** L118
[40] Perri S and Zimbardo G 2009b *Advances in Space Research* **44** 465
[41] Potgieter M S and Nndanganeni R R 2013 *Astrophysics and Space Science* **345** 33
[42] Potgieter M S, Vos E E, Boezio M, De Simone N, Di Felice V and Formato V 2014 *Solar Physics* **289** 391
[43] Qin G and Zhang L-H 2014 *Astrophys. J.* **787** 12
[44] Qin G and Shalchi A 2015 *Physics of Plasmas* **22** 012905
[45] Qin G and Shalchi A 2016 *Astrophys. J.* **823** 23
[46] Rechester A B and Rosenbluth M N 1978 *Phys. Rev. Lett.* **40** 38
[47] Schlickeiser R 2002 *Cosmic Ray Astrophysics* (Berlin: Springer)
[48] Shalchi A, Bieber J W and Matthaeus W H 2004 *Astrophys. J.* **615** 805
[49] Shalchi A 2006 *Astronomy & Astrophysics Letters* **453** L43
[50] Shalchi A and Kourakis I 2007 *Astronomy & Astrophysics* **470** 405
[51] Shalchi A 2010 *Astrophys. J. Lett.* **720** L127
[52] Shalchi A 2011a *Advances in Space Research* **47** 1147
[53] Shalchi A 2011b *Contributions to Plasma Physics* **51** 10
[54] Shalchi A, Rempel T J and Rempel T J 2011 *Plasma Phys. Control. Fusion* **53** 105016
[55] Shalchi A 2013 *Astrophysics and Space Science* **344** 187
[56] Shalchi A 2014 *Advances in Space Research* **53** 1024
[57] Shalchi A 2015 *Physics of Plasmas* **22** 010704
[58] Shalchi A and Hussein M 2015 *Astrophysics and Space Science* **355** 243
[59] Shalchi A 2017 *Physics of Plasmas* **24** 050702
[60] Shalchi A 2019 Field line random walk, field line separation, and particle transport in turbulence with weak transverse complexity *Advances of Space Research* **in press**
[61] Shen Z-N and Qin G 2018 *Astrophys. J.* **854** 137
[62] Spatschek K H 2008 *Plasma Phys. Control. Fusion* **50** 12402
[63] Sun P and Jokipii J R 2011 *32nd international cosmic ray conference, Beijing 2011*
[64] Taylor G I 1922 *Proceedings of the London Mathematical Society* **20** 196
[65] Wawrzynczak A and Alania M V 2010 *Advances in Space Research* **45** 622
[66] Webb G M, Zank G P, Kaghashvili E Kh and le Roux J A 2006 *Astrophys. J.* **651** 211
[67] Zank G P, Li G, Florinski V, Matthaeus W H, Webb G M, and Le Roux J A 2004 *Journal of Geophysical Research* **109** A04107
[68] Zwillinger D 2012 *Standard Mathematical Tables and Formulae* (Boca Raton: CRC Press)
[69] Zybin K P and Istomin Y N 1985 *Zh. Eksp. Teor. Fiz.* **89** 836

Comparisons Between the Field Lines Using an Accelerating and a Constant Solar Wind model

S. Tasnim[1], **Iver H. Cairns**[2], **M. S. Wheatland**[2], **B. Li**[2], **and Gary P. Zank**[1]

[1] Center for Space Plasma and Aeronomic Research (CSPAR), University of Alabama in Huntsville, Huntsville, AL 35805, USA

[2] School of Physics, The University of Sydney, NSW-2006, Australia

E-mail: samira.tasnim@uah.edu, iver.cairns@sydney.edu.au, michael.wheatland@sydney.edu.au, garyp.zank@gmail.com

Abstract. Magnetic field line mapping between the Sun and the Earth is important to trace the nonthermal particles. We generalize a recently developed mapping approach (**B** stepping) where this approach allows us to map the magnetic field lines by stepping along the local magnetic field direction. We employ an advanced solar wind model which includes acceleration, angular momentum conservation, and intrinsic non-radial velocities and magnetic fields at the inner boundary /source surface. We map the field lines using Wind spacecraft data for two solar rotation periods: one near a solar minimum between CR2118 and CR2119 and other CR1992 near a solar maximum. Maps using the accelerating solar wind model are compared with the maps using a constant solar wind model. On a broad scale, maps using two solar wind models for the same solar rotation periods are very similar. However, in a small scale, there are significant differences, e.g. differences are evident in connectivities, paths, and winding angles. In addition, field lines using the accelerating solar wind model are more azimuthally oriented for during the solar maximum. These differences demonstrate the significance of inclusion of the accelerating radial speed profile, intrinsic azimuthal velocity and magnetic field components.

1. Introduction

The heliospheric magnetic field structure and configuration help us to study and understand the transport of the nonthermal particles and solar wind's source regions (Feldman et al. 1975, Rosenbauer *et al.* 1977, Owens & Forsyth 2013). Nonthermal particles, e.g. suprathermal Strahl electrons with energy $\approx 80\,\mathrm{eV} - 2\,\mathrm{keV}$ (Rosenbauer *et al.* 1977), solar energetic particles (SEPs) with energy $1\,\mathrm{MeV}$ (Kutchko et al. 1982, Richardson et al. 1991, Ruffolo *et al.* 2006), and the subrelativistic to relativistic electrons that produce type III solar radio bursts (Suzuki & Dulk 1985, Reiner et al. 1995)), are much faster than convective velocity $\mathbf{E} \times \mathbf{B}$ and the thermal particle. This results in the nonthermal particle to trace easily the magnetic field line structure (Schatten et al. 1968, Owens & Forsyth 2013, Li et al. 2016a,b).

Parker first explained that the rotation of the Sun naturally produces spiral like magnetic field structure (Parker 1958). The magnetic field lines are observed to follow Parker spiral when we averaged them over many solar rotation periods. However, in the short time-scale, they show significant deviations from the Parker spiral (Forsyth et al. 1996, Borovsky 2010, Schulte in den Bäumen et al. 2011, 2012, Tasnim & Cairns 2016, Tasnim et al. 2018). Moreover, complex

Content from this work may be used under the terms of the Creative Commons Attribution 3.0 licence. Any further distribution of this work must maintain attribution to the author(s) and the title of the work, journal citation and DOI.
Published under licence by IOP Publishing Ltd

magnetic field structures are often observed by tracing the solar energetic particles, relativistic electrons, and protons (Suzuki & Dulk 1985, Owens & Forsyth 2013).

Various solar wind models are proposed to explain the deviations from Parker spiral (Fisk 1996, Gosling & Skoug 2002, Schwadron et al. 2010, Smith 1979, Schulte in den Bäumen et al. 2011, 2012, Tasnim & Cairns 2016, Tasnim et al. 2018). Besides, a number of mapping approaches are developed to predict the field lines from close to the Sun to the Earth's location (Schatten et al. 1968, Nolte & Roelof 1973*b*,*a*, Li et al. 2016a,b). Ecliptic magnetic field was first mapped on a large scale by Schatten et al. (1968), where they employed the near-Earth spacecraft data. The field lines using the solar maximum data were observed to have more structures, whereas these lines are observed to be more open near the solar minimum (Schatten et al. 1968). Later, Nolte & Roelof (1973*b*,*a*) showed that the longitudes of the source surface are related to the longitudes of the observed magnetic field data. However, these approaches are failed to distinguish between the open and closed field lines (Gosling & Roelof 1974).

A recently developed mapping approach Li et al. (2016a) employ an equatorial two dimensional solar wind model of Schulte in den Bäumen et al. (2012) and 1 AU near Earth data to predict the field lines from the source surface to 1 AU. Note that Schulte in den Bäumen et al. (2012)'s model includes an intrinsic azimuthal magnetic field component. The predicted field lines agreed 90% with the observed pitch angle distribution data (Li et al. 2016a). In further research, they extended the work to predict the correlation between the path of subrelativistic electrons of the Type III radio burst and the path of the magnetic field lines. They also compare them with the observed field line inversion in the equatorial plane. However, all the above mapping approaches employed solar wind models where the solar wind's radial speed is constant along the radial distance.

In this paper, we employ an advanced generalized solar wind model which includes an accelerating radial wind profile, intrinsic azimuthal magnetic field, and angular momentum conservation. In addition, the model includes a deviation from corotation, which results in an intrinsic azimuthal velocity (Tasnim et al. 2018). We improve **B**-step algorithm developed by Li et al. (2016a) and Li et al. (2016b) and map the field lines which are not only biased along the 1 AU. This model also uses 1 AU near Earth data from the Wind spacecraft. We predict the field lines for two solar rotations, one near the solar maximum and another the solar minimum. Note that Tasnim et al. (2018)'s model does not consider dynamical variations between the source surface and 1 AU. However, the non-Parker like magnetic field structure and the large values of the observed azimuthal velocity at 1 AU can be due to dynamic effects as stream-stream interactions (SIRs) at corotating interaction regions (CIRs) (Hu & Habbal 1992, Hu 1993, Odstřcil 1994, Richardson et al. 1996).

This paper is structured as follows. Section 2.1 briefly expresses and describes the magnetic field components using the accelerating solar wind model of Tasnim et al. (2018) whereas Section 2.2 describes the magnetic field components using the constant solar wind model of Schulte in den Bäumen et al. (2012). The predicted magnetic field components are used to extract the magnetic field vectors for both of the models. Section 3 provides a short description of the modified **B**-stepping ((Li et al. 2016a), (Li et al. 2016b)) approach. Section 4 qualitatively and quantitatively describes magnetic field lines using two solar wind models. Section 5 gives a brief discussion of the results and concludes the paper.

2. Solar Wind Models and Predictions

2.1. Magnetic Field Components using an Advanced Data Driven Accelerating Solar Wind Model

We use Tasnim et al. (2018)'s data driven equatorial accelerating solar wind model to extract the radial and azimuthal magnetic field components. The above model includes Parker's accelerating radial speed profile using an implicit method along with intrinsic non-radial velocities and magnetic fields at the inner boundary/source surface. The model uses several magnetodyhrodynamic (MHD) equations: (a) frozen-in field equation, (b) angular momentum conservation equation $\partial L_\star/\partial t + \rho \mathbf{r} \times (\mathbf{v} \cdot \nabla)\mathbf{v} = (1/\mu_0)\mathbf{r} \times (\mathbf{B} \cdot \nabla)\mathbf{B} = 0$ by assuming a radially symmetric thermal and magnetic pressure, (c) conservation of mass $\rho(r)v_r(r)r^2 =$ constant, (d) Gauss's Law $\nabla \cdot \mathbf{B} = 0$, and (e) Faraday's Law $\nabla \times \mathbf{E}' = 0$ assuming a time stationary electric field in the rotating frame. Here ρ is the mass density, t is the time, $\mathbf{v}$ is the velocity, $\mathbf{E}$ is the electric field, and $\mathbf{B}$ is the magnetic field.

The accelerating solar wind model of Tasnim et al. (2018) considers that the solar wind pattern is constant over a solar rotation period. This assumption allows the fixed pattern to rotate with the Earth and to vary the wind radially and longitudinally. We assume the magnetic field and velocity have radial and azimuthal components in the equatorial plane, i.e. $\mathbf{B} = (B_r, B_\phi)$ and $\mathbf{v} = (v_r, v_\phi)$. Here v_r and v_ϕ are the radial and azimuthal components of the velocity whereas B_r and B_ϕ are the radial and azimuthal components of the magnetic field. We assume the non-radial component of the magnetic field is constant along the longitudinal direction locally, i.e. $\partial B_\phi/\partial \phi = 0$. Detailed description and derivation is provided in Tasnim et al. (2018). We use primed variables to present the plasma quantities and magnetic field components in the corotating frame. Above assumptions yield the following expression for the radial magnetic field:

$$r'^2 B'_r(r', \phi') = r^2 B_r(r, \phi, t) = r'^2_{\rm s} B'_r(r'_{\rm s}, \phi'_{\rm s}). \tag{1}$$

Similarly, we can derive the expression for the azimuthal magnetic field at any location (r, ϕ) which provides

$$\begin{aligned} B_\phi(r', \phi') = B_\phi(r, \phi, t) = & \frac{r_{\rm s} v_r(r_{\rm s}, \phi_{\rm s}, t) B_\phi(r_{\rm s}, \phi_{\rm s}, t)}{r v_r(r, \phi, t)} - \frac{r_{\rm s} \delta v_\phi(r_{\rm s}, \phi_{\rm s}, t) B_r(r_{\rm s}, \phi_{\rm s}, t)}{r v_r(r, \phi, t)} \\ & + \frac{v_\phi(r, \phi, t) B_r(r, \phi, t)}{v_r(r, \phi, t)} - \frac{\Omega r_{\rm s}^2 B_r(r_{\rm s}, \phi_{\rm s}, t)}{r v_r(r, \phi, t)}, \end{aligned} \tag{2}$$

where $r_{\rm s}$ defines the radial distance of the source surface or the inner boundary of the solar wind. Note that we define the source surface/inner-boundary as a locus where the solar wind originates from. Here $\Omega = 2\pi/T$ presents the Sun's rotation frequency, $T = 27\,\text{days} = 648\,\text{hours}$ is synodic rotation period of the Sun. In addition, $v_r(r_{\rm s}, \phi_{\rm s}, t)$ is the radial velocity at the source surface, $\delta v_\phi(r_{\rm s}, \phi_{\rm s}, t)$ is the deviation from corotation, $v_\phi(r, \phi, t)$ and $v_r(r, \phi, t)$ are the azimuthal and radial velocities at any location (r, ϕ) at any time t of the solar rotation period, $B_r(r, \phi, t)$ is the radial magnetic field at (r, ϕ) and time t.

2.2. Magnetic Field Components for a Constant Radial Speed Solar Wind Model

We use Schulte in den Bäumen et al. (2012)'s model which was previously employed by Li et al. (2016a) and Li et al. (2016b) to map the magnetic field lines. This model assumes a constant radial speed and includes an intrinsic azimuthal magnetic field component at the inner boundary / source surface. We re-write the original equations in a corotating frame to make the magnetic field components consistent with those from Section 2.1. Note that Schulte in den Bäumen et al. (2012)'s model does not include the intrinsic azimuthal velocity or deviation from the corotation.

The radial magnetic field component for the Schulte in den Bäumen et al. (2012) model is same as for the Tasnim et al. (2018)'s model. In addition, the expression for the azimuthal component of the magnetic field is:

$$B_\phi(r', \phi') = B_\phi(r, \phi, t) = \frac{B_\phi(r_{\rm s}, \phi_{\rm s}, t) r_{\rm s}}{r} - \frac{B_r(r, \phi, t)\Omega(r - r_{\rm s})}{v_r(\phi, t)}. \tag{3}$$

2.3. Extraction of the Plasma Quantities and Magnetic Field Components

We use 1 AU data from the Wind Spacecraft and fit them with the model predictions to extract $B'_\phi(r'_s, \phi'_s)$, $B'_r(r'_s, \phi'_s)$, and $\delta v_\phi(r'_s, \phi'_s)$. Then, we predict $B'_r(r', \phi') = B_r(r, \phi, t)$ and $B'_\phi(r', \phi') = B_\phi(r, \phi, t)$ using the above expressions and 1 AU data at any (r, ϕ) location of the equatorial plane.

3. B-step Mapping Approach

We improve Li et al. (2016a,b)'s mapping algorithm to display and provide a "global" and unbiased magnetic field map where the field lines are not only biased around 1 AU. The updated algorithm is applicable to map the field lines at any radial and longitudinal location instead of field lines around 1 AU. We also name it as the B-step algorithm since we map the lines by stepping along the local magnetic field, i.e. the lines step along $\mathbf{B}$ and $-\mathbf{B}$ from a starting point $\mathbf{r}_0 = (x_0, y_0)$ in the equatorial plane between the Sun and the Earth.

The field line locations are calculated by averaging over four neighbouring local points to move toward a new point $r = (x, y)$ which can be expressed as

$$x = x_0 + \Delta x = x_0 + \sum_{i=1}^{4} \frac{B_{xi}}{B_i} dl \ , \quad (4)$$

$$y = y_0 + \Delta y = y_0 + \sum_{i=1}^{4} \frac{B_{yi}}{B_i} dl \ , \quad (5)$$

where x and y components of the magnetic field vectors are defined by B_{xi} and B_{yi} with $i = 1, 2, 3,$ and, 4 for the four nearest neighbouring points. These field lines start from an initial starting point $\mathbf{r}_0$ and step toward a new point $\mathbf{r}_i$. The process continues still the field lines reach the chosen end point. The mapping procedure continues in the opposite direction by the conversions $(\Delta x, \Delta y) \rightarrow (-\Delta x, -\Delta y)$ along with $dl \rightarrow -dl$ in Equations 4 - 5. These equations completely satisfy the Gauss's Law, but do not preserve the direction $\mathbf{B}$ at 1 AU which is due to fact that above equations use a local average instead of using an interpolated data.

4. Field Line Maps for the Accelerating and the Constant Radial Speed Wind

This section describes the detailed structures and orientations of the magnetic field lines for an accelerating and a constant solar wind and discusses the qualitative correlation between observed data at 1 AU using the Wind Spacecraft data for the solar rotation period 6 December 2011 and 1 January 2012 (between CR 2118 – CR 2119) near a solar minimum and another solar rotation period near a solar maximum from 1 July 2002 to 27 July 2002 (CR1992). We define the antisunward direction with angles $90^\circ \leq \Phi_B \leq 270^\circ$ and the remaining angles as (within 360° rotation) sunward ($0^\circ \leq \phi_B < 90^\circ$ or $270^\circ < \phi_B \leq 360^\circ$) direction, where ϕ_B is the in-ecliptic angle and $\phi_B = \tan^{-1}(B_\phi/B_r)$. Parker spiral model restricts ϕ_P between $90^\circ < \Phi_B < 180^\circ$ or $270^\circ < \phi_P < 360^\circ$, where ϕ_P is the Parker spiral angle. Mapping algorithm also allows the field lines to orient approximately radial and even sometimes to orient anticlockwise direction at a large heliocentric distance (an explanation is given in the following subsection). Identifications of magnetic field orientations strongly depend on the definition of the field orientations (Owens & Forsyth 2013, Li et al. 2016a,b).

4.1. Predicted Magnetic Maps Between CR2118 – CR2119: 6 December 2011 to 1 January 2012

Magnetic field line starts at 000 UT for each map in Figures 1 and 2, where the mapping lines are spaced at 8 hours interval in the Earth's orbit over the full solar rotation period. More

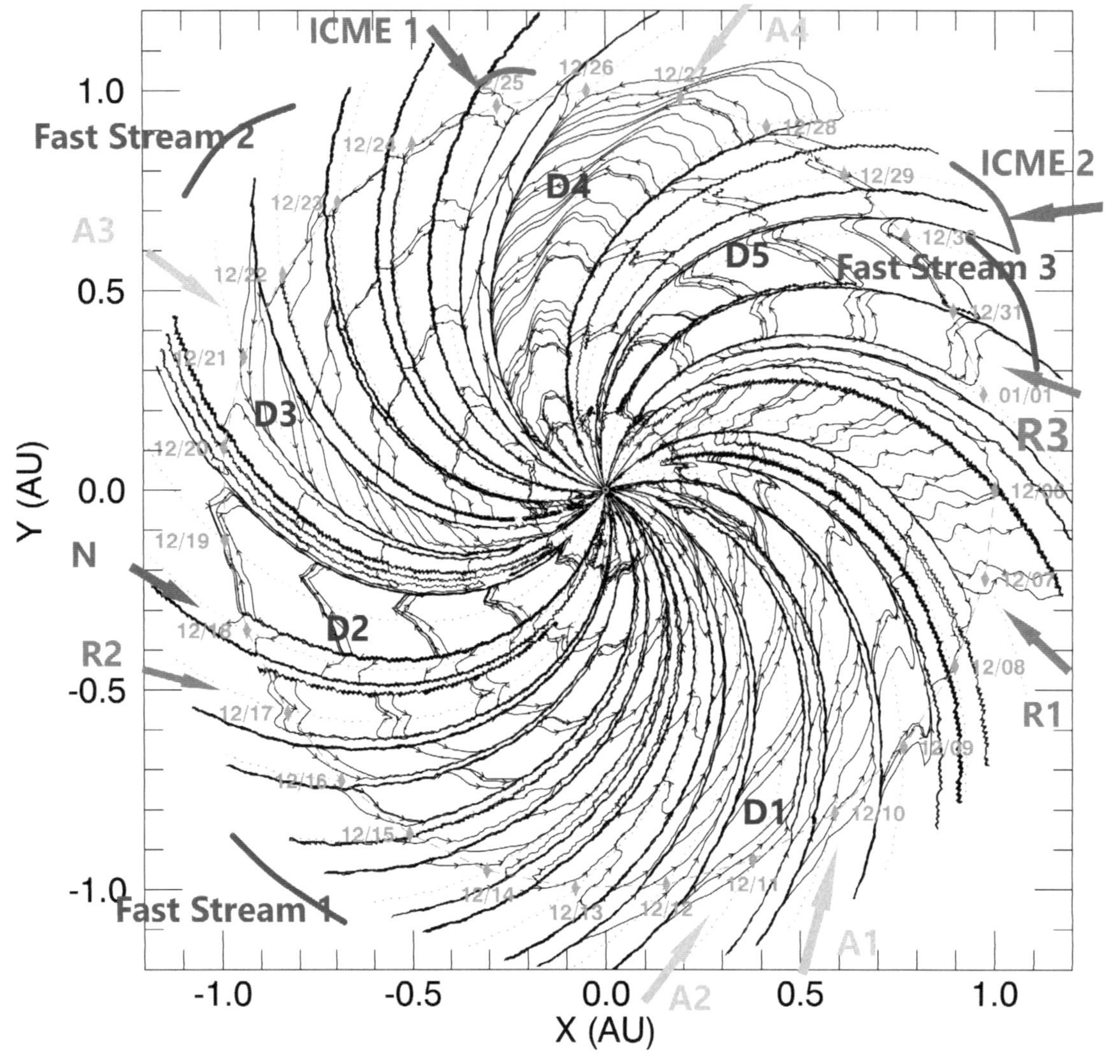

Figure 1. A map of magnetic field lines predicted using the accelerating solar wind model of Tasnim et al. (2018) in the equatorial plane for the solar rotation period between CR2118 and CR2119 (6 December 2011 and 1 January 2012). The map keeps the Sun at the center while the Earth moves around the Sun in the clockwise direction with increasing time. Here the green diamond markers show Earth's longitudinal locations every day at the universal time 0000. Inward arrows represent the inward field line, whereas outward arrows represent the opposite. Symbols Ai identify azimuthal field lines for $i = 1, 2, 3...$, and similarly Di identify the periods with low B, radial field lines as Ri, and field lines in the anticlockwise direction as N.

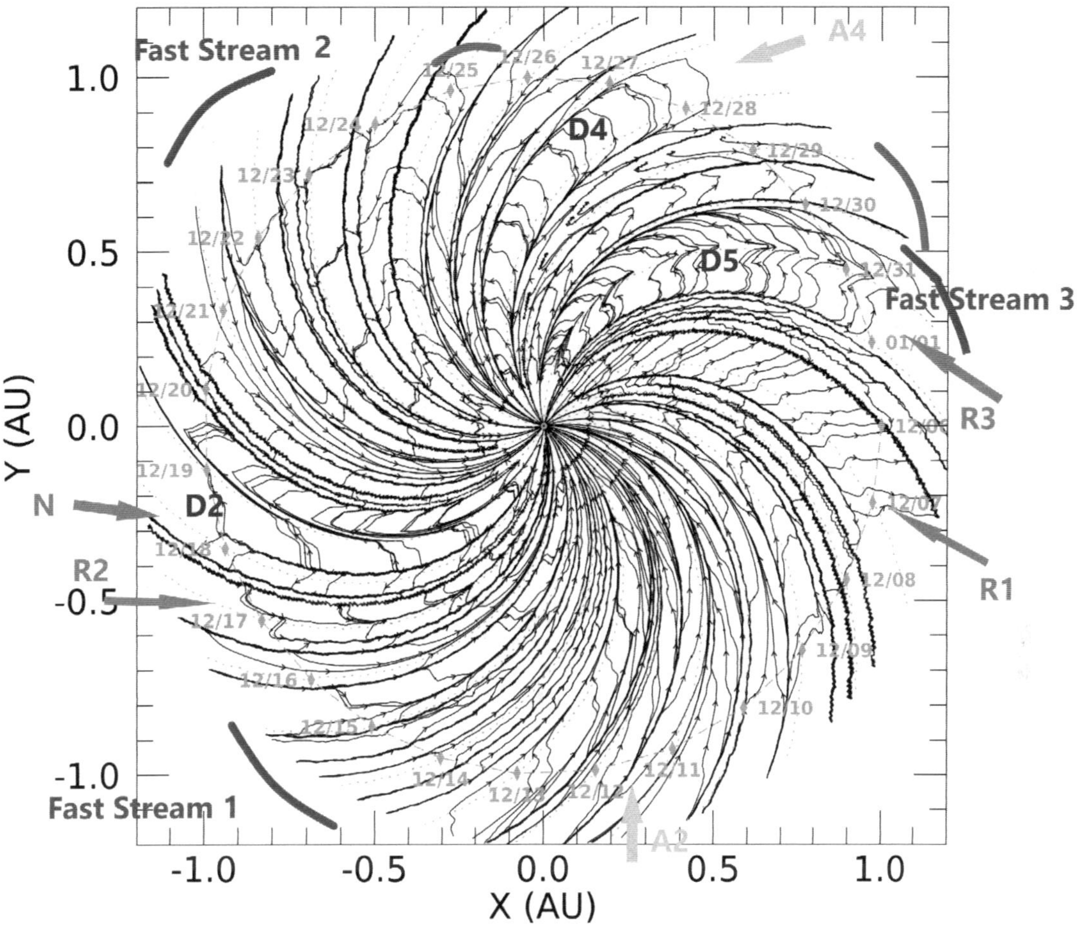

Figure 2. Magnetic field lines using the constant solar wind model of (Schulte in den Bäumen et al. 2012) for the solar rotation period between CR2118 and CR2119. Symbols have the same meaning as Figure 1.

precisely, we map the field lines at 0000 UT, 0800 UT, 1600 UT of each day. The choice of other starting points or time intervals will provide us with field lines of the remaining hours. We use definitions from Li et al. (2016a), Li et al. (2016b), and Li et al. (2016c) to describe the field lines of the accelerating $v_r(r, \phi)$ and constant v_r model (Schulte in den Bäumen et al. 2011, 2012, Li et al. 2016a,b). Several features of a solar rotation between CR2118 – CR2119 are noted below:

(i) The solar rotation contains two ICME events: ICME 1 on 25 December (0200 UT to 1700 UT) and ICME 2 between 29 to 30 December (2200 UT to 0900 UT) (Richardson & Cane 2015), and three fast stream regions: Fast Stream 1, 2, and 3 in Figures 1 and 2. These events modify the magnetic structure of the solar wind.

(ii) Figures 1 and 2 show open field lines within 24 to 25 December during the ICME event (marked as ICME 1). Note that curved red lines indicate the ICME regions. Open field lines also present in 18, 20, 21, 28, and 31 December. These lines are all sector boundaries which falsely linked as this mapping algorithm does not identify the SBs. Artificially connected lines also exist between the start and end of the solar rotation period in Figures 1 and 2.

Therefore, we assume the field lines are open when it crosses SBs and within the start and end of the solar rotation period for further analysis.

(iii) Field lines also experience the local field inversion. Figure 1 and 2 show line inversions between 21 to 22 December and between 29 to 30 December of 2011. We already mentioned that a CME event is present between 29 to 30 December (marked as ICME 2). In contrast, the inversion near 21 December is due to the presence of SIR event, which is due to the interaction between fast (labeled as Fast Stream 2) and slow streams.

(iv) Observed data and data-driven field lines magnetic show that occasionally, field lines are aligned approximately radial at large heliocentric distances (labeled as R1, R2, R3). We also see some field lines are oriented in anticlockwise direction instead of orienting in a clockwise direction like Parker spirals (labeled as N).

We now focus on similarities and dissimilarities between field lines predicted by Tasnim et al. (2018) and Li et al. (2016a,b,c). On a broad scale, Figures 1 and 2 show that magnetic maps using two solar wind models between CR2118 and CR2119 have strong similarities and some differences. Qualitative comparisons between Figure 1 and Figure 2 for the solar rotation are listed below:

(i) The field line densities are not always the same. Some days have a higher field line population than the other days. Both Figures 1 and 2 show low populated areas between 19 – 20, 26 – 28, and 30 – 31 December (labeled as D2, D4, and D5). An interpretation is that both of the solar wind models (employed to map the field lines) include intrinsic azimuthal field components $[B_\phi(r_s, \phi_s)]$. Azimuthally oriented fields at the small heliocentric distances in both Figures 1 and 2 demonstrate the inclusion of $B_\phi(r_s, \phi_s)$.

(ii) In addition, the map on Figure 1 shows more low populated regions than Figure 2, e.g., within 11 to 12, and 20 to 22 December (labeled as D1, and D3). The field lines are mostly azimuthal at 1 AU during this period. Additional low populated regions in the accelerating solar wind model indicates implication of the combined effect of intrinsic azimuthal velocities $\delta v_\phi(r_s, \phi_s) \neq 0$ and the accelerating radial wind speed $v_r(r, \phi)$. On the other hand, the constant speed model allowed a strong corotation with the Sun $\delta v_\phi = 0$ and restricted the radial speed to be constant with radial distance $v_r = v_r(\phi)$ along the velocity streamlines.

(iii) Field lines are azimuthal between 10 and 11 December and between 26 and 28 December (labeled as A1 and A4) for both of these models. However, the field lines are more azimuthal for the accelerating model. Additional azimuthally oriented field lines (marked by A2 and A3 on Figure 1) demonstrate the effect of intrinsic azimuthal velocity in Tasnim et al. (2018)'s model.

(iv) The magnetic field lines using accelerating $v_r(r, \phi)$ are sometimes disconnected close to Sun, which is due to the absence of the field vector at the Alfvénic critical radius (r_A). Tasnim et al. (2018)'s model excludes magnetic field vectors at $r_A(\phi_A)$ to remove the singularities at $r = r_A(\phi_A)$ when $M_A(r, \phi) = 1$. Put in other words, the accelerating solar wind excludes some points due to the presence of singularities at the Alfvénic critical point where the radial Mach number becomes 1 where the denominator in the analytic expression of the magnetic field contains a factor $(M^2_{Ar}(r, \phi) - 1)$ [detailed description is available in Tasnim et al. (2018)]. Therefore, field line mapping stops at $r = r_A(\phi_A)$ for predicted $\mathbf{B}$ using Tasnim et al. (2018).

(v) Variation of $r_A(\phi_A)$ with heliolongitudes ϕ results the predicted field lines discontinuities at different radial locations [e.g. disconnected field lines $r \leq 0.2$ AU]. Note that singularities in accelerating solar wind models affect both of the algorithms. Contrarily, Schulte in den Bäumen et al. (2011, 2012)'s constant solar wind model is much simpler and only considers frozen-in field conditions. Therefore, the model does not deal with singularity. As a result, Figure 2 shows connectivity with the source surface.

(vi) Field line connections in maps of Figures 1 and 2 between 6 December of 2011 and 1 January of 2012 indicate that **B**-step approach artificially connects field lines at the start and end of the solar rotation period.

Figure 3 shows field lines between 8 to 10 December (top two panels) and 24 and 29 December (bottom two panels) 2011. Field lines on the left-hand side are using the accelerating solar wind model, and the lines on the right-hand side are using the constant solar wind model. These lines explicitly show a notable path and connectivity differences. Dissimilarities in field-line connections and traveled paths indicate that the inclusion of acceleration and intrinsic non-radial velocity significantly changes the field line structures, even though their orientations at 1 AU remain very similar. As a result, field line orientations at 1 AU are categorized in the same PAD classes for both of the solar wind models.

4.2. Maps Between 1 July and 27 July 2002 (CR1992)

Figures 4 and 5 show the maps for a solar rotation period CR1992 near a solar maximum. The Wind Spacecraft data for the period shows the existence of of two ICMEs (ICME 1 between 18 July to 19 July and ICME 2 between 20 July to 22 July) (Richardson & Cane 2015) and three fast streams.

Field lines for the accelerating and constant solar wind models have strong similarities as like the corresponding maps for the solar minimum. However, Figure 4 shows more azimutally oriented field lines for accelerating solar wind models during the period near solar maximum than for the solar minimum. One interpretation is probably partly due to the existence of solar activities between CR1992 over a longer period than the solar rotation period between CR2118 and CR2119. For instance, field lines between 9 – 10, 16 – 19, and 20 – 23 July of 2002. Yellow arrows labeled as A1, A2, A3, A4, and A5 also indicate that the influence of deviation of corotation or intrinsic azimuthal velocities is more dominating during the solar maximum. Note that all the symbols have the same meaning as like Figure 1.

Figure 5 shows few filed lines between 6 to 11 July (top panels) and 13 to 17 July (bottom panels) of 2002 for the accelerating (left panels) and the constant (right panels) solar wind models near 1 AU. These fields lines clearly show that intrinsic azimuthal velocity strongly affects the path of the field lines during the solar maximum than during the solar minimum.

5. Discussion and Conclusions

This paper improves the **B**-stepping mapping algorithm of Li et al. (2016a) to provide an unbiased and complete magnetic map for any starting point at any radial and longitudinal location from the source surface to 1 AU and even beyond.

The primary achievement of this paper is the application of an advanced solar wind model, which includes acceleration of the solar wind, intrinsic azimuthal velocity by allowing a deviation from the strong corotation along with a non-zero azimuthal intrinsic magnetic field. More precisely, this work generalizes the mapping approach of Li et al. (2016a,b) with a more advanced solar model. Mapped field lines are very similar for both of these models; notably, they have the same orientation at 1 AU.

Another importance of this work apart from generalising the analyses of Li et al. (2016a,b) to a new solar wind model, is that the very similar results in Figures 1 and 2 suggest that the comprehensive testing of Li et al. (2016a,b) against observational data will apply with minimal changes to use of the accelerating wind model. Thus, further work or mapping lines in the solar wind can confidently use Tasnim et al. (2018) new accelerating wind model instead of the simpler Schulte in den Bäumen et al. (2011, 2012) model used by Li et al. (2016a,b).

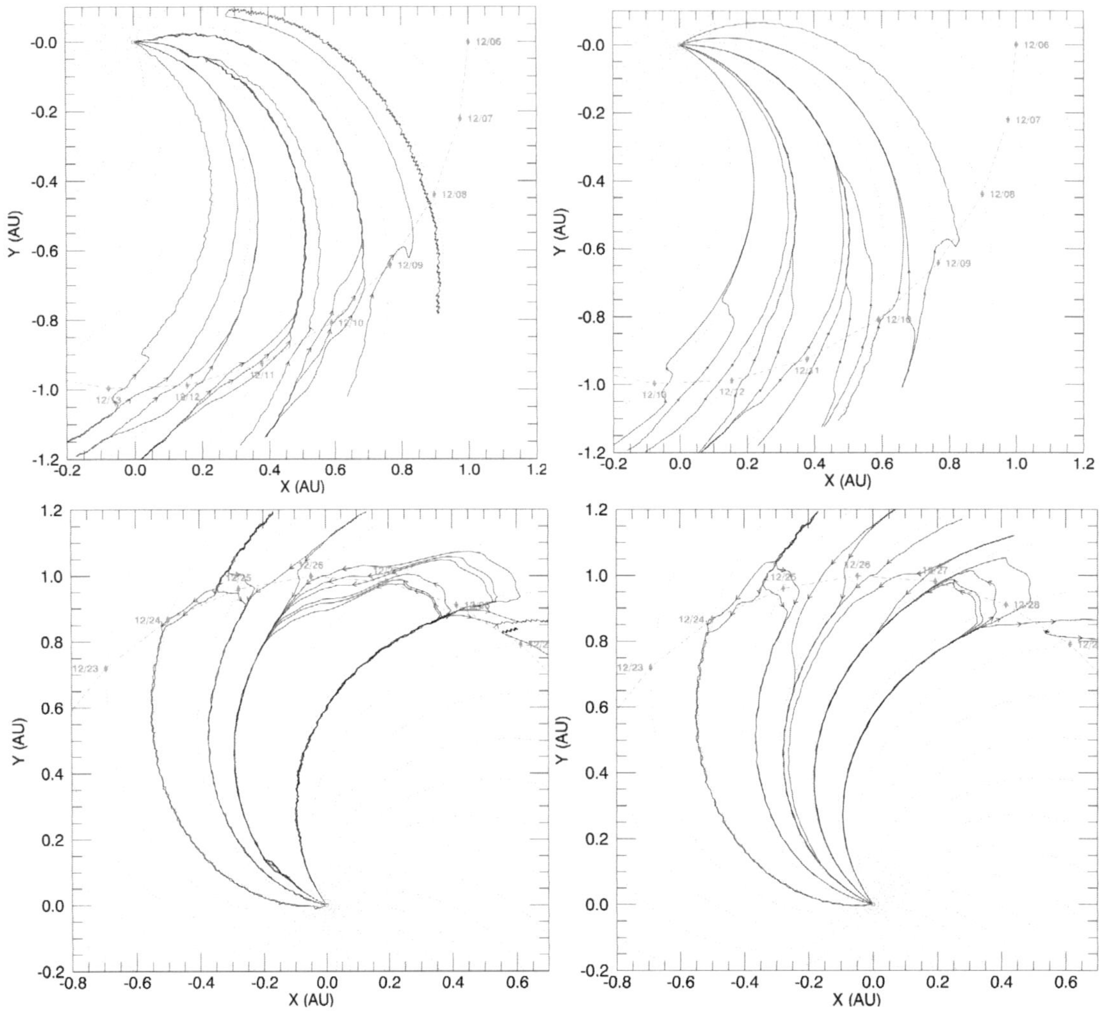

Figure 3. The predicted magnetic field lines between 8 to 13 December 2011 (top panels) for the accelerating (left) and the constant (right) wind models. Similarly, the lines between 24 to 26 December 2011 are shown in bottom panels for the accelerating (left) and the constant (right) solar wind models.

In summary, maps from two solar wind models are similar in a broad-scale. However, close observations of these maps show that the magnetic field lines for the accelerating radial speed model (Figure 1) and the constant radial speed wind model (Figure 2) have a notable path and connectivity differences. See, for instance, the almost horizontal field lies near 1 AU at the top of each map. These dissimilarities suggest that the inclusion of acceleration of the radial speed and intrinsic non-radial velocity have significant effects on magnetic field lines. This result agrees with the analytic expression of the magnetic field components where field components are correlated to the velocity components.

Maps in Figure 1 and 2 also demonstrate that the field lines do not always follow velocity streams. The success rate for predicting magnetic field lines for Tasnim et al. (2018)'s model is little different than the constant radial speed model from Li et al. (2016a) and Li et al. (2016b).

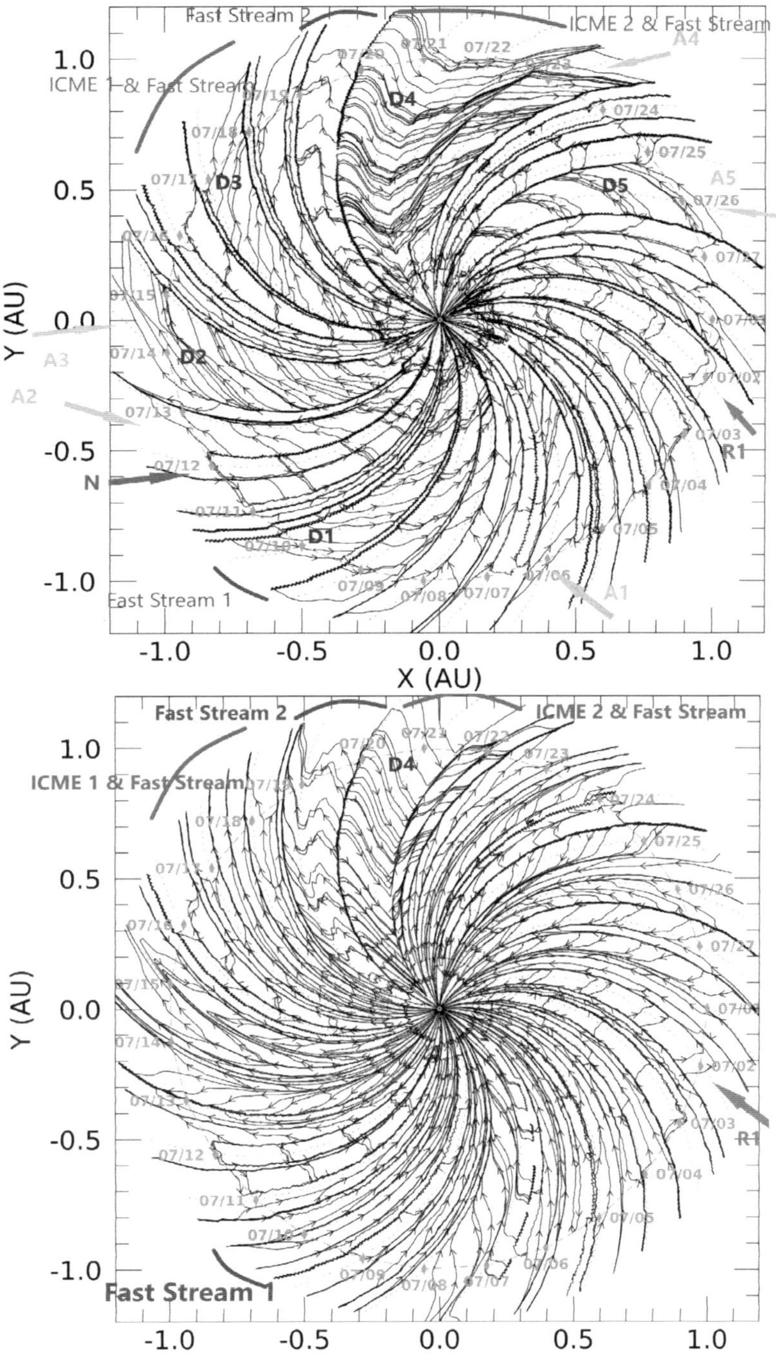

Figure 4. Predicted magnetic field line maps for the solar rotation period 1 July to 27 July 2002 (CR 1992) using the accelerating solar wind model of Tasnim et al. (2018) (top) and the constant solar wind model of Schulte in den Bäumen et al. (2012) (bottom).

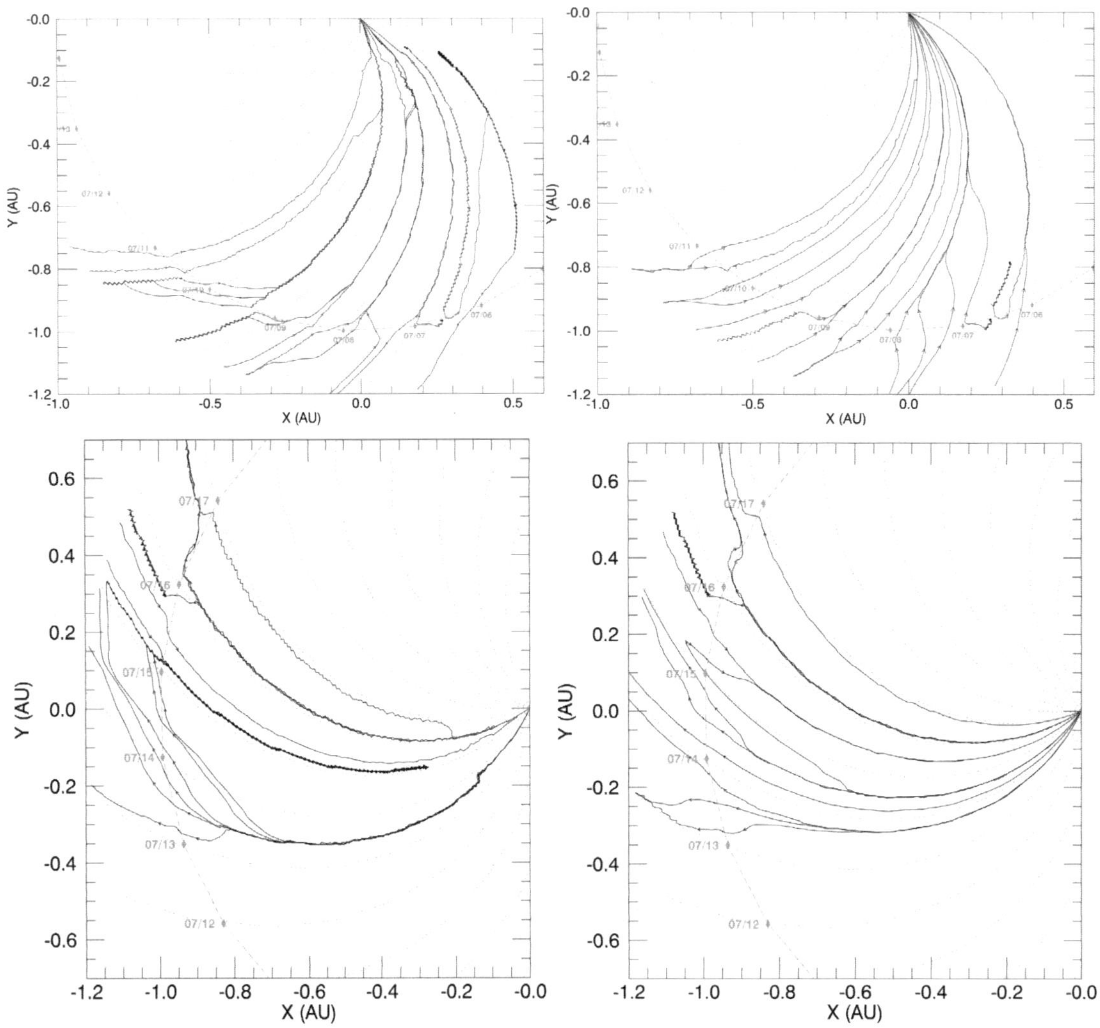

Figure 5. The predicted field lines between 6 to 11 July (top panels) and 13 to 17 July of 2002 (bottom panels) where the left two panels show the lines for the accelerating solar wind model and the right two panels shows the lines for a constant solar wind model.

Acknowledgments
G.P.Z. acknowledge the partial support of an NSF DOE grant PHY - 1707247 and a NASA grant SV4 - 84017. This material is based also in part upon work supported by the NSF EPSCoR RII-Track-1 Cooperative Agreement OIA-1655280. G.P.Z. acknowledges the generosity of the International Space Science Institute (ISSI) in supporting him through the 2017 Johannes Geiss Fellowship. This work is also partly supported by the International Space Science Institute (ISSI) in the framework of International Team 504 entitled Current Sheets, Turbulence, Structures and Particle Acceleration in the Heliosphere. We thank Jakobus le Roux for the useful discussion. The simulations are performed at Los Alamos National Laboratory.

References
Borovsky, J. E. (2010), 'On the variations of the solar wind magnetic field about the Parker spiral direction', *J. Geophys. Res.* **115**, A09101. doi: 10.1029/2009JA015040.

Feldman, W. C., Asbridge, J. R., Bame, S. J., Montgomery, M. & Gary, S. P. (1975), 'Solar wind electrons', *J. Geohys. Res.* **80**, 4181–4196. doi: 10.1029/JA080i031p04181.

Fisk, L. A. (1996), 'Motion of the footpoint of the heliospheric magnetic field lines at the sun: Implications for recurent energetic particle events at high heliographic latitude', *J. Geophys. Res.* **101**, 547. doi: 10.1029/96JA01005.

Forsyth, R. J., Balogh, A., Horbury, T. S., Erdos, G., Smith, E. J. & Burton, M. E. (1996), 'The heliospheric magnetic field at solar minimum: Ulysses observations from pole to pole', *Astron. Astrophys.* **316**, 287.

Gosling, J. T. & Roelof, E. C. (1974), 'A comment on the detection of closedmagnetic structures in the solar wind', *Sol. Phys.* **39**, 405408. doi: 10.1007/BF00162433.

Gosling, J. T. & Skoug, R. M. (2002), 'On the origin of radial magnetic fields in the heliosphere', *J. Geophys. Res.* **107(A10)**, 1327. doi: 10.1029/2002JA009434.

Hu, Y. Q. (1993), 'Evolution of corotating stream structures in the heliospheric equatorial plane', *J. Geophys. Res.* **98**, 13,201–13,214.

Hu, Y. Q. & Habbal, S. R. (1992), 'Double shock pairs in the solar wind', *J. Geophys. Res.* **98**, 3551.

Kutchko, F. J., Briggs, P. R. & Armstrong, T. P. (1982), 'The bidirectional particle event of October 12, 1977, possibly associated with a magnetic loop', *J. Geophys. Res.* **82**, 14191431. doi: 10.1029/JA087iA03p01419.

Li, B., Cairns, I. H., Gosling, J. T., Malaspina, D., Neudegg, D., Steward, G. & Lobzin, V. (2016b), 'Comparisons of mapped magnetic field lines with the source path of the 7 April 1995 type III solar radio burst', *J. Geophys. Res.* **121**, 61416156. doi: 10.1002/2016JA022756.

Li, B., Cairns, I. H., Gosling, J. T., Steward, G., Francis, M., Neudegg, D., in den Bumen, H. S., Player, P. R. & Milne, A. (2016a), 'Mapping magnetic field lines between the Sun and Earth', *J. Geophys. Res.* **121**, 925–948. doi: 10.1002/2015JA021853.

Li, B., Cairns, I. H., Owens, M. J., Neudegg, D., Lobzin, V. V. & Steward, G. (2016c), 'Magnetic field inversions at 1 AU: Comparisons between mapping predictions and observations', *J. Geophys. Res.* **121**, 1072810743. doi: 10.1002/2016JA023023.

Nolte, J. T. & Roelof, E. C. (1973*a*), 'Large-scale structure of the interplanetary medium i: High coronal source longitude of the quite-time solar wind', *Sol. Phys.* **33**, 241–257. doi: 10.1007/BF00152395.

Nolte, J. T. & Roelof, E. C. (1973*b*), 'Large-scale structure of the interplanetary medium II: Evolving magnetic configurations deduced from multi-spacecraft observations', *Sol. Phys.* **33**, 483–504. doi: 10.1007/BF00152435.

Odstřcil, D. (1994), 'Interactions of solar wind streams and related small structures', *J. Geophys. Res.* **99**, 17,653–17,671.

Owens, M. J. & Forsyth, R. J. (2013), 'The heliospheric magnetic field', *Living Rev. Solar Phys.* **10**, 5. doi: 10.12942/lrsp-2013-5.

Parker, E. N. (1958), 'Dynamics of the interplanetary gas and magnetic fields', *Z. Astrophys* **29**, 274. doi: 10.1086/146579.

Reiner, M. J., Fainberg, J. & Stone, R. G. (1995), 'Large-scale interplanetarymagnetic field configuration revealed by solar radio bursts', *Science* **270**, 461464. doi: 10.1126/science.270.5235.461.

Richardson, I. G. & Cane, H. V. (2015), 'Near-Earth interplanetary coronal mass ejections since january 1996', http://www.srl.caltech.edu/ACE/ASC/DATA/level3/icmetable2.htm. [Updated September 22, 2015].

Richardson, I. G., Cane, H. V. & von Rosenvinge, T. T. (1991), 'Prompt arrival of solar energetic particles from far eastern events The role of large-scale interplanetarymagnetic field structure', *J. Geophys. Res.* **96**, 78537786. doi: 10.1029/2011GL049578.

Richardson, J. D., Belcher, J. W., Lazarus, A. J., Paularena, K. I., Steinberg, J. T. & Gazis, P. R. (1996), 'Non-radial flows in the solar wind', *AIP Conf. Proc.* **382**, 479.

Rosenbauer *et al.*, H. (1977), 'A survey on initial results of the helios plasma experiment', *J. Geohys.* **42**, 561580.

Ruffolo *et al.*, D. (2006), 'Relativistic solar protons on 1989 October 22: Injection and transport along both legs of a closed interplanetary magnetic loop', *Astrophys. J.* **639**, 1186. doi: 10.1086/499419.

Schatten, K. H., Ness, N. F. & Wilcox, J. M. (1968), 'Influence of a solar active region on the interplanetary magnetic field', *Sol. Phys.* **5**, 240–256. doi: 10.1007/BF00147968.

Schulte in den Bäumen, H., Cairns, I. H. & Robinson, P. A. (2011), 'Modeling 1 AU solar wind observations to estimate azimuthal magnetic fields at the solar source surface', *Geophys. Res. Lett.* **38**, L24101. doi: 10.1029/2011GL049578.

Schulte in den Bäumen, H., Cairns, I. H. & Robinson, P. A. (2012), 'Nonzero azimuthal magnetic fields at the solar source surface: Extraction, model, and implications', *J. Geophys. Res.* **117**, 104. doi: 10.1029/2012JA017705.

Schwadron, N. A., Connick, D. E. & Smith, C. W. (2010), 'Magnetic flux balance in the heliosphere', *Astrophys. J. Lett.* **722**, L132. doi: 10.1088/2041-8205/722/2/L132.

Smith, E. J. (1979), 'Interplanetary magnetic fields', *Rev. Geophys* **17**, 610. doi: 10.1029/RG017i004p00610.

Suzuki, S. & Dulk, G. A. (1985), *Bursts of type III and type V, in Solar Radiophysics*, Cambridge (U. K.).

Tasnim, S. & Cairns, I. H. (2016), 'An equatorial solar wind model with angular momentum conservation and non-radial magnetic fields and flow velocities at an inner boundary', *J. Geophys. Res.* **121**, 4966–4984. doi: 10.1002/2016JA022725.

Tasnim, S., Cairns, I. H. & Wheatland, M. S. (2018), 'A generalized equatorial model for the accelerating solar wind', *J. Geophys. Res.* **123**, 1061–1085. doi: 10.1002/2017JA024532.

Journal of Physics: Conference Series **1332** (2019) 012016 doi:10.1088/1742-6596/1332/1/012016

Characteristics of a Langmuir Soliton Observed in a Solar Type III Burst

G. Thejappa[1], **R. J. MacDowall**[2]
[1] Department of Astronomy, University of Maryland, College Park, MD 20742, USA
[2] NASA/Goddard Space Flight Center, Greenbelt MD 20771, USA

Abstract. We present the high time resolution in situ observations of a Langmuir wave packet obtained by the STEREO WAVES experiment in the source region of a solar type III radio burst. This wave packet is unique in the sense that it occurs as an intense localized one-dimensional magnetic field aligned wave packet with peak intensity well in excess of the threshold for the oscillating two stream instability (OTSI), soliton formation and related strong turbulence processes. Most importantly, the measured half-width of this wave packet is approximately equal to the expected half-width of a Langmuir soliton with the same peak intensity as that of the wave packet. Furthermore, a density cavity is created by the ponder-motive force of this wave packet as expected of a Langmuir soliton. These findings indicate that the observed wave packet is most likely a Langmuir soliton formed as a result of the balance between the non-linearity related self-compression and the dispersion related spreading. We discuss the implication of these observations for theories of solar type III radio bursts.

1. Introduction

In a recent paper [14], we have reported the results of our search for Langmuir solitons in the high time resolution in situ wave data obtained by the STEREO WAVES experiment [2] in the source regions of 10 solar type III radio bursts. This search has yielded the identification of several localized magnetic field aligned one dimensional strongly turbulent wave packets as Langmuir solitons. In this paper, we describe the characteristics of one of those events.

In the subsonic limit, the electric field $E(Z,T)$ and the corresponding density perturbation $\frac{\delta n_e}{n_e}$ of the one dimensional Langmuir soliton can be described as [5]:

$$E(Z,t) = \frac{E_t}{\cosh[k_0(Z-ut)]}\exp[i(k_L Z - \omega t)] \quad (1)$$

$$\frac{\delta n_e}{n_e} = \frac{W_L}{n_e T_e} \quad (2)$$

Content from this work may be used under the terms of the Creative Commons Attribution 3.0 licence. Any further distribution of this work must maintain attribution to the author(s) and the title of the work, journal citation and DOI.
Published under licence by IOP Publishing Ltd

where Z is the longitudinal coordinate, t is the time, k_0 is the wave number and u is the velocity of the soliton. The frequency and wave numbers are defined as [5]

$$\omega = \omega_{pe} + \frac{3}{2}\lambda_{De}^2(k_L^2 - k_0^2)\omega_{pe} \tag{3}$$

$$k_0 = \lambda_{De}^{-1}(\frac{W_L}{6n_eT_e})^{1/2}. \tag{4}$$

Here $\frac{W_L}{n_eT_e} = \frac{\epsilon_0 E_t^2}{4n_eT_e}$ is the normalized peak energy density (E_t is the peak amplitude of the wave packet, ϵ_0 is the dielectric constant, n_e and T_e are the electron density and temperature, respectively), λ_{De} is the Debye length, k_L is the wave number of the Langmuir wave and $f_{pe} = \frac{\omega_{pe}}{2\pi}$ is the electron plasma frequency. The velocity of the soliton u is defined as

$$u = \frac{\partial\omega}{\partial k_L} = 3\omega_{pe}\lambda_{De}^2 k_L. \tag{5}$$

The half-power width L_E is related to the peak amplitude E_t of the soliton as

$$L_E \approx \frac{2.6}{k_0} \approx \lambda_{De}(\frac{40n_eT_e}{W_L})^{1/2}, \tag{6}$$

which is obtained from the relation

$$\frac{E}{E_t} = \frac{1}{\cosh(k_0 Z)} = \frac{1}{2}. \tag{7}$$

Thus, any observed localized wave packet of peak amplitude E_t can be identified as the Langmuir soliton, if its half-power width L_τ is comparable to L_E as given in equation (6).

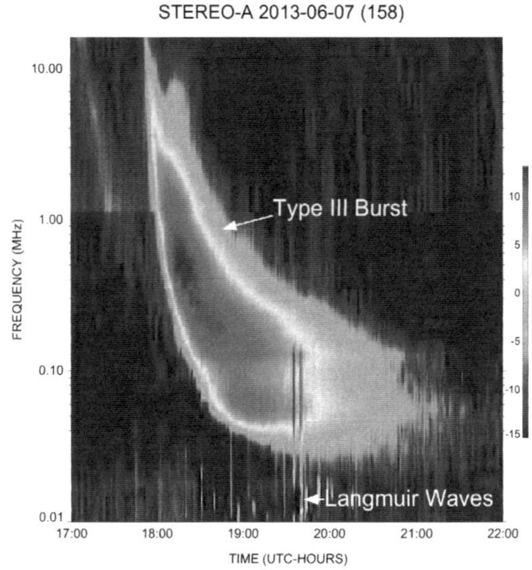

Figure 1. A local type III radio burst and associated Langmuir waves

2. Observations

In Figure 1, we present the dynamic spectrum of a local type III radio burst and its associated in situ waves, observed by the STEREO WAVES experiment [2]. The emission drifting fast from $\approx 14MHz$, all the way to $\approx$ 26 kHz is the type III radio burst. Using its fast negative frequency drift, we estimate the velocity of the electron beam responsible for this event as $v_b \approx 0.37c$ (c is the velocity of light). We identify the non-drifting emissions seen in the 11-15 kHz interval as the Langmuir waves. The time domain sampler (TDS) of the STEREO WAVES experiment has resolved these Langmuir waves into several wave packets. In the left panel of Figure 2, we present the X, Y and Z electric field components of one of such wave packets in the spacecraft frame. We have transformed these field components into the magnetic field aligned coordinate system, whose X-, Y- and Z-axes are aligned along the $\vec{b}$, $\vec{b} \times \vec{v}$,

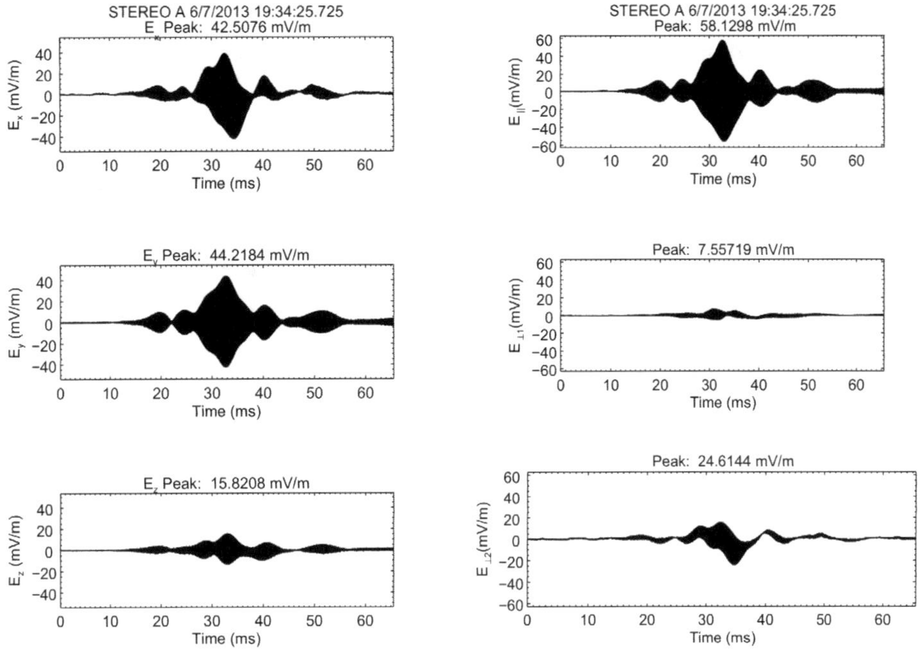

Figure 2. (left) The X- Y- and Z- components of one of the type III associated Langmuir wave packets in spacecraft frame, and (right) the parallel and perpendicular electric field components with respect to the magnetic field. The unit vectors of magnetic field and solar wind velocity used in the coordinate transformation are $\vec{b}$ = [0.43176, 0.81756, -0.37836] and $\vec{v}$ = [0.98834, -0.049404, 0.14406], respectively.

and $\vec{b} \times (\vec{v} \times \vec{b})$, respectively. Here $\vec{b}$ and $\vec{v}$ are the unit vectors of the magnetic field ($\vec{B}$) and solar wind velocity ($\vec{v_{sw}}$), respectively. These are provided by the STEREO IMPACT magnetic field [1] and STEREO PLASTIC [3] experiments, respectively. Here, we use $\vec{b}$ = [0.43176, 0.81756, -0.37836] and $\vec{v}$ = [0.98834, -0.049404, 0.14406] as given in aten.igpp.ucla.edu/forms/stereo/. In the right panel of Figure 2, we present the field components in the B-aligned coordinate system. Since the peak amplitude of the parallel component $\approx$ 58.1298 mVm^{-1} is larger than those of the perpendicular components $\approx$ 7.6 mVm^{-1} and $\approx$ 24.6 mVm^{-1}, we consider this wave packet as the one-dimensional field aligned wave packet. We have used TDS data to study non-linear processes in solar radio bursts [9, 10, 11, 12, 13, 14, 15]. Previously, we have used the Ulysses observations for similar studies [6, 7, 8].

3. Discussion

The threshold condition for the strong turbulence processes, such as the oscillating two stream instability (OTSI), soliton formation and spatial collapse is [16].

$$\frac{W_{\rm L}}{n_{\rm e}T_{\rm e}} > (k_L\lambda_{\rm De})^2. \tag{8}$$

If the wave packet corresponds to a Langmuir condensate formed as a result of weak turbulence processes, threshold for the supersonic collapse can be written as [16]

$$\frac{W_{\rm L}}{n_{\rm e}T_{\rm e}} \geq \frac{m_{\rm e}}{m_{\rm i}}, \tag{9}$$

where $m_{\rm e}$ and $m_{\rm i}$ are the electron and ion masses, respectively. If these conditions are satisfied, the weak turbulence approximation is not valid, and the Langmuir oscillations get trapped in the self-generated density cavities. As a result, the turbulence becomes highly inhomogeneous by consisting of localized collapsing soliton-caviton pairs. The energy transfer in this case occurs in the direction of higher wave numbers.

In Figure 3a, we plot the time profile of total electric field of the wave packet $E_t = \sqrt{E_X^2 + E_Y^2 + E_Z^2}$, which shows that it is a localized wave packet with peak intensity of $E_t \sim 60.5$ mVm^{-1}. Using this value, we can estimate its normalized peak energy density as $\frac{W_{\rm L}}{n_{\rm e}T_{\rm e}} = \frac{\epsilon_0 E_t^2}{4n_e T_e} \approx 3.5 \times 10^{-3}$. Here, for electron temperature ($T_{\rm e}$), we have assigned a typical value of 10^5 K, since the measurements of $T_{\rm e}$ are not available, and by assuming that the intense peak in the spectrum of the parallel component of the wave packet corresponds to electron plasma frequency, $f_{\rm pe} \approx 11.54$ kHz, we have estimated the electron density as $n_e \approx 1.64 \times 10^6$ m^{-3} using the relation $f_{\rm pe}[Hz] = 9n_{\rm e}^{1/2}$. These values yield Debye length, $\lambda_{\rm De} = 69T_{\rm e}^{1/2}n_{\rm e}^{-1/2} \approx 17$ m. As far as the wave number of Langmuir waves is concerned, we estimate it as $k_{\rm L} = \frac{2\pi f_{\rm pe}}{v_{\rm b}} \approx 6.5 \times 10^{-4}$ m^{-1}, and $(k_{\rm L}\lambda_{\rm De})^2 \approx 1.2 \times 10^{-4}$ for $v_{\rm b} \approx 0.37c$, and $\frac{m_e}{m_i} \sim 5.5 \times 10^{-4}$. Thus, the threshold conditions (8) and (9) are easily satisfied, which implies that the current wave packet is in the strong turbulence regime.

By assuming that the observed wave packet is the Langmuir soliton, we can estimate its expected half-width as $L_E = \lambda_{De}(\frac{40n_eT_e}{W_L})^{1/2} \sim 107\lambda_{De}$ for $\frac{W_{\rm L}}{n_{\rm e}T_{\rm e}} \approx 3.5 \times 10^{-3}$. As seen in Figure 3a, the half-power temporal width of the wave packet is $\tau_{0.5} \approx 8.08$ ms. By assuming that the observed wave packet is stationary in the solar wind, $\tau_{0.5}$ measured in the spacecraft frame can be translated into half-width L_τ of the wave packet in the solar wind frame using the relation

$$L_\tau = \tau_{0.5}v_{sw}\cos\theta. \tag{10}$$

The STEREO/PLASTIC experiment [3] has measured the solar wind speed as $v_{\rm sw} \approx 628.6$ kms^{-1}, and the angle between the solar wind and magnetic field θ as $\sim 70°$. These values yield the half-width of the wave packet as $L_\tau \approx \tau_{\rm e}v_{\rm sw}\cos\theta \approx 100\lambda_{\rm De}$. Thus, the measured half-width $L_\tau \approx 100\lambda_{De}$ is very close to the expected half-width $L_E = \lambda_{De}(\frac{40n_eT_e}{W_L})^{1/2} \sim 107\lambda_{De}$. This indicates that the observed wave packet most

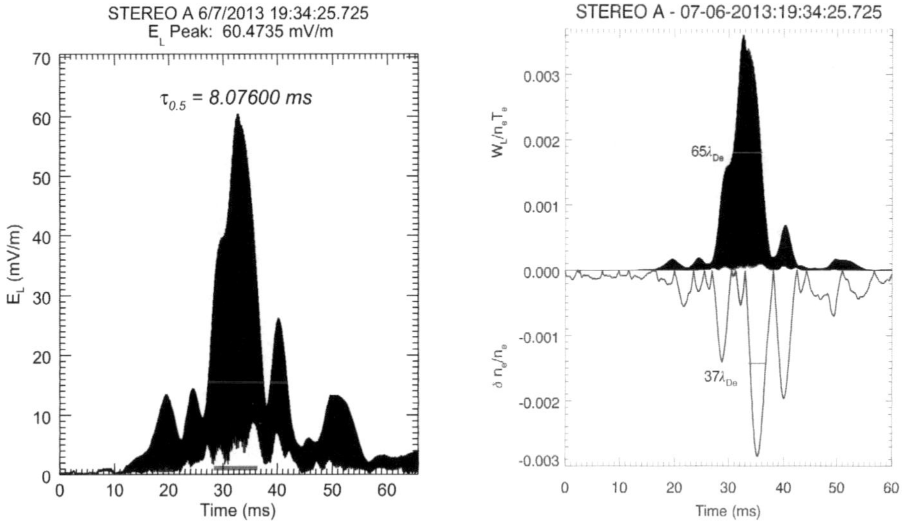

(a) (b)

Figure 3. (a) The time profile of the total electric field $E_t = \sqrt{E_X^2 + E_Y^2 + E_Z^2}$ (the $\frac{1}{2}$-power duration of 8.076 ms is equivalent to the spatial scale of $\sim 100\lambda_{De}$), and (b) the observed time profiles of the $\frac{W_L}{n_eT_e}$ and its associated density cavity $\frac{\delta n_e}{n_e}$, which show that $\frac{1}{e}$-power spatial scales are $\sim 65\lambda_{De}$ and $\sim 35\lambda_{De}$, respectively.

likely is the Langmuir soliton formed as a result of cancellation of non-linearity related compression by the dispersion related spreading. If this is true, the ponderomotive force of this wave packet should create a density cavity. As shown in equation (2) the depth of this cavity $\frac{\delta n_e}{n_e}$ should be equal to $\sim \frac{W_L}{n_eT_e}$.To verify this, in Figure 3b, we present the observed time profiles of the $\frac{W_L}{n_eT_e}$ and its associated density cavity $\frac{\delta n_e}{n_e}$. The density fluctuations are measured using the spacecraft potential as shown in [4, 14, 15]. Figure 3b shows that $\frac{1}{e}$-power spatial scales of the $\frac{W_L}{n_eT_e}$ wave packet and density cavity $\frac{\delta n_e}{n_e}$ are $\sim 65\lambda_{De}$ and $\sim 35\lambda_{De}$, respectively. Furthermore, as seen in Figure 3b, the $\frac{W_L}{n_eT_e} \sim 3.5 \times 10^{-3}$ agrees reasonable well with $\frac{\delta n_e}{n_e} \sim 2.8 \times 10^{-3}$. Thus, the observed density cavity is most likely created by the ponderomotive force of the observed wave packet. In [14], we have identified 17 Langmuir wave packets as Langmuir solitons. In Figure 4, we plot the measured half-widths (L_T) versus their expected half-widths (L_E), which shows an excellent correlation with correlation

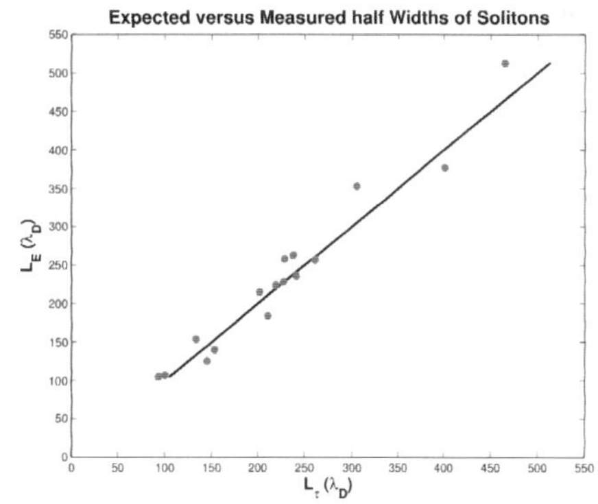

Figure 4. The measured versus expected widths of solitons

coefficient of ~ 0.98. This provides unambiguous evidence for Langmuir solitons in type III solar bursts.

4. Conclusions

We have presented the observations of an intense localized one dimensional magnetic field aligned Langmuir waveform obtained by the STEREO WAVES experiment in the source region of a local solar type III radio burst. The theories predict that the bump-on-tail distributions of electron beams propagating along the magnetic field excite one-dimensional field aligned Langmuir wave packets. The observations presented in this study represent one of such wave packets. Our analysis has revealed that (1) the peak intensity and short spatial scale of this wave packet easily satisfy the threshold condition for excitation of OTSI and related strong turbulence processes, and (2) the observed wave packet represents the Langmuir soliton formed as a result of OTSI. Thus, the observations presented in this study well confirm that the OTSI and related strong turbulence processes play significant roles in the stabilization of the electron beam as well as excitation of solar radio emissions at the fundamental and higher harmonics of the electron plasma frequency, f_{pe}.

5. Acknowledgment

The SWAVES instruments include contributions from the Observatoire of Paris, University of Minnesota, University of California, Berkeley, and NASA/GSFC. G. T. acknowledges the NASA STEREO and Wind projects for the financial support.

6. References

[1] Acuna, M. H. et al., 2008, *Space Sci. Rev.*, 136, 203
[2] Bougeret, J. L. et al. 2008, *Space Sci. Rev.*, **136**, 487
[3] Galvin, A. B. *et al.* 2008, *Space Sci. Rev.*, 136, 437
[4] Kellogg, P. J., Goetz, K., Monson, S. J., Bale, S. D., Reiner, M. J., and Maksimovic, M. 2009, *J. Geophys. Res.*, 114, A02107, doi:10.1029/2008JA013566
[5] Nezlin, M. V. 1993, (in) Physics of intense beams in plasmas, p. 186-249, Institute of Physics Publications, Philadelphia.
[6] Thejappa, G., Lengyel-Frey, D., Stone, R. G., and M. L. Goldstein, M. L. 1993, *Astrophys. J.*, 416, 831
[7] Thejappa, G., and MacDowall, R. J. 1998, *Astrophys. J.*, 498, 465
[8] Thejappa, G., Goldstein, M. L., MacDowall, R. J., Papadopoulos, K., and Stone R. G. 1999, *J. Geophys. Res.*, 104, 28279
[9] Thejappa, G., MacDowall, R. J., Bergamo, M., and Papadopoulos, K. 2012a, *Astrophys. J.* , 747, L1
[10] Thejappa, G., MacDowall, R. J., and Bergamo, M. 2012b, *Geophys. Res. Lett.*, 39, LXXXX, doi:10.1029/2012GL051017
[11] Thejappa, G., MacDowall, R. J., and Bergamo, M. 2012c, *J. Geophys. Res.* , 117, A08111, doi:10.1029/2012JA017695
[12] Thejappa, G., MacDowall, R. J., and Bergamo, M. 2013a, *Ann. Geophys.*, 31, 1417-1428
[13] Thejappa, G., MacDowall, R. J., and Bergamo, M. 2013b, *J. Geophys. Res.* , 118, 114, doi:10.1002/jgra.50441
[14] Thejappa, G., and MacDowall, R. J. 2018a, *Astrophys. J.*, 864:122, http://doi.org/10.3847/1538-4357/aad5e4
[15] Thejappa, G., and MacDowall, R. J. 2018b, *Astrophys. J.*, 862:75, https://doi.org/10.3847/1538-4357/aaca3b
[16] Zakharov, V. E. 1972, *Sov. Phys.-JETP*, 35, 908

Isotopic Fractionation in ^{3}He-rich SEP Events

M E Wiedenbeck[1], G M Mason[2], B Klecker[3]

[1] Jet Propulsion Laboratory, California Institute of Technology, Pasadena, CA 91109 USA
[2] Applied Physics Laboratory, Johns Hopkins University, Laurel, MD 20723 USA
[3] Max-Planck-Institut für extraterrestrische Physik, Garching, Germany

E-mail: mark.e.wiedenbeck@jpl.nasa.gov

Abstract. A well established characteristic of ^{3}He-rich solar energetic particle events is a pattern of abundance enhancements relative to normal solar system values that increase approximately monotonically with atomic number (or mass), leading to enhancements that can exceed 100× for the heaviest elements relative to oxygen. In a 2018 paper, Mason and Klecker (M&K) suggested that the heavy element enhancements could be due to characteristics of the stopping powers of low-energy nuclei penetrating thin layers of H gas or plasma. We present results of a new calculation of enhancements in a H gas and show them to be in reasonable agreement with the M&K results. Studies of heavy-element isotopic composition in ^{3}He-rich events using data from ACE/SIS have also shown that heavier isotopes of an element typically exhibit significant abundance enhancements relative to lighter isotopes of the same element. We applied the new calculation to investigate whether the observed isotopic fractionation can be explained using the M&K H-gas model with the same parameters that account for the heavy-element abundances and found that the measured isotopic enhancements are significantly greater than the predicted values. A search of a larger parameter space was also not able to reconcile the elemental and isotopic enhancement values. However, this search did reveal trends in the dependence on the model parameters that merit further investigation.

1. Introduction

Impulsive solar energetic particle (SEP) events are characterized by enhancements of ^{3}He/^{4}He by several orders of magnitude over ths solar wind value of $\sim 4 \times 10^{-4}$ [1] and a pattern of heavy element enhancements that increases approximately monotonically with the atomic number (or the mass) of the element, with Fe/O typically enhanced by $\sim$ 10× and with enhancements of the heaviest elements relative to O that can reach ~100× or more [2, 3]. Studies have shown [4] that the heavy element enhancements have little, if any, correlation with the enhancements of ^{3}He/^{4}He in the same events, leading to the strong suspicion that two different mechanisms are responsible. A number of authors argued that the ^{3}He enhancement results from the resonant interaction of this nuclide with some type of plasma waves that heat only this ion, which has a mass-to-charge ratio, M/Q, far from that of all other nuclides commonly observed in these events (e.g., [5]). The mechanism responsible for the heavy-element fractionation has lacked a compelling, quantitative explanation for more than four decades.

Mason & Klecker [6] (henceforth M&K) showed that the pattern of heavy element enhancements can be produced by a simple mechanism associated with the effective atomic charge that low-energy, heavy elements attain when slowing down in matter, taking into account both ionizing and non-ionizing energy losses. As a result of the relatively larger number of orbital

Content from this work may be used under the terms of the Creative Commons Attribution 3.0 licence. Any further distribution of this work must maintain attribution to the author(s) and the title of the work, journal citation and DOI.
Published under licence by IOP Publishing Ltd

electrons attached to an Fe nucleus than to an O nucleus at equal values of their energy per nucleon (E/M) below $\sim$1 MeV/nuc, Fe is more penetrating that O.

M&K showed that if power-law energy spectra of the elements are incident on a thin layer of hydrogen gas and only those particles that penetrate the layer are further accelerated, the layer can act as a filter that modifies the relative abundances of the elements. Values were found for the power-law spectral index, γ, and the material thickness ("grammage", defined as thickness $\times$ mass density) of the layer, L, such that this filter did a reasonable job of reproducing the observed pattern of heavy-element enhancements [2, 3], at least over the mass range $4 \leq M \lesssim 100$ (corresponding to elements with $2 \leq Z \lesssim 50$).

M&K also reported that the same mechanism operating in the more realistic environment of hot hydrogen plasma, such as would be encountered in the solar corona, can produce a similar enhancement pattern, albeit with different parameters and larger uncertainties in the calculation.

Measurements of heavy-element isotopic composition in ^{3}He-rich SEP events and their correlations with one another and with the Fe/O enhancement [7, 8, 9, 10] provide additional probes of the fractionation mechanism. Characteristically, the heavier isotopes of an element are enhanced relative to the lighter isotopes of that element. For the heavy isotope ratio, ^{22}Ne/^{20}Ne, enhancement factors exceeding $\sim 3\times$ are not unusual.

In this paper, we further test the M&K model for the case of a medium consisting of neutral H gas. In Section 2, we summarize the key features of the model and describe a new calculation. In Section 3, we apply the model to the calculation of the elemental abundance enhancement pattern using parameters favored by M&K and show that this new calculation does a reasonable job of reproducing results found by M&K. We then apply the model with these same parameters to the calculation of isotope ratio enhancements and compare the calculated values with measurements from ACE/SIS [9, 10]. We find that these parameters do not account for isotope ratio enhancements as large as have been observed. We then discuss the search of a larger range of parameter space to find parameters that can account for the isotopic and elemental enhancements. We show that more extreme power-law spectra are required to account for the observations of ^{22}Ne/^{20}Ne, particularly if the model is to simultaneously account for measured Fe/O values. In Section 4 we discuss the model predictions for the correlations between enhancements of ^{22}Ne/^{20}Ne and ^{26}Mg/^{24}Mg and between ^{22}Ne/^{20}Ne and Fe/O. We find that the enhancement factors for these ratio pairs are nearly proportional and are only weakly dependent on the actual values of the model parameters used to account for the observations. In Section 5 we discuss the need to extend the the isotope calculations to the case of energy loss in a hot plasma, as done by M&K for element enhancements, and the complications that need to be addressed to do so. In Section 6 we summarize the key results of this study.

2. Calculation of Abundance Ratio Enhancements

We used version 10.3 of the Geant4 simulation package[1] [11, and references therein] to replicate calculations reported by M&K, which were based on the widely used "The Stopping and Range of Ions in Matter" (SRIM) package[12]. Primary nuclei with specified atomic number (Z), mass number (M), and initial kinetic energy per nucleon (E/M) were injected at the center of a large volume of hydrogen gas at standard temperature and pressure. The software followed the energy loss and interactions (e.g, knock-on electron production) in a series of steps until each nucleus came to rest. The total distance travelled while slowing down and also the projection of that distance on the direction of the particle's incident velocity were recorded. For each primary beam, a large number of nuclei were followed in order to derive the range and projected-range distributions. Figure 1 shows the range–energy relations for ^{16}O and ^{56}Fe (righthand panel) and the corresponding dependence of the specific ionization per nucleon on E/M obtained by

[1] The Geant4 physics option `emstandard_opt3` was used.

differentiation (lefthand panel). The crossover at low energies, which is evident in both plots, is the basis for the fractionation effect proposed by M&K.

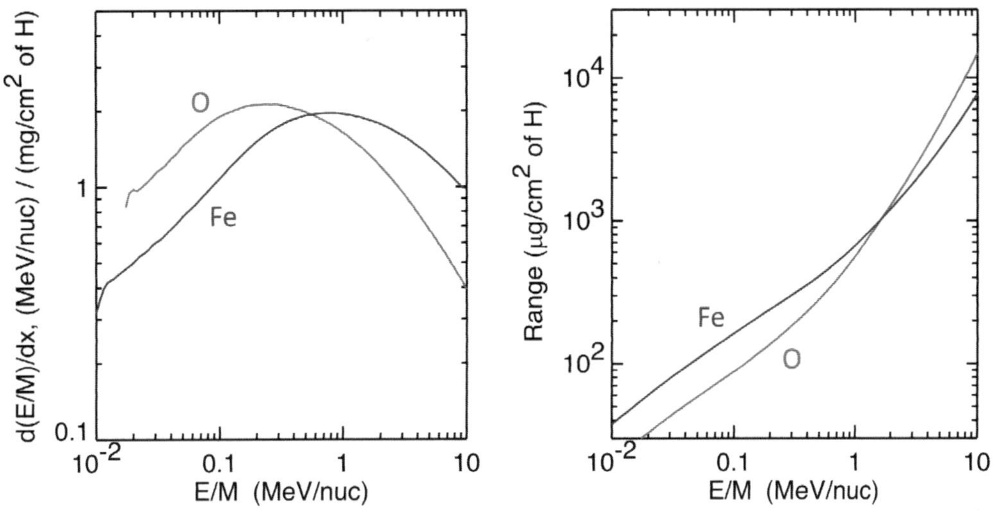

Figure 1. Dependence of $d(E/M)/dx$ (left) and R (right) on E/M obtained from the Geant4 model.

For selected nuclides, the upper panel of Figure 2 shows with solid bars the E/M intervals over which essentially all particles incident on a H gas layer of thickness $200\,\mu\text{g/cm}^2$ would penetrate the layer and with hatched bars the E/M intervals over which range straggling would cause only a fraction of the particles to penetrate. At still lower energies, all of the particles would stop in the layer. For comparison, the lower panel of Figure 2 illustrates power-law energy spectra that could be incident on the H gas layer for several values of the spectral index, γ. As a result of energy losses in the H layer, the energy spectra of the exiting particles (not shown) are significantly distorted from the incident spectra and extend all the way down to zero energy. However, if all of the transmitted particles subsequently undergo acceleration independent of the energies they had when exiting from the absorber, this distortion of the spectral shape will not affect the fractionation. This is the case considered by M&K and by us. On the other hand, if there is an energy threshold below which particles exiting the layer are not accelerated, then the shape would matter.

Comparing ^{56}Fe with ^{16}O, the partial-transmission regions are well separated from one another. For the power laws that are shown, the selection of only penetrating particles for acceleration results in a significant excess of Fe relative to O. Comparing the isotopes ^{20}Ne and ^{22}Ne, the partial-transmission regions are much closer together, so to obtain a large enhancement of ^{22}Ne/^{20}Ne requires a steep (soft) power law.

It is notable that for two isotopes of the same element, the heavier isotope should always have the longer range at any specified E/M. Thus, one expects that the fractionation will always favor the heavier isotopes. This is the pattern observed in the data [8, 9, 10].

When the incident power law is very steep, there can be a significant variation of the incident flux even over the rather narrow region of partial transmission. For this reason, we mapped out the partial-transmission region in considerable detail, finding the E/M values at which 1%, 2%, ..., 99% of the incident particles make it through the layer. Let $R_f^{Z,M}(E/M)$ denote the grammage for which a fraction f of the particles of a particular nuclide Z, M and incident energy E/M are transmitted. We simulated 1000 particles for each of a series of incident beams with

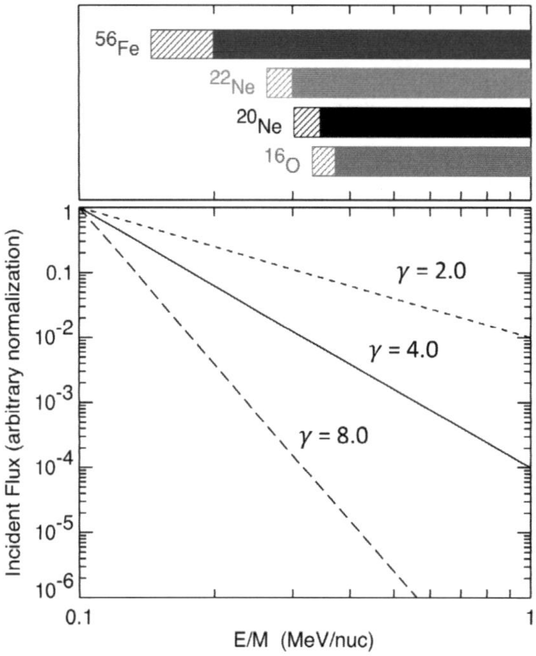

Figure 2. Upper panel: E/M intervals over which beams of selected nuclides penetrate a layer of H of grammage $200\,\mu\text{g/cm}^2$ either completely (solid section) or partially (hatched section). Lower panel: Example power-law incident spectra with selected values of the spectral index, γ.

300 logarithmically spaced E/M values per decade (step size of 0.77%) from 0.01 to 10 MeV/nuc and used the resulting projected ranges to determine $R_f^{Z,M}(E/M)$ for $f = 0$ to 1 in steps of 0.01 for each of the simulated E/M values. To find $R_f^{Z,M}(E/M)$ for values of E/M that were not simulated, we interpolated between adjacent E/M values at equal f.

The relative numbers of transmitted particles for two different nuclides with a given incident power-law spectrum, $(E/M)^{-\gamma}$, was calculated by adding the analytically calculated integral over the region for 100% transmission to the numerically calculated integral over the region of partial transmission and taking the ratio of the integrals for the two nuclides. This ratio gives the relative enhancement of the two nuclides. It is noteworthy that for steep incident spectra the population of transmitted particles is dominated by those with energies close to the threshold for penetration. Conclusions about the enhancements of abundance ratios should be applicable as long as the spectrum can be reasonably approximated by a power law between the penetration thresholds for the two species.

3. Elemental and Isotopic Enhancement Patterns

The lefthand panel of Figure 3 shows the pattern of elemental abundance enhancements that we obtained and compares them to measured values [2]. They are generally consistent with the calculations of M&K for the same parameters. The righthand panel of the figure shows our calculation of various isotope-ratio enhancements for the same set of parameters. The comparison with measurements from the 2002 Aug 20 ^{3}He-rich event [10] shows that the observed isotope enhancements significantly exceed those that would be expected from the model that reproduces the elemental values.

We examined a wider range of spectral indices for the same grammage ($200\,\mu\text{g/cm}^2$) to see if

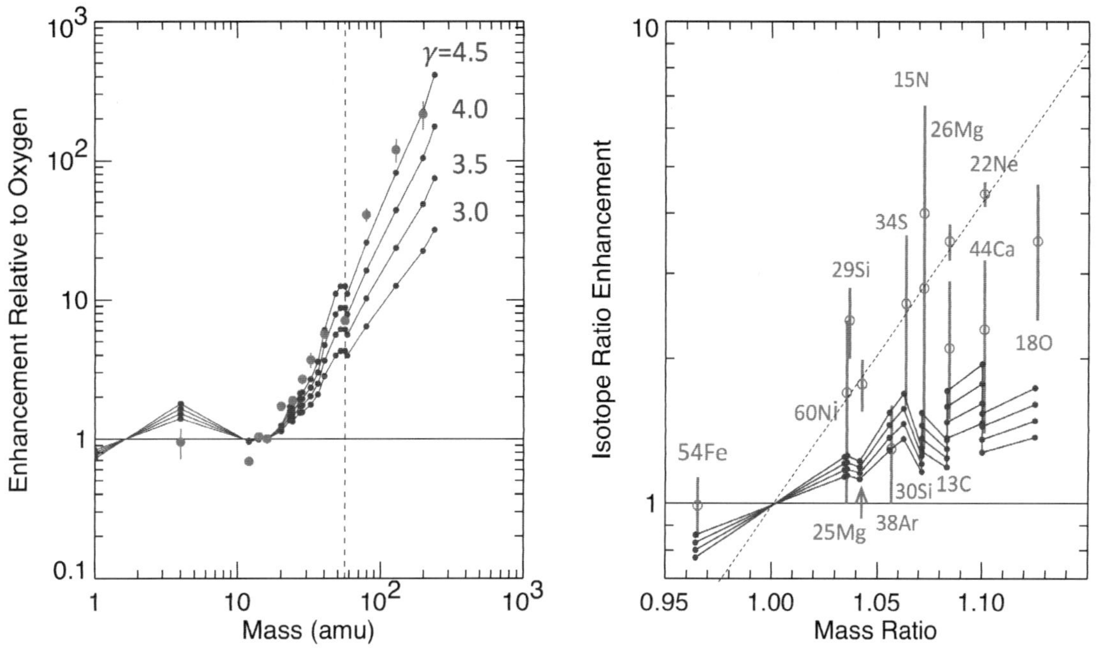

Figure 3. Left panel: Pattern of elemental abundance enhancements calculated for the grammage of 200 μg/cm^2 of H gas used by M&K and four values of γ, the index of the incident power-law spectrum (blue curves). Red points show measured values from [2]. Dashed vertical line indicates the Fe point. Right panel: Isotope ratio enhancements versus mass ratio. Blue curves calculated using the same model parameters as used for the left panel. Red points show measured values from [10] labeled with the nuclide in the numerator of the ratio. The denominator in each case is the most abundant isotope of the same element. Dashed line shows power-law fit to the isotope data from the 2002 Aug 20 event [10].

the isotope enhancements could be accounted for, independent of the elemental enhancements. The lefthand panel of Figure 4 shows the resulting comparison for a spectral index $\gamma = 9.0$. The model results account for the observations at least as well at the simple power-law dependence on the mass ratio that was presented by [10], which is shown by the dashed line in Figures 3 and 4. The deviations from a simple power law are an interesting feature of the model. However, it should be noted that in two cases where values are available for identical mass ratios (^{22}Ne/^{20}Ne and ^{44}Ca/^{40}Ca with a ratio of 1.100 and ^{13}C/^{12}C and ^{26}Mg/^{24}Mg with a ratio of 1.083), the isotope ratio that shows a larger enhancement in the data is predicted to have a smaller enhancement by the model. Thus, it is quite possible that these features in the data are fortuitous.

Using the Fe/O and ^{22}Ne/^{20}Ne ratios as indicative of the elemental and isotopic enhancements that can be obtained from the model, we carried out a more through investigation of the model predictions as a function of grammage and spectral index. The righthand panel of Figure 4 shows contours of constant enhancements of the two ratios as functions of γ versus L. The enhancements of these ratios have significant correlations, particularly for grammages $\lesssim$300 μg/cm^2, where the contours of constant Fe/O and ^{22}Ne/^{20}Ne are approximately parallel. For large values of the grammage, the contours of constant Fe/O become nearly vertical and are no longer parallel to the isotope contours. This suggests that it may be possible to account for the observed enhancements using $500 \lesssim L \lesssim 800$ μg/cm^2 and $\gamma > 10$. To date, this region of

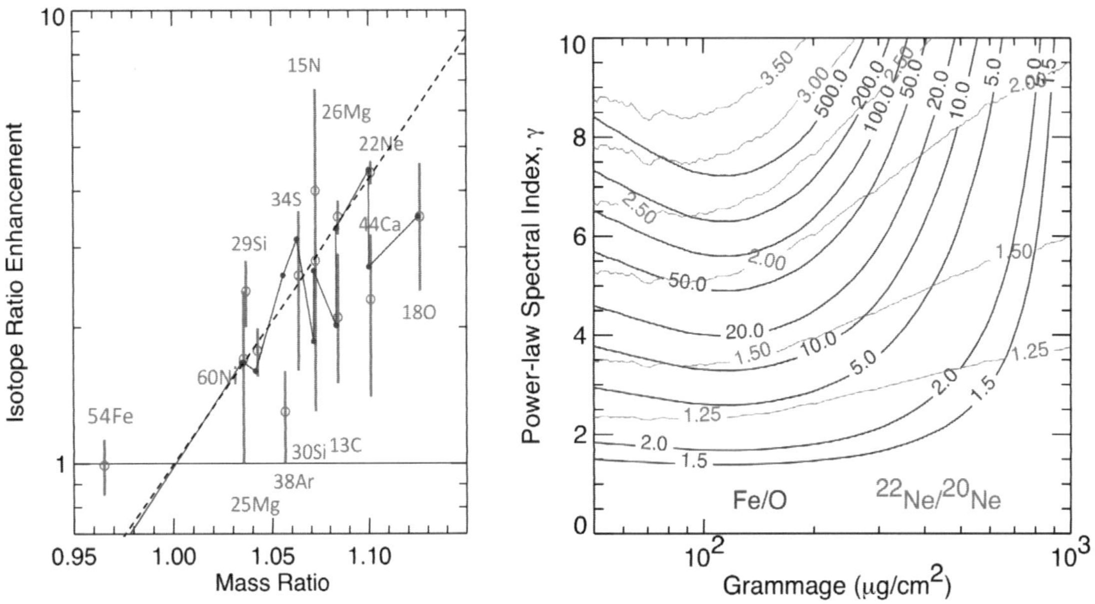

Figure 4. Left panel: Same as the righthand panel in Fig.3 except with $\gamma = 9.0$. Right panel: Contours of constant enhancement factors for the calculated ratios Fe/O (blue) and ^{22}Ne/^{20}Ne (red) as a function of grammage (L) and spectral index (γ).

parameter space has not been explored.

4. Enhancement Correlations

Strong correlations have been reported between the event-to-event variations of ^{22}Ne/^{20}Ne and ^{26}Mg/^{24}Mg [9] and also between ^{22}Ne/^{20}Ne and Fe/O [13]. We have investigated whether the model can account for these correlations. The isotope measurements from [9] are shown in the lefthand panel of Figure 5 and are compared with calculated values obtained from the model for a variety of different values of L (distinguished by line color) and γ (indicated by points along the lines). The dash-dot line shows the fit to the ^{3}He-rich SEP event data points from [9]. The model does a reasonable job of accounting for the correlation slope between the two isotope ratios, with only a weak dependence on the exact choice of the model parameters. The righthand panel of Figure 5 shows the ^{22}Ne/^{20}Ne enhancement versus the Fe/O enhancement for ^{3}He-rich events (yellow points) measured with ACE/SIS [13] and for a set of gradual SEP events [14, × symbols]. The slope of the correlation between ^{22}Ne/^{20}Ne and Fe/O that is predicted by the model is distinctly shallower than observed in the data, but is comparable to the slope found in the gradual events. In addition, as noted above, the largest values of the ^{22}Ne/^{20}Ne enhancement cannot be obtained using the parameters found by M&K to account for the pattern of elemental abundance enhancements. Further investigation of this correlation in the high-L, high-γ region of the contour plot in Figure 4 appears warranted.

5. Discussion

Although the model provides some interesting insights into the sort of fractionation effects that can occur when heavy nuclei penetrate a thin H gas layer before undergoing the final step in their acceleration, the model does not include a realistic description of the material that the pre-accelerated particles would encounter in the solar corona where temperatures greatly exceed the few $\times 10^4$ K above which H should be fully ionized. M&K discussed the expected effects on the elemental enhancements if the stopping medium is a hot H plasma rather than a neutral H gas. They found that qualitatively similar fractionation can occur as a result of the effective charge states acquired by heavy ions as they slow in the plasma, but with a smaller grammage required and with a Z dependence that is farther from being monotonic. There are significant uncertainties in the calculation of heavy-ion slowing in a hot plasma. In addition to the effects on the effective charge of the energetic ions, the energy loss rate for a given effective charge should be greater than in a gas because energy loss in distant collisions with free electrons can be significantly greater than if the electrons were bound in H atoms [15]. A proper treatment of the fractionation in a hot H plasma is beyond the scope of the present study. Thus the results presented here are, at best, an indication of the types of isotope ratio enhancements that might be expected from a more complete model.

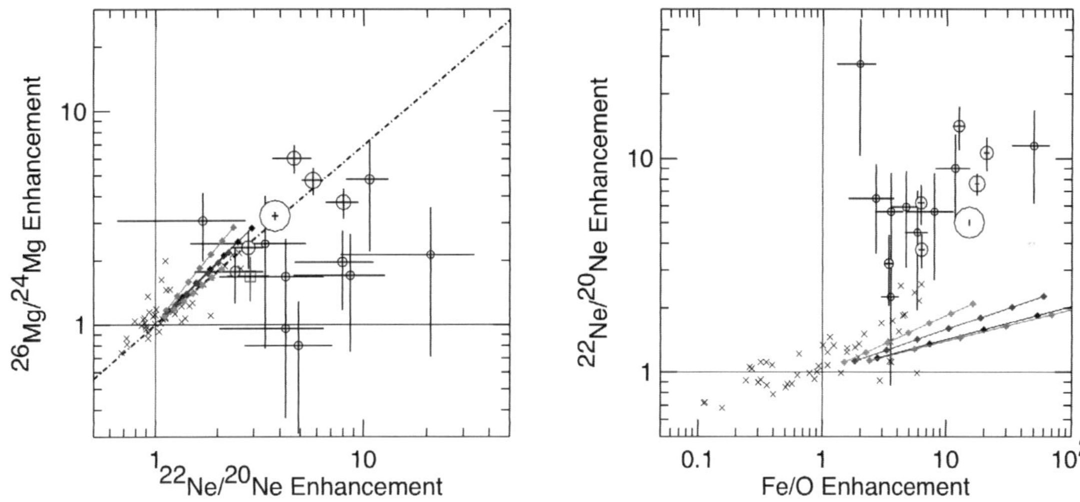

Figure 5. Left panel: Measured Ne and Mg isotope ratios reported in ^{3}He-rich SEP events [9] shown as yellow circles with areas inversely proportional to the measurement uncertainty to emphasize the highest-precision measurements. Yellow square shows the average enhancement in a number ^{3}He-rich events too small to include individually. Points indicated by $\times$ are from large, gradual SEP events [14]. Calculated enhancements are shown as diamonds for grammages of 100 (black), 200 (red), 300 (blue), and 400 μg/cm^2 (magenta) and γ=2, 3, ...8 with the enhancements increasing with increasing γ. Dash-dot line shows the correlation expected for fractionation proportional to a power of Q/M [16]. Right panel: Correlation between the Ne isotope ratio enhancement and the Fe/O enhancement. Symbols are the same as in the left panel.

6. Summary

We were able to successfully reproduce the M&K calculation of heavy element fractionation resulting from the dependence of specific ionization in a neutral H gas on atomic number and energy per nucleon. Carrying out the calculations using Geant4 and using the same grammage and spectral index of the incident power-law energy spectrum gave enhancements that are quantitatively very similar to those obtained by M&K. We then applied this calculation to investigate isotopic fractionation effects. To account for observed isotope ratio enhancements, we required significantly more extreme power-law spectra than obtained from the elemental enhancements. A more extensive search of the parameter space of grammage versus spectral index did not establish conditions under which observed enhancements of both Fe/O and ^{22}Ne/^{20}Ne could be obtained simultaneously using a neutral H gas as the stopping medium, but did identify a trend that suggests that further investigation is warranted of the region with $500 \lesssim L \lesssim 800\,\mu\text{g/cm}^2$ and $\gamma > 10$. The model did yield correlations between enhancements of different isotope ratios that are similar to those that have been measured, but did not reproduce the more extreme enhancement factors that have been observed. An improved assessment of the model will require an accurate treatment of heavy ion stopping in a hot, fully-ionized plasma.

Acknowledgments
This study benefitted from discussions within the International Space Science Institute (ISSI) team on "Origins of ^{3}He-Rich Solar Energetic Particles". We thank D. Low, N. Lal and D. Haggerty for advice on the use of Geant4. MEW's work was supported by the National Aeronautics and Space Administration at the Jet Propulsion Laboratory and the California Institute of Technology.

References

[1] Gloeckler G and Geiss J 1998 Measurement of the Abundance of Helium-3 in the sun and in the local interstellar cloud with SWICS on Ulysses *Space Sci. Rev.* **84**, 275
[2] Mason G M, Mazur J E, Dwyer J R, Jokipii J R, Gold R E, and Krimigis S M 2004 Abundances of heavy and ultraheavy ions in ^{3}He-rich solar flares *Astroph. J.* **606** 555
[3] Reames D V and Ng C K Heavy-element abundances in solar energetic particle events *Astroph. J.* **610** 510
[4] Dwyer J R, Mason G M, Mazur J E, Gold R E, Krimigis S M, Möbius E and Popecki M 2001 Isotopic composition of solar energetic particle events measured by Advanced Composition Explorer/ULEIS *Astroph. J.* **563** 403
[5] Fisk L A 1978 ^{3}He-rich flares: a possible explanation *Astroph. J.* **224** 1048
[6] Mason G M and Klecker B 2018 A possible mechanism for enriching heavy ions in ^{3}He-rich solar energetic particle events *Astroph. J.* **862** 7
[7] Cohen C M S, Mewaldt R A, Leske R A, Cummings A C, Stone E C, Wiedenbeck M E, Christian E R and von Rosenvinge T T 1999 New observations of heavy-ion-rich solar particle events from ACE *J. Geoph. Res.* **26** 2697
[8] Cohen C M S, Leske R A, Christian E R, Cummings A C, Mewaldt R A, Slocum P L, Stone E C, von Rosenvinge T T and Wiedenbeck M E 2000 The isotopic composition of solar energetic particles *Acceleration and Transport of Energetic Particles Observed in the Heliosphere* ed R A Mewaldt, J R Jokipii et al (Melville, NY: American Institute of Physics) CP528 55
[9] Wiedenbeck M E, Leske R A, Cohen C M S, Cummings A C, Mewaldt R A, Stone E C and von Rosenvinge T T 2008 Elemental and isotopic fractionation in ^{3}He-rich solar energetic particle events *Particle Acceleration and Transport in the Heliosphere and Beyond–7th Annual Astrophysics Conf.* ed G Li, Q Hu et al (Melville, NY: American Institute of Physics) CP1039 149
[10] Wiedenbeck M E, Cohen C M S, Leske R E, Mewaldt R M, Cummings A C, Stone E C and von Rosenvinge T T 2010 Heavy-ion fractionation in the impulsive solar energetic particle event of 2002 August 20: elements, isotopes, and inferred charge states *Astroph. J.* **719** 1212
[11] Allison J et al. 2016 Recent developments in Geant4 *Nucl. Instr. and Meth. in Phys. Res.* A **845** 186
[12] Ziegler J F, Biersack J P and Ziegler M D 2008 *SRIM–The Stopping and Range of Ions in Matter* (Morrisville, NC: Lulu Press)
[13] Wiedenbeck M E, Leske R A, Cohen C M S, Cummings A C, Mewaldt R A, Stone E C and von Rosenvinge T T 2008 Neon and magnesium isotopic composition in ^{3}He-rich solar energetic particle events *AGU Spring Meeting Abstracts* SH43C-06
[14] Leske R A, Mewaldt R A, Cohen C M S, Cummings A C, Stone E C, Wiedenbeck M E and von Rosenvinge T T 2007 Solar isotopic composition as determined using solar energetic particles *Space Sci. Rev.* **130**, 195
[15] Hoffmann D H H *et al.* 1994 Energy loss of fast heavy ions in plasmas *Nucl. Instr. & Meth. in Phys. Res.* B **90** 1
[16] Brenman H and Stone E C 1985 Solar coronal and photospheric abundances from solar energetic particle measurements *Astroph. J.* **299** L57

Journal of Physics: Conference Series **1332** (2019) 012018 doi:10.1088/1742-6596/1332/1/012018

Spreading protons in the heliosphere: a note on cross-field diffusion effects

N. Wijsen[1,2], A. Aran[2], J. Pomoell[3] and S. Poedts[1]

[1] Department of Mathematics/Centre for mathematical Plasma Astrophysics, KU Leuven, Belgium
[2] Departament Física Quàntica i Astrofísica, Institut de Ciències del Cosmos (ICCUB), Universitat de Barcelona (IEEC-UB), Spain
[3] Department of Physics, University of Helsinki, Finland

E-mail: nicolas.wijsen@kuleuven.be

Abstract. We study how a high-speed solar wind stream embedded in a slow solar wind affects the transport and energy changes of solar energetic protons in interplanetary space, assuming different levels of cross-field diffusion. This is done using a particle transport model that computes directional particle intensities and first order parallel anisotropies in a background solar wind generated by the magnetohydrodynamic model EUHFORIA. In particular, we consider a mono-energetic 4 MeV proton injection over an extended region located at a heliographic radial distance of 0.1 AU. By using different values for the perpendicular proton mean free path, we study how cross-field diffusion may affect the energetic particle spread and intensity profiles near a high-speed solar wind stream and a corotating interaction region (CIR). We find that both a strong cross-field diffusion and a solar wind rarefaction region are capable of dispersing SEPs efficiently, producing overall low particle intensities which can in some cases prevent the SEPs from being detected in-situ, since their intensity may drop below the detected pre-event intensity levels. We also discuss how accelerated particle populations form on the reverse and forward shock waves, separated by the stream interface inside the CIR. Under strong levels of cross-field diffusion, particles cross the SI and hence both accelerated particle populations merge together.

1. Introduction

Solar energetic particles (SEPs) are electrons, protons and heavy ions that have been accelerated to supra-thermal energies (up to a hundred of MeV or even $\sim$ GeVs, in the case of ions), typically during solar eruptive events. These charged particles may for example be energised at solar flare sites during magnetic reconnection events, or at shock waves driven by coronal mass ejections (CMEs). After escaping from their acceleration site, SEPs propagate through the inner heliosphere, guided by the interplanetary magnetic field (IMF). On their journey through space, a non-nominal solar wind as well as turbulent fluctuations may strongly affect the SEP trajectories, altering both the spatial and temporal dependencies of energetic particle distributions in the heliosphere. Non-nominal solar wind conditions may for example be caused by interplanetary CMEs, or by solar wind streams of different speeds interacting with each other. When a fast solar wind stream is followed by a slow stream, a rarefaction region forms between the two, whereas in the opposite case, the high-speed stream compresses the slow solar wind, which may eventually result in the formation of a corotating interaction region (CIR) [1]. CIRs may accelerate particles at large heliocentric radial distances, where one or both of the shock waves bounding the CIR are present, via diffusive shock acceleration [2], and at smaller radial distances due to the compressional waves bounding the transition region between the slow and the fast solar winds, e.g., [3, 4].

Content from this work may be used under the terms of the Creative Commons Attribution 3.0 licence. Any further distribution of this work must maintain attribution to the author(s) and the title of the work, journal citation and DOI.
Published under licence by IOP Publishing Ltd

To study the influence of a non-nominal IMF on SEP events, Wijsen et al. [5] and [6] used a three-dimensional particle transport model to propagate particles in a solar wind generated by the EUropean Heliospheric FORecasting Information Asset (EUHFORIA) [7], which is a three-dimensional magnetohydrodynamic (MHD) model of the solar wind and CME evolution in the inner heliosphere. Using EUHFORIA, these authors modelled a fast solar wind stream embedded in a slow solar wind, generating self-consistently a CIR from ~1.3 AU onward. By considering impulsive injections of 4 MeV protons from different extended injection regions located at 0.1 AU, Wijsen et al. [6] illustrated how the longitudinal width of the particle intensity distributions can strongly increase or decrease as a function of radial distance, depending on the underlying IMF topology. Moreover, by performing simulations both with and without cross-field diffusion, it was shown that cross-field diffusion widens the region where particles propagate, i.e., the particle streaming zone, yet it does not influence the qualitative behaviour of the longitudinal width as a function of radial distance [6]. **We want to point out that we leave it for future work to study the three-dimensional spread of particles in the early phase of SEP events, that is, where particles propagate closer to the Sun than our inner boundary at 0.1 AU. Both observational and modelling studies indicate that particles may be accelerated by shocks associated with CMEs already $< 10\,\mathbf{R}_\odot$ (e.g., [8, 9, 10, 11, 12, 13, 14, 15], and references therein). In addition, shock waves associated with CMEs contribute to the spread of SEPs in interplanetary space (e.g., [16]). Recently, Hu et al. [17] used the two-dimensional iPATH model ([18]) to model both diffusive shock acceleration and interplanetary transport of protons starting from $10\,\mathbf{R}_\odot$ during a gradual SEP event, as seen by virtual observers located at different radial distances in the solar equatorial plane, and covering a 60° azimuthal range. Assuming nominal solar wind conditions upstream of the CME driven shock wave, these authors used the Nonlinear Guiding Center (NLGC) theory [19, 20, 21] to illustrate how cross-field diffusion can alter the intensity-time profiles for 4.4 MeV and 46.4 MeV particles, especially for observers magnetically not well-connected with the travelling shock. The intensity profiles measured by these observers showed more prompt enhancements for simulations with larger cross-field diffusion coefficients.**

In this work, we extend the studies performed in [5] and [6] **by applying different cross-field diffusion conditions to the particles propagating in non-nominal solar wind conditions in the interplanetary space.** In particular, we investigate how a larger mean free path perpendicular to the magnetic field might alter the results obtained in these previous studies. After briefly introducing our energetic particle and solar wind models in Section 2, we show how the longitudinal width of the particle intensity distribution varies as a function of radial distance, time, and level of cross-field diffusion (Section 3). Next, we investigate how the different levels of cross-field diffusion used in our simulations affect the particle acceleration at the shock waves bounding the CIR (Section 4). We end in Section 5 with a summary and discussion of the obtained results.

2. Simulation of SEP transport within a CIR

To study SEPs in the inner heliosphere, we model the evolution of the directional particle intensity $j(\mathbf{x}, p, \mu, t)$ using the focused transport equation (FTE) (see e.g. [22, 23, 24]):

$$\begin{aligned}\frac{\partial j}{\partial t} + \frac{\partial}{\partial \mathbf{x}} \cdot \left[\left(\frac{d\mathbf{x}}{dt} + \frac{\partial}{\partial \mathbf{x}} \cdot \kappa_\perp\right) j\right] + \frac{\partial}{\partial \mu}\left[\left(\frac{d\mu}{dt} + \frac{\partial D_{\mu\mu}}{\partial \mu}\right) j\right] + \frac{\partial}{\partial p}\left(\frac{dp}{dt} j\right) \\ = \frac{\partial^2}{\partial \mu^2}\left[D_{\mu\mu} j\right] + \frac{\partial}{\partial \mathbf{x}} \cdot \left[\frac{\partial}{\partial \mathbf{x}} \cdot (\boldsymbol{\kappa}_\perp j)\right],\end{aligned} \tag{1}$$

with

$$\frac{d\mathbf{x}}{dt} = \mathbf{V}_{\rm sw} + \mu v \mathbf{b} + \mathbf{V}_d \tag{2}$$

$$\frac{d\mu}{dt} = \frac{1-\mu^2}{2}\left(v\nabla\cdot\mathbf{b} + \mu\nabla\cdot\mathbf{V}_{\rm sw} - 3\mu\mathbf{bb}:\nabla\mathbf{V}_{\rm sw} - \frac{2}{v}\mathbf{b}\cdot\frac{d\mathbf{V}_{\rm sw}}{dt}\right) \tag{3}$$

$$\frac{dp}{dt} = \left(\frac{1-3\mu^2}{2}(\mathbf{bb}:\nabla\mathbf{V}_{\rm sw}) - \frac{1-\mu^2}{2}\nabla\cdot\mathbf{V}_{\rm sw} - \frac{\mu}{v}\mathbf{b}\cdot\frac{d\mathbf{V}_{\rm sw}}{dt}\right)p. \tag{4}$$

In these equations, $\mathbf{x}$ denotes the phase-space spatial coordinate and t the time, both measured in an inertial frame, whereas the cosine of the pitch angle μ and the momentum magnitude p or speed v are expressed in a frame co-moving with the solar wind. Furthermore, $\mathbf{V}_{\rm sw}$ is the solar wind velocity, and $\mathbf{b}$ the unit vector in the direction of the mean magnetic field, $D_{\mu\mu}$ is the pitch-angle diffusion coefficient, $\kappa_\perp$ the spatial cross-field diffusion tensor, and $\mathbf{V}_d$ the drift velocity due to the gradient and curvature of the mean magnetic field. Our particle transport model solves Eq. (1) by integrating the equivalent set of Itô stochastic differential equations forward in time [25]. The solar wind velocity and magnetic field variables appearing in Eqs. (2)–(4) are obtained from the three-dimensional MHD model EUHFORIA. For details regarding the numerical procedures used by our model, we refer to [5].

As in [5] and [6], we use EUHFORIA to model a slow solar wind with an embedded fast solar wind stream. This is done by prescribing everywhere at the inner boundary of EUHFORIA a solar wind speed of 330 km/s, except for a region satisfying (longitude − 75°)^2 + (latitude − 5°)^2 < (20°)^2, where we prescribe a solar wind of speed 660 km/s. Moreover, this high-speed region is surrounded by a transition zone of width 6°. This transition eventually evolves into a CIR, bounded by reverse and forward MHD shock waves. As discussed in [5], the precise location where the large-amplitude compression waves bounding the transition region evolve into shock waves, and hence marking the formation of the CIR, is difficult to pinpoint. This is because the finite resolution of our MHD simulation produces inevitably shock waves that are wider than observed interplanetary CIR shocks. However, we observe in our particle simulations that particles start to get accelerated from a heliographic radial distance of ~1.3 AU onward, indicating that the large-amplitude compression waves have steepened sufficiently such that their characteristic length scale is considerably smaller than the particle mean free path that we assume (see below).

In contrast to the EUHFORIA simulation used in [5] and [6], we assume in the current study a monopolar positive magnetic field at the inner boundary of the model, such that there is no current sheet present in our solar wind. This is done to avoid any effects of the current sheet on the simulated particle intensities, which is necessary due to the larger spread of the particles in this work compared to the simulations performed in [5] and [6]. We want to note that the guiding centre approximation, used in Eq (1), breaks down near the current sheet [26]. Apart from the magnetic field polarity, all other MHD variables prescribed at the inner boundary of the model are the same as in [5], and hence we refer to that paper for further details.

The simulated source of particles is located at the inner boundary of the model, at a heliographic radial distance of 0.1 AU, and centred around the equatorial plane with a latitudinal width of 10°. The source has a longitudinal width of 30° such that it covers a region extending from the slow to the fast solar wind. As a consequence, the particle injection region is magnetically connected to both the forward and reverse compression/shock waves, hence also encompassing the SI inside the CIR, where the shocked slow and fast solar wind plasma meet. We note that the injection region considered in this work is identical to the one of [5] and to case 4 of [6] (see Fig 1 of [6]). Likewise, we also consider a mono-energetic injection of 4 MeV protons, allowing us to easily track the energy changes of the particles during their propagation in the solar wind. Unlike in [5] and [6], in order to mimic a prompt injection of particles, we inject the protons according to a Reid-Axford time profile [27]:

$$j(t) \propto \frac{1}{t}\exp\left(-\frac{\beta}{t} - \frac{t}{\tau}\right),$$

where we choose the parameters $\beta = 1.5$ hours and $\tau = 0.2$ hours (like in [28]).

The particle scattering conditions are determined by the pitch-angle diffusion coefficient $D_{\mu\mu}$ and the cross field diffusion tensor $\boldsymbol{\kappa}_{\perp}$. As detailed in [5], we use quasi-linear theory to prescribe the functional form of $D_{\mu\mu}$. The strength of the magnetic turbulence acting as pitch-angle scattering centres is quantified by assuming a constant radial mean free path $\lambda_r^{\parallel} = 0.3$ AU for 4 MeV protons. For the cross-field diffusion tensor we assume an expression similar to the one used by e.g. [29, 30]:

$$\boldsymbol{\kappa}_{\perp} = \alpha \frac{\pi}{12} v \lambda_{\parallel} \frac{B_0}{B} \left(\mathbb{I} - \mathbf{bb}\right), \tag{5}$$

where $\mathbb{I}$ is the unit tensor, **bb** the dyadic product of the magnetic unit vector, and $\lambda_{\parallel}$ the parallel mean free path. B_0 is a reference magnetic field strength value fixed to the maximum value of B at 1 AU, $B_0 = 9.7$ nT, which is attained in the shocked slow solar wind, adjacent to the SI.

We note that we used rather simple functional forms for the pitch-angle and cross-field diffusion coefficients. However, the significant uncertainties regarding the turbulent conditions and hence the particle diffusion properties in large scale structures like CIRs, make it difficult to establish more realistic assumptions for the particle diffusion coefficients [31, 32, 33]. Nonetheless, in Wijsen et al. [5, 6] it is demonstrated how our choice for the pitch-angle and the cross-field diffusion coefficients can reproduce different features of the energetic particle intensity profiles associated with CIRs, like e.g. the particle acceleration near the shock waves and the intensity dip near the SI. This is also discussed in more detail in Section 5. We remark that in reality the parallel and perpendicular mean free path of the particles might vary rather strongly across the CIR due to the presence of different turbulent regimes [31]. The increase in turbulence near the CIR shocks (see e.g, [34] and Section 5 for a discussion) would most likely decrease the parallel mean free path and amplify some of the effects that we describe in this work. A smaller radial mean free path would e.g., imply that the particles will scatter across the CIR shock waves more often, resulting in more particle acceleration. On the other hand, using a large constant radial mean free path everywhere in our simulations allows us to more easily trace the effects of the underlying non-nominal IMF topology and the different levels of cross-field diffusion. In addition, we note that more advanced theories of the cross-field diffusion, like e.g., NLGC theory, require the specification of the magnetic turbulence level δB in the heliosphere. For nominal solar wind conditions, one typically assumes that this quantity decreases as a power-law with radial distance from the Sun (see e.g., [35, 36, 18]). However, such a singular dependence on the radial coordinate is not valid in a solar wind containing a CIR, as the turbulent conditions in a CIR are significantly different from those encountered in the fast or slow solar wind (see e.g., [31, 32, 33]). Therefore, since the behaviour of δB in a CIR is not well understood, we instead adopt the phenomenological form (5) **for the cross-field diffusion, as its effects are more easily traceable in the simulations.**

By defining the perpendicular mean free path as $\lambda_{\perp} = 3\kappa_{\perp}/v$, the parameter α determines the ratio $\lambda_{\perp}/\lambda_{\parallel}$ at 1 AU up to a factor of order unity. We consider three different values for the parameter α, namely 2×10^{-4}, 2×10^{-3}, and 2×10^{-2}. In the remainder of the text, we will refer to the simulations performed with these different values of α as case 1, case 2 and case 3, respectively. These values of α help us to classify the 'level' or 'strength' of the cross-field diffusion employed in our simulations. In this way, the case with the strongest cross-field diffusion corresponds to case 3 ($\alpha = 2 \times 10^{-2}$).

3. Evolution of the particle spread in the solar equatorial plane

In this section, we analyse how the different levels of cross-field diffusion affect the longitudinal width of the particle intensity distributions in the solar equatorial plane. For this purpose, we define the particle streaming zone as the region where the omni-directional intensity, integrated over energy, is larger than the limiting value $10^{-8.5}$ (in arbitrary units). This threshold is set three orders of magnitude below the peak-intensity attained at 1 AU for a simulation without cross-field diffusion. Defining the longitudinal coordinate φ in anticlockwise direction, we obtain at each heliographic radial distance, r, and time, t, the

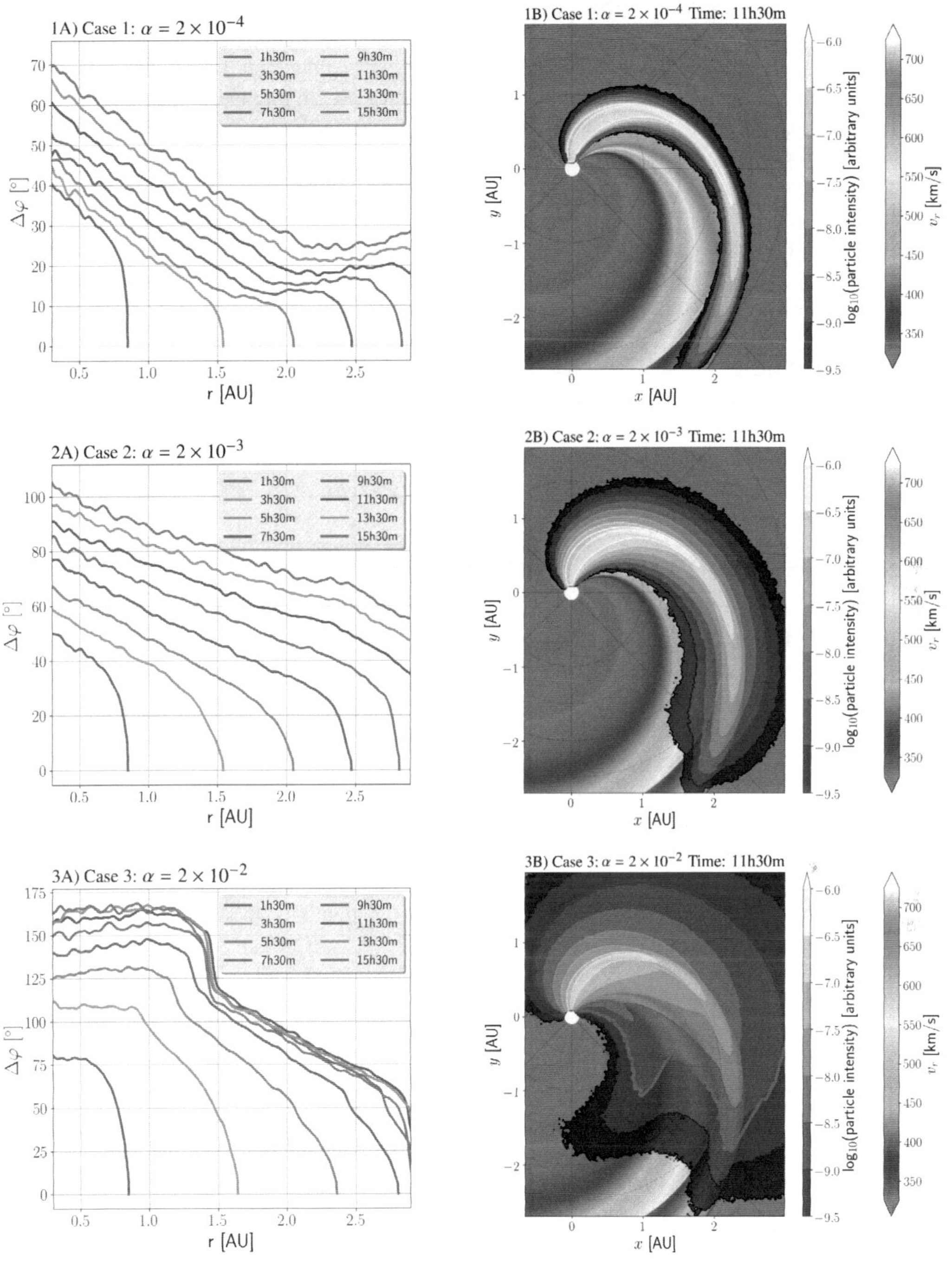

Figure 1. Column A (left column) shows the longitudinal width of the omni-directional intensity distribution as a function of radial distance for the different indicated times. Column B (right column) shows a snapshot at 11.5 hours of the omni-directional particle intensities, drawn in grey shades on top of the radial velocity profile of the solar wind. The green and blue curves represent, respectively, the lower and upper longitudinal boundaries of the streaming zone, as determined by the prescribed intensity threshold value $10^{-8.5}$ (arbitrary units). The first row is for case 1, the second row for case 2, and the third row for case 3.

lower, $\varphi_L(t,r)$, and upper, $\varphi_U(t,r)$, edge longitudes of the particle streaming zone, such that we can define the longitudinal width of the particle streaming zone as $\Delta\varphi(r,t) = \varphi_U(t,r) - \varphi_L(t,r)$. The left column of Fig. 1 shows for the three cases the radial dependence of $\Delta\varphi(r,t)$ at different times. The right column of Fig. 1 shows for each case the corresponding snapshots of the particle intensities at 11.5 hours. In these figures, the lower longitudinal boundary is indicated by a green line, whereas the upper longitudinal boundary is shown in blue.

From the first column of Fig. 1, we see that for all cases, $\Delta\varphi(r,t)$ generally decreases with increasing radial distance, reflecting the IMF compression at the shock waves bounding the CIR (see also Fig. 6 of [5]). However, apart from this common decreasing trend, there are several differences between the radial evolution of $\Delta\phi$ when comparing the different cases.

For case 1, panel 1A shows that at large radial distances ($r \gtrsim 2$ AU), the width of the particle streaming zone remains approximately constant and even starts to increase slightly. At these distances, the particle streaming zone is located entirely within the CIR, meaning that all the IMF lines on which particles are travelling have crossed the shock waves at distances $r < 2$ AU. In contrast, we see in panel 2A that for case 2 the longitudinal width of the particle streaming zone does not become constant from $r = 2$ AU onward. This is because the stronger cross-field diffusion has brought particles to IMF lines that cross the shock waves at radial distances larger than 2 AU, and hence, contrary to case 1, there are still particles outside the CIR at these large radial distances. However, since the forward and reverse shock waves propagate into the slow and fast solar wind, respectively, we expect that from a certain radial distance onward, all particles will again reside inside the CIR, and hence the behaviour of the longitudinal width will probably become qualitatively similar to what is observed for case 1.

Panel 3A shows that the strong cross-field diffusion of case 3 introduces several features to the radial dependence of the longitudinal width that are not seen in the two other cases. At small radial distances, the width of the particle streaming zone remains approximately constant, and even increases slightly, which is because some particles have been transported to the unperturbed slow solar wind behind the high-speed stream. However, panel 3A shows that the constant longitudinal width of the particle intensity distribution is followed by a sudden drop at larger radial distances. For times later than 9.5 hours, this drop occurs at approximately the same radial distance, $r = 1.4$ AU. Looking at panel 3B, we see that this decrease is because the omni-directional intensity drops below the $10^{-8.5}$ threshold in the rarefaction region behind the high-speed solar wind stream (the pronounced indent in the green intensity threshold contour in Figure 3B). This is a result of the strong expansion of magnetic flux tubes inside the rarefaction region, which causes the energetic particle density to get rarefied as well. In addition, the large cross-field diffusion for case 3 spreads the energetic particles over a much larger region than for the two other cases, hence rarefying the particle densities everywhere in the heliosphere. This is especially true in the fast solar wind stream and the rarefaction region, since the low magnetic field strength implies a more efficient cross-field diffusion (see Eq. (5)). As a consequence, the intensities for case 3 are everywhere relatively low as compared to the previous cases. These low intensities might imply that a large part of the SEP event may go undetected if the pre-event intensity levels are high enough or the measuring instrument is not sensitive enough. In fact, such a situation is illustrated in panel 3B, where we see that a large region is characterised by particle intensities below the $10^{-8.5}$ threshold, which is therefore not taken into account when determining the width of the intensity distributions. The low intensity values due to the efficient particle spreading also explains why in panel 3A the longitudinal width does not increase any further after 9.5 hours, and why for a radial distance larger than 1.5 AU, the longitudinal width at time 15.5 hours is even smaller than at the three previous time instances.

From the first column of Fig. 1, we note that for all three cases the longitudinal width of the particle streaming zone at small radial distances and at 1.5 hours is already significantly larger than the longitudinal width of the injection region, which was 30°. A rapid spread of particles at early times can be understood by considering the diffusive cross-field flux, which can, for small radial distances and in the solar equatorial plane, be approximated as in [37]:

$$\left|\mathcal{F}_{\perp}^{\mathrm{dif}}\right| \sim \kappa_{\perp}\left|\frac{\partial j}{\partial \varphi}\right|.$$

From this expression, we see that the cross-field diffusion will be more efficient when large longitudinal gradients are present in the directional intensity function. Since our particles are injected with a uniform intensity from a source region of limited spatial extend, we have in fact that, upon injection, $|\partial j/\partial \varphi| \longrightarrow +\infty$ at the boundaries of our injection region in the solar equatorial plane. We note that this infinite derivative does not produce numerical problems due to the stochastic nature of our model. However, it explains the efficient cross-field diffusion early on in the event (see also [37] for a discussion).

Moreover, we note that the above argument is more generally valid at any location in the heliosphere when considering the gradient of the intensity in the direction perpendicular to the magnetic field. More specifically, for an isotropic $\kappa_{\perp}$, the diffusive cross-field flux in the direction of a unit vector $\mathbf{e}_{\perp}$, with $\mathbf{e}_{\perp} \perp \mathbf{b}$, can be defined as

$$\left|\mathcal{F}_{\perp}^{\mathrm{dif}}\right| \sim \kappa_{\perp}\left|\mathbf{e}_{\perp} \cdot \nabla j\right|. \tag{6}$$

Hence, cross-field diffusion will be most effective where $\kappa_{\perp}$ is large or where there are large particle intensity gradients perpendicular to the magnetic field. Such large gradients are induced by the IMF topology inside the CIR, where the magnetic field is compressed. As discussed above, this compression of the magnetic field lines leads to a compression of the particle intensity distribution, and hence a decreasing longitudinal width. However, the same compression will also produce larger particle intensity gradients perpendicular to the magnetic field, hence increasing the effectiveness of cross-field diffusion according to Eq. (6). At the same time, we note that the inverse scaling of our cross-field diffusion with the magnetic field strength implies that the perpendicular mean free path decreases inside the CIR. However, as discussed in [5] and [6], a small cross-field motion of the particles inside the CIR where the IMF lines are closely compressed may transport the particles on field lines that are significantly separated outside the CIR, i.e., at smaller radial distances. The observed cross-field diffusion in our simulations results thus from a non-trivial interplay between the magnetic field dependence of the cross-field diffusion tensor and the intensity gradients perpendicular to the magnetic field. Both mechanisms are strongly influenced by non-nominal solar wind conditions.

Finally, we remark that by adding the cross-field diffusion term in the spatial part of the FTE, i.e., in Eq. (2), the energetic particles are given an additional velocity component on top of their actual velocity v. As a consequence, particles with $\mu \sim 1$ might spread faster than their speed allows for, hence violating causality [38]. Following a similar procedure as in [38] to estimate the impact of this effect in our simulations, we found it to be negligible since the intensities resulting from such particles are more than two orders of magnitude below the selected threshold value. This non-physical behaviour is a consequence of treating the motion of particles perpendicular to the mean magnetic field as a spatial diffusion process, and would not appear if one would instead e.g., model cross-field motions by stochastically varying only the propagation direction of the particles. Such an approach boils down to varying the magnetic unit vector, which is similar to the method implemented by Laitinen et al. [39] to model the motion of particles along random-walking magnetic field lines.

4. The effect of cross-field diffusion on shock accelerated particle populations

We now examine the particle intensity and anisotropy profiles in the solar equatorial plane at a fixed heliographic radial distance of 1 AU. We focus hereby on energy channels above the initial 4 MeV injection energy of the particles, meaning that we only consider particles that have been accelerated at the shock waves bounding the CIR. **We note that the acceleration of the protons and ions in shock waves depends on several factors, including the scattering conditions (e.g., [40, 41, 42, 43]), the shock strength (e.g., [44, 45, 46, 47, 48]), and the shock obliquity (e.g., [49, 50, 51]), among others. In Section 5 we provide a discussion on these factors in the context of CIR shocks, and how they influence the results presented in this section.**

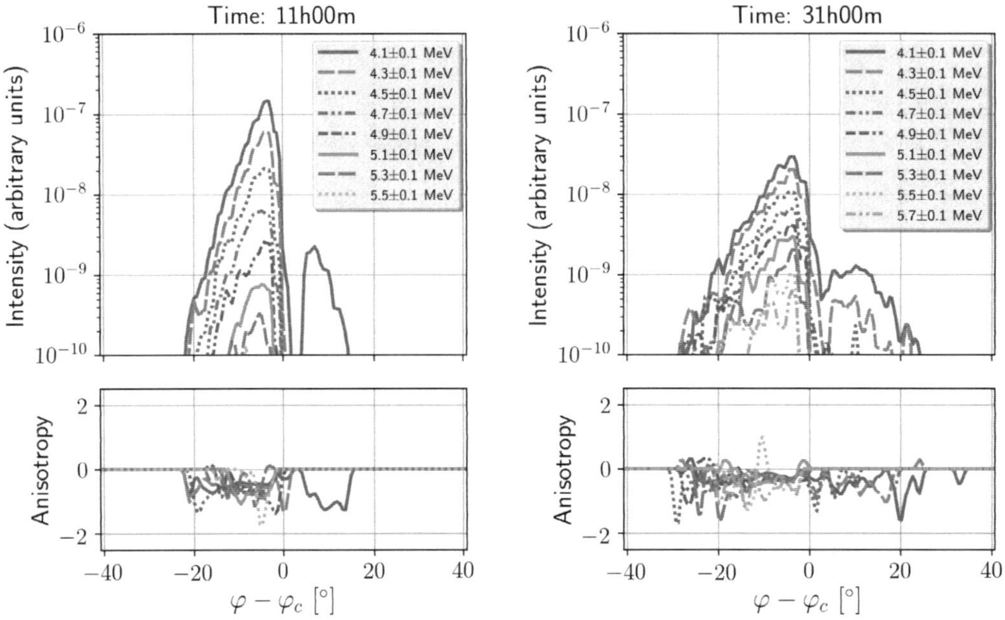

Figure 2. Snapshot of the proton intensities (upper panels) and anisotropies (lower panels) for case 1 at 11.0 hours (left panels) and 31.0 hours (right panels). The profiles are measured in the solar equatorial plane at a fixed heliocentric radial distance of 1 AU, and shown as a function of the longitudinal separation from the centre of the injection region.

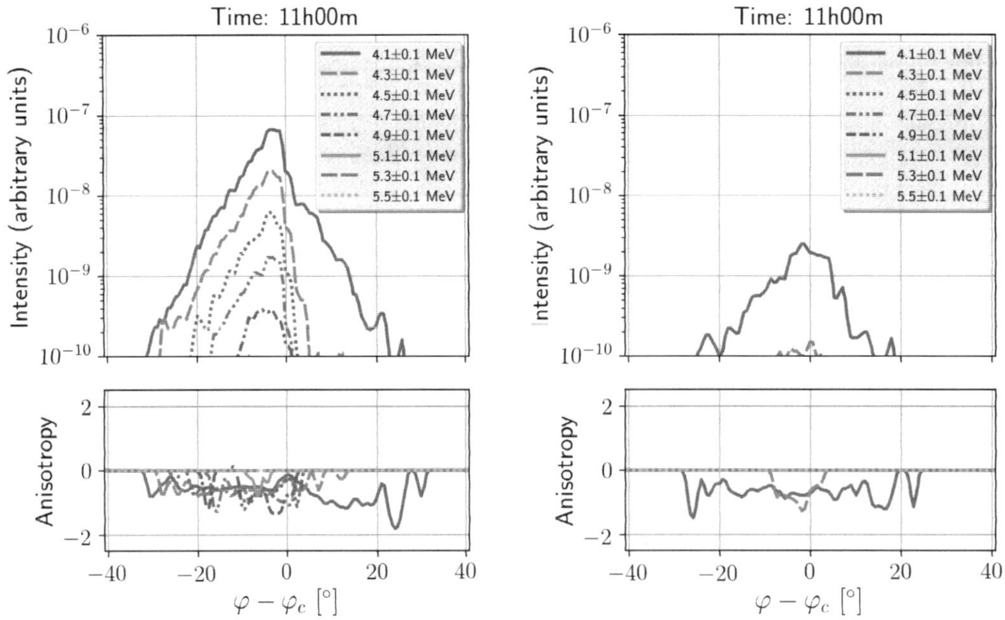

Figure 3. Snapshot of the proton intensities (upper panels) and anisotropies (lower panels) at time 11.0 hours for case 2 (left panel) and case 3 (right panel). The profiles are measured in the solar equatorial plane at a fixed heliocentric radial distance of 1 AU, and shown as a function of the longitudinal separation from the centre of the injection region.

Figure 2 shows two snapshots, at 11.0 hours (left) and at 31.0 hours (right), of the intensity and anisotropy profiles for case 1, as a function of $\varphi - \varphi_c$, where φ_c denotes the longitude of the point at 1 AU that connects magnetically to the centre of the particle injection region at 0.1 AU. The left panel of Fig. 2 shows two distinct intensity enhancements in the 4.1 ± 0.1 MeV energy channel, resulting from particle acceleration at the two shock waves bounding the CIR. The intensity enhancement at $\varphi - \varphi_c < 0$ is produced by particle acceleration at the reverse shock, whereas the intensity enhancement at $\varphi - \varphi_c > 0$ is produced by particle acceleration at the forward shock. The negative anisotropies in the lower panels of Fig. 2 indicate that the acceleration of the particles occurs at radial distances larger than 1 AU. This is in accordance with the findings of Wijsen et al. [5], where we observed the generation of accelerated particles at the shocks for $r \gtrsim 1.3$ AU. In our simulations, the reverse shock is a considerably more efficient particle accelerator than the forward shock. For the former, we see that after 11 hours there are particles accelerated up to ~1.0 MeV above their initial energy, whereas the latter only accelerates particles slightly above 4 MeV.

The two intensity enhancements in the left panel of Fig. 2 can co-exist without merging partly because the SI separating them is characterised by a strong magnetic field, and hence by a weak cross-field diffusion since the latter scales inversely proportional to the magnetic field strength (see Eq.(5)). However, since the cross-field diffusion is not exactly zero at the SI, the separation between the two intensity enhancements will slowly diffuse away. The latter might be aided by the large particle intensity gradients perpendicular to the magnetic field observed at the SI. Looking at the right panel of Fig. 2, which shows the intensity and anisotropy profiles of case 1 at 31.0 hours, we realise that the intensity dip has indeed increased to a higher intensity level. By this time, the reverse shock has accelerated protons above 5.3 MeV, although the intensities and hence the statistics for these particles are low.

Figure. 3 shows the longitudinal intensity and anisotropy profiles at 1 AU for case 2 and 3 at 11.0 hours. Focusing on the left panel of Fig. 3, we see that the stronger cross-field diffusion of case 2 has merged the two 4.1 ± 0.1 MeV intensity enhancements of case 1 into one wide enhancement. Moreover, we see that there is less particle acceleration when compared to the previous case, since the 5.1 ± 0.1 MeV and 5.3 ± 0.1 MeV energy channels are no longer populated by particles. The peak intensities of the other channels have also decreased, yet we note that these channels cover now a larger longitudinal range. The drop in accelerated particle intensities with increasing cross-field diffusion is even more notable when looking at the right panel of Fig. 3, which shows the intensity profiles for case 3. This panel illustrates that the shock waves are barely capable of accelerating particles due to the large cross-field diffusion assumed in case 3. This is because the cross-field diffusion efficiently moves particles away from the shock waves by transporting the particles to IMF lines that either cross the shock wave at much larger radial distances, or that enter the CIR before the shock waves have formed.

5. Summary and discussion

In this work, we have extended the studies performed in [5] and [6] that investigate SEP transport near and within a high-speed solar wind stream that eventually evolves into a CIR. For this, we used a three dimensional particle transport model that computes particle directional intensities and anisotropies in a solar wind generated by EUHFORIA. In particular, we considered an injection of 4 MeV protons at the inner boundary of our model, from a source region that extends both in the slow and fast solar wind regimes, hence also covering the transition region between both plasma environments. Assuming a cross-field diffusion coefficient that scales inversely proportional to the magnetic field strength, we performed simulations with three different levels of cross-field diffusion. This allowed us to study how the strength of cross-field diffusion may alter the particles spread and the energy content of SEPs in a solar wind that deviates from a nominal solar wind configuration, under the assumptions of our model.

From our simulations, we found that the longitudinal widths of the particle intensity distributions show a general decreasing trend with increasing radial distance, reflecting the underlying IMF topology near and within the CIR. Such a decreasing width is in contrast with what is observed in nominal solar wind configurations, where the width of the particle intensity distribution remains constant in the absence of cross-field diffusion or increases with radial distance if cross-field diffusion is present (see [6]). Although our three simulations with different levels of cross-field diffusion showed overall a similar decreasing trend, there were some notable differences. For the case with the strongest cross-field diffusion, there was a sudden drop in the longitudinal width of the particle intensity distributions at ~1.4 AU, because the particle intensity falls below our prescribed threshold. This intensity drop occurred in the rarefaction region behind the high-speed solar wind stream, where the magnetic flux tubes are strongly expanding, hence rarefying the SEP densities. In addition, the strong cross-field diffusion in our simulation dispersed particles over such a large region in the heliosphere that the intensities decreased everywhere significantly in comparison with the simulations that use a weaker cross-field diffusion. As a consequence, a substantial amount of particles resided in regions of the solar wind showing intensities below our prescribed threshold, and were therefore not taken into account when calculating the longitudinal width. A similar situation may happen with actual in-situ observed SEP events, i.e., when particles propagate either under strong diffusive transport conditions or in solar wind rarefaction regions, they might be dispersed efficiently such that their intensities drop below the background level measured by a given particle instrument, as similarly noticed by Pacheco et al. [52] for electron events simulated under diffusive parallel transport conditions.

Similar to what Strauss et al. [37] obtained for electron propagation in a nominal IMF, we find that the cross-field diffusion is very effective early in the event, when large intensity gradients are present. However, in contrast to a Parker solar wind, the compressed IMF topology inside the CIR also gives rise to large intensity gradients perpendicular to the magnetic field, increasing the effectiveness of cross-field diffusion. In contrast, the strongly diverging magnetic field topology in the rarefaction region might produce very low particle intensity gradients, hence decreasing the diffusive cross-field flux as defined in Eq. (6). We find thus that for a cross-field diffusion coefficient that scales inversely proportional to the magnetic field strength, the effectiveness of the cross-field diffusion is determined by an intricate interplay between the variations in the magnetic field strength and the particle intensity gradients, both modulated by the IMF topology and hence by the converging or diverging properties of the solar wind flow.

In Section 4, we showed the intensity profiles of the shock-accelerated particles at 1 AU, as a function of longitude. For case 1, i.e., the case with the weakest cross-field diffusion, two separate intensity enhancements were visible, reflecting the particle acceleration occurring at the two different shock waves. The negative anisotropies indicate that the acceleration of the particles happens at larger radial distances, where the shocks have formed. This agrees with in-situ observations at 1 AU, showing that the flow of energetic ions accelerated at a CIR shock is typically towards the Sun [53].

Moreover, the double structure in the intensity profiles is a feature that is also commonly detected with in-situ observations at different radial distances from the Sun [54, 55, 56, 57, 58]. Like in our simulations, observations measure typically more accelerated particle fluxes originating from the reverse shock than from the forward shock [56, 3, 59]. This difference might be partly due to the fact that the relative speed difference between the unperturbed solar wind and the shock is higher for the reverse shock than for the forward shock, since the reverse shock propagates radially inwards in the solar wind frame (see also [60]).

Another important factor that might contribute to the observed differences in accelerated particle fluxes originating from the different shocks is the angle between the shock normal and the upstream magnetic field, i.e. the shock angle, θ_{Bn}. In our simulation, the reverse shock of the CIR is more oblique than the forward shock, which follows from the difference in speed between the fast and slow solar wind. Tsurutani et al. [56] demonstrated that proton fluxes near CIRs are correlated with the shock angle, with the most intense proton events occurring for shocks that are quasi-perpendicular. This is in accordance with the results obtained by Zank et al. [50], which illustrated that perpendicular shock can be very efficient particle accelerators given that there is a sufficient amount of turbulence present in the solar wind to diffuse the particle across magnetic field lines. Interesting to note is that CIR shocks are indeed typically characterised by increased

amounts of turbulence (see e.g. [31]) as compared to the unperturbed solar wind, despite the fact that wave excitation by the streaming instability is quenched for highly oblique shocks. The latter is a consequence of the proportionality of the wave growth to the shock angle [41, 61, 45].

The dip between the two intensity enhancements at the reverse and forward shocks coincides with the SI, and can be explained by noting that the field lines adjacent to the SI do not cross the reverse shock wave, nor the forward shock wave. Hence the only way for shock-accelerated particles to reach these field-lines is through cross-field transport. Using observations made by the Pioneer 10 and 11 spacecraft of a CIR lasting for 14 rotations, Tsurutani et al. [56] showed that the intensity dip increases with a factor of 10 over 11 solar rotations, likely due to cross-field diffusion. Similarly, we see in the simulations of case 1 that the intensity minimum increases in time. Figure 2 illustrated that the dip increased by a similar factor of ~10, yet in only 20 hours, i.e., much faster than the observations in [56]. This difference can have various causes, as discussed in the following. It may suggest that the cross-field diffusion used in case 1, for which $\lambda_{\perp}/\lambda_{\|} \sim 10^{-4}$ at 1 AU, might still be too strong inside a CIR. However, Dwyer et al. [57] obtained values for $\lambda_{\perp}/\lambda_{\|}$ of the order of unity for three CIRs using 44 – 313 keV/nucleon helium measurements from the Wind spacecraft. Important to note here is that the results of Tsurutani et al. [56] regarding the increase in the intensity dip near the SI were done when the Pioneer 10 and 11 spacecraft were at a radial distance of ~4.4 and 4.9 AU, whereas the Wind spacecraft is located at ~1 AU, and hence the properties of the solar wind turbulence might be different. Other non-negligible factors that affect the acceleration of particles and hence also the evolution of the intensity dip are the plasma normal velocity jump and the compression ratio across the shocks as well as the obliquity of the shocks. In fact, Desai et al. [3] showed that the proton intensity at CIRs is well correlated with the compression ratio, even if there is no shock present, at least when the particle seed remains constant. In our simulations the particles accelerated at the CIR shocks originate from a flare-like impulsive event, with the particle source region magnetically connected to a limited part of the CIR shocks. However, studies have shown that the seed particle population of most CIR events is likely the supra-thermal tail of the solar wind (see e.g. [62, 63]). This means that the seed population consists of charged particles with energies around ~ 10 keV/nucleon, i.e., much lower than the 4 MeV injection energy that we consider. Pitch-angle scattering, cross-field diffusion and the acceleration processes in these non-nominal solar wind conditions are dependent of the momentum of the particles. A continuous reservoir of low-energy seed particles at the CIR shocks may thus also help explaining the differences observed between our model and the observations in [56]. **However, the observation that CIRs accompanied by the largest energetic particle enhancements are often associated with quasi-perpendicular shocks [56], raises the question how low energy protons would overcome the associated injection problem [64, 65]. Possible explanations can be found in the high levels of turbulence found near CIR shocks, or by considering a seed population originating from the supra-thermal tail of a shock-heated solar wind plasma. Prinsloo et al. [48] illustrated that the prior heating of the solar wind distribution is important in reproducing the intensities of typical energetic storm particle events observed at shock waves driven by CMEs, especially when the injection energy is high like is typically the case for quasi-perpendicular shocks [66, 50]. A similar conclusion may hold for the shock waves bounding the CIR. We note that in our simulations we avoid any injection problem by considering protons of sufficiently high energy only.** We leave it for future work to study how our results change when e.g., continuously injecting the low-energy particles from the supra-thermal tail of the solar wind.

The double-enhancement intensity structure never forms at 1 AU in our case 2 and 3 simulations, where we use a cross-field diffusion magnitude comparable to that of Dwyer et al. [57]. This may be due to strong turbulence variations near and within the CIR that are not captured by our cross-field diffusion model. More specifically, the plasma near the reverse shock is often characterised by significant magnetic fluctuations [34, 31], while inside the CIR, the turbulent fluctuation of the magnetic field typically decreases towards the SI [56, 57]. More recently, Strauss et al. [67] showed the decrease in turbulence at the SI by computing the variance of the in-situ measured magnetic field components for a CIR. In addition, these authors argue that a tangential magnetic discontinuity like the SI may produce an anisotropic cross-field diffusion tensor, by strongly damping magnetic field fluctuations in the direction perpendicular to the discontinuity. As a consequence, the SI would act as an efficient diffusion barrier, while still allowing for strong cross-field diffusion near the CIR boundaries. Hence, the model proposed by Strauss et al. [67] seems to explain both the slow increase of the intensity dip in the CIR shown in [56] and the large values of $\lambda_\perp/\lambda_\parallel$ reported in [57] near the reverse shock. In our simulations, a strong cross-field diffusion near the shocks leads to less particle acceleration, owing to the efficiency of the cross-field diffusion at spreading particles away from the shock wave.

As mentioned above, in addition to different shock properties, this apparent inconsistency between our simulation results and in-situ observations may be due to the fact that we are considering 4 MeV protons, whereas Dwyer et al. [57] obtain their results for helium with lower energies per nucleon. The ion parallel mean free path typically decreases with decreasing rigidity [68, 69], and therefore low-energy particles would be more efficiently trapped near the shock waves. In addition, the strong levels of turbulence observed near CIR shocks might reduce the parallel mean free path of the particles significantly, hence both increasing $\lambda_\perp/\lambda_\parallel$ and facilitating multiple shock encounters. Moreover, we note that cross-field diffusion might have a non-negligible pitch-angle dependence, as discussed in [38], which could also influence the scattering of the particles near the shock waves. Further investigation of these issues is required, as it will eventually lead to a better understanding of particle transport in the complex turbulent environments that characterise large scale interplanetary structures like CIRs. The strongly changing turbulence properties found in the different parts of a CIR, and the possibility to have in-situ measurements of these regions, make CIRs the ideal laboratories to study the transport of particles perpendicular to the mean magnetic field.

Acknowledgments
N. Wijsen is supported by a PhD Fellowship of the Research Foundation Flanders (FWO). The computational resources and services used in this work were provided by the VSC (Flemish Supercomputer Center), funded by the Research Foundation Flanders (FWO) and the Flemish Government, department EWI. The work at KU Leuven was done in the framework of the projects GOA/2015-014 (KU Leuven), G.0A23.16N (FWO-Vlaanderen) and C 90347 (Prodex). Funding for this work at University of Barcelona was partially provided by the Spanish MINECO under project MDM-2014-0369 of ICCUB (Unidad de Excelencia 'María de Maeztu'). The work at University of Helsinki was carried out in the Finnish Centre of Excellence in Research of Sustainable Space (Academy of Finland grant numbers 312390 and 312351).

References
[1] Smith E J and Wolfe J H 1976 *Geophys. Res. Lett.* **3** 137–140
[2] Bell A R 1978 *MNRAS* **182** 147–156
[3] Desai M I, Marsden R G, Sanderson T R, Balogh A, Forsyth R J and Gosling J T 1998 *J. Geophys. Res.* **103** 2003
[4] Giacalone J, Jokipii J R and Kóta J 2002 *ApJ* **573** 845–850
[5] Wijsen N, Aran A, Pomoell J and Poedts S 2019 *A&A* **622** A28
[6] Wijsen N, Aran A, Pomoell J and Poedts S 2019 *A&A* **624** A47 (*Preprint* 1903.09072)
[7] Pomoell J and Poedts S 2018 *J. Space Weather Space Clim.* **8** A35
[8] Roussev I I, Sokolov I V, Forbes T G, Gombosi T I, Lee M A and Sakai J I 2004 *ApJ* **605** L73–L76
[9] Lee M A, Mewaldt R A and Giacalone J 2012 *Space Sci. Rev.* **173** 247–281
[10] Rouillard A P, Sheeley N R, Tylka A, Vourlidas A, Ng C K, Rakowski C, Cohen C M S, Mewaldt R A, Mason G M and Reames D 2012 *ApJ* **752** 44
[11] Pomoell J, Aran A, Jacobs C, Rodríguez-Gasén R, Poedts S and Sanahuja B 2015 *J. Space Weather Space Clim.* **5** A12
[12] Lario D, Kwon R Y, Vourlidas A, Raouafi N E, Haggerty D K, Ho G C, Anderson B J, Papaioannou A, Gómez-Herrero R, Dresing N and Riley P 2016 *ApJ* **819** 72
[13] Schwadron N A et al. 2017 *Space Sci. Rev.* **212** 1069–1106
[14] Afanasiev A, Vainio R, Rouillard A P, Battarbee M, Aran A and Zucca P 2018 *A&A* **614** A4 (*Preprint* 1806.05497)
[15] Kozarev K A, Dayeh M A and Farahat A 2019 *ApJ* **871** 65
[16] Cane H V and Lario D 2006 *Space Sci. Rev.* **123** 45–56
[17] Hu J, Li G, Fu S, Zank G and Ao X 2018 *ApJ* **854** L19

[18] Hu J, Li G, Ao X, Zank G P and Verkhoglyadova O 2017 *J. Geophys. Res.(Space Physics)* **122** 10,938–10,963
[19] Matthaeus W H, Qin G, Bieber J W and Zank G P 2003 *ApJ* **590** L53–L56
[20] Zank G P, Li G, Florinski V, Matthaeus W H, Webb G M and le Roux J A 2004 *J. Geophys. Res.(Space Physics)* **109**
[21] Shalchi A, Li G and Zank G P 2009 *Astrophysics and Space Science* **325** 99 ISSN 1572-946X
[22] Roelof E C 1969 *Lectures in High-Energy Astrophysics* ed Ögelman H and Wayland J R p 111
[23] Isenberg P A 1997 *J. Geophys. Res.* **102** 4719–4724
[24] le Roux J A and Webb G M 2009 *ApJ* **693** 534–551
[25] Gardiner C W 2004 *Handbook of Stochastic Methods for Physics, Chemistry and the Natural Sciences* 3rd ed (*Springer Series in Synergetics* vol 13) (Berlin: Springer-Verlag) ISBN 3-540-20882-8
[26] Speiser T W 1965 *J. Geophys. Res.* **70** 4219–4226
[27] Reid G C 1964 *J. Geophys. Res.* **69** 2659–2667
[28] Lario D, Aran A, Agueda N and Sanahuja B 2007 *Advances in Space Research* **40** 289–294
[29] Zhang M, Qin G and Rassoul H 2009 *ApJ* **692** 109–132
[30] Wang Y, Qin G and Zhang M 2012 *ApJ* **752** 37
[31] Horbury T and Schmidt J 1999 *Space Science Reviews* **89** 61–76 ISSN 1572-9672
[32] Crooker N U et al. 1999 *Space Sci. Rev.* **89** 179–220
[33] Richardson I G 2018 *Living Reviews in Solar Physics* **15** 1 ISSN 1614-4961
[34] Tsurutani B T, Ho C M, Arballo J K, Goldstein B E and Balogh A 1995 *Geophys. Res. Lett.* **22** 3397–3400
[35] Mason G M, Li G, Cohen C M S, Desai M I, Haggerty D K, Leske R A, Mewaldt R A and Zank G P 2012 *The Astrophysical Journal* **761** 104
[36] Zhao L, Li G, Ebert R W, Dayeh M A, Desai M I, Mason G M, Wu Z and Chen Y 2016 *J. Geophys. Res.(Space Physics)* **121** 77–92 (*Preprint* 1706.04322)
[37] Strauss R D T, Dresing N and Engelbrecht N E 2017 *ApJ* **837** 43
[38] Strauss R D and Fichtner H 2015 *ApJ* **801** 29
[39] Laitinen T, Dalla S and Marsh M S 2013 *ApJ* **773** L29
[40] Lee M A 1983 *J. Geophys. Res.* **88** 6109–6120
[41] Gordon B E, Lee M A, Möbius E and Trattner K J 1999 *J. Geophys. Res.* **104** 28263–28278
[42] Vainio R and Laitinen T 2007 *ApJ* **658** 622–630
[43] Afanasiev A, Battarbee M and Vainio R 2015 *A&A* **584** A81 (*Preprint* 1603.08857)
[44] Lario D, Sanahuja B and Heras A M 1998 *ApJ* **509** 415–434
[45] Rice W K M, Zank G P and Li G 2003 *J. Geophys. Res.(Space Physics)* **108** 1369
[46] Rodríguez-Gasén R, Aran A, Sanahuja B, Jacobs C and Poedts S 2011 *Advances in Space Research* **47** 2140–2151
[47] Rodríguez-Gasén R, Aran A, Sanahuja B, Jacobs C and Poedts S 2014 *Sol. Phys.* **289** 1745–1762 (*Preprint* 1310.4651)
[48] Prinsloo P L, Strauss R D and Le Roux J A 2019 *ApJ* **878** 144-144 (*Preprint* 1905.08458)
[49] Zank G P, Rice W K M, le Roux J A, Cairns I H and Webb G M 2001 *Physics of Plasmas* **8** 4560–4576
[50] Zank G P, Li G, Florinski V, Hu Q, Lario D and Smith C W 2006 *J. Geophys. Res.(Space Physics)* **111** A06108
[51] Battarbee M, Vainio R, Laitinen T and Hietala H 2013 *A&A* **558** A110 (*Preprint* 1309.2062)
[52] Pacheco D, Agueda N, Aran A, Heber B and Lario D 2019 *A&A* **624** A3 (*Preprint* 1902.06602)
[53] Marshall F E and Stone E C 1978 *J. Geophys. Res.* **83** 3289–3298
[54] Barnes C W and Simpson J A 1976 *ApJ* **210** L91–L96
[55] Decker R B, Pesses M E and Krimigis S M 1981 *J. Geophys. Res.* **86** 8819–8831
[56] Tsurutani B T, Smith E J, Pyle K R and Simpson J A 1982 *J. Geophys. Res.* **87** 7389–7404
[57] Dwyer J R, Mason G M, Mazur J E, Jokipii J R, von Rosenvinge T T and Lepping R P 1997 *ApJ* **490** L115–L118
[58] Mason G M and Sanderson T R 1999 *Space Sci. Rev.* **89** 77–90
[59] Mason G M et al. 1999 *Space Sci. Rev.* **89** 327–367
[60] Kocharov L, Kovaltsov G A, Torsti J, Anttila A and Sahla T 2003 *J. Geophys. Res.(Space Physics)* **108** 1404
[61] Li G, Zank G P and Rice W K M 2003 *J. Geophys. Res.(Space Physics)* **108** 1082
[62] Chotoo K, Schwadron N A, Mason G M, Zurbuchen T H, Gloeckler G, Posner A, Fisk L A, Galvin A B, Hamilton D C and Collier M R 2000 *J. Geophys. Res.* **105** 23107–23122
[63] Mason G M, Desai M I and Li G 2012 *ApJ* **748** L31
[64] Jokipii J R 1987 *ApJ* **313** 842
[65] Zank G P, Pauls H L, Cairns I H and Webb G M 1996 *J. Geophys. Res.* **101** 457–478
[66] Tylka A J, Cohen C M S, Dietrich W F, Lee M A, Maclennan C G, Mewaldt R A, Ng C K and Reames D V 2005 *ApJ* **625** 474–495
[67] Strauss R D, le Roux J A, Engelbrecht N E, Ruffolo D and Dunzlaff P 2016 *ApJ* **825** 43
[68] Dröge W 2000 *ApJ* **537** 1073–1079
[69] Dröge W 2003 *ApJ* **589** 1027–1039

Journal of Physics: Conference Series **1332** (2019) 012019 doi:10.1088/1742-6596/1332/1/012019

Synchrotron spectra of GRB prompt emission and pulsar wind nebulae

Siyao Xu

Department of Astronomy, University of Wisconsin, 475 North Charter Street, Madison, WI 53706, USA; Hubble Fellow

E-mail: sxu93@wisc.edu

Abstract. Particle acceleration is a fundamental process in many high-energy astrophysical environments and determines the spectral features of their synchrotron emission. We have studied the adiabatic stochastic acceleration (ASA) of electrons arising from the basic dynamics of magnetohydrodynamic (MHD) turbulence and found that the ASA acts to efficiently harden the injected electron energy spectrum. The dominance of the ASA at low energies and the dominance of synchrotron cooling at high energies result in a broken power-law shape of both electron energy spectrum and photon synchrotron spectrum. Furthermore, we have applied the ASA to studying the synchrotron spectra of the prompt emission of gamma-ray bursts (GRBs) and pulsar wind nebulae (PWNe). The good agreement between our theories and observations confirms that the stochastic particle acceleration is indispensable in explaining their synchrotron emission.

1. Introduction

Many cosmic accelerators are huge reservoirs of magnetic energy. The dissipated magnetic energy is converted to energies of particles, accounting for the observed synchrotron emission [1, 2, 3]. In both turbulent and magnetized medium, a proper description of magnetohydrodynamic (MHD) turbulence is crucial for studying the interaction between particles and turbulent magnetic fields and the resulting particle acceleration [4, 5, 6]. Recent theoretical [7, 8] and numerical [9, 10, 11] studies reveal a critical balance between the turbulent motions in the direction perpendicular to the local magnetic field and magnetic wave-like motions parallel to the local magnetic field in MHD turbulence. Accordingly, the turbulent dynamo with magnetic field lines stretched by turbulent motions [12] and the turbulent reconnection of stochastic magnetic fields [8] are also in dynamical balance in MHD turbulence. As a new acceleration mechanism proposed by [13], particles entrained on turbulent magnetic field lines undergo cycles of deceleration in turbulent dynamo regions and acceleration in turbulent reconnection regions, leading to a globally diffusive acceleration process, which we term as "adiabatic stochastic acceleration (ASA)". The ASA becomes the dominant acceleration process in MHD turbulence when the non-adiabatic resonant scattering of particles by anisotropic MHD turbulence is inefficient [14]. In our recent studies, we applied the ASA to interpreting the Band function spectrum [15] of the prompt emission of gamma-ray bursts (GRBs) [16, 17] and the synchrotron spectra of pulsar wind nebulae (PWNe) [18].

Content from this work may be used under the terms of the Creative Commons Attribution 3.0 licence. Any further distribution of this work must maintain attribution to the author(s) and the title of the work, journal citation and DOI.
Published under licence by IOP Publishing Ltd

2. Electron energy spectrum resulting from the ASA

The time evolution of the electron energy spectrum $N(E,t)$ is described by

$$\frac{\partial N}{\partial t} = a_2 \frac{\partial}{\partial E}\Big(E\frac{\partial (EN)}{\partial E}\Big) + \beta\frac{\partial (E^2 N)}{\partial E} + Q(E). \tag{1}$$

The three terms on the RHS correspond to the ASA, radiation losses, and particle injection. The acceleration rate of ASA is

$$a_2 = \xi \frac{u_{\text{tur}}}{l_{\text{tur}}}, \tag{2}$$

where l_{tur} and u_{tur} are the correlation length and speed of turbulence, $\xi = \Delta E/E$ is the cumulative fractional energy change during the trapping of particles within individual turbulent eddies during the eddy turnover time $\tau_{\text{tur}} = l_{\text{tur}}/u_{\text{tur}}$. ξ is of order unity for non-relativistic turbulence and of order γ_{tur}^2 for relativistic turbulence with γ_{tur} as the turbulence Lorentz factor. In the case of both synchrotron and synchrotron-self-Compton (SSC) losses, β is expressed as

$$\beta = \frac{\sigma_T c B^2 (1+Y)}{6\pi (m_e c^2)^2}, \tag{3}$$

where Y is the ratio between the powers of SSC and synchrotron radiation and is zero when only synchrotron is considered. Besides, B is the magnetic field strength, σ_T is the Thomson cross section, γ_e and m_e are the electron Lorentz factor and the electron rest mass, and c is the speed of light. The third term $Q(E) = CE^{-p}$ represents a steady injection of power-law electron spectrum with a power-law index p, accounting for other possible instantaneous acceleration processes, e.g., shock acceleration, reconnection acceleration, that generate power-law electron spectra.

In [16, 17], we analytically solved Eq. (1) in the energy range where the ASA dominates over radiation losses. We found that $N(E,t)$ evolves from a spectral shape governed by the injected particle distribution

$$N(E,\tau) = \frac{C'\sqrt{\tau}}{\sqrt{\pi}} E^{-p} \exp\Big(-\frac{E}{E_{\text{cf}}}\Big), \tag{4}$$

to a hard spectrum under the effect of ASA,

$$N(E,\tau) = \frac{C'(E_u^{-p+1} - E_l^{-p+1})\sqrt{\tau}}{(-p+1)\sqrt{\pi}} E^{-1} \exp\Big(-\frac{E}{E_{\text{cf}}}\Big), \tag{5}$$

where $C' = C/a_2$, $\tau = a_2 t$, and E_l and E_u are the lower and upper limits of the energy range of injected electrons. Here $E_{\text{cf}} = a_2/\beta$ is the cutoff energy of the ASA, where the acceleration rate equalizes with the cooling rate. This hardening of electron energy spectrum due to the ASA is also illustrated in Fig. 1. It shows that irrespective of the injected spectral form, the ASA leads to diffusive particle acceleration and thus a hard electron energy spectrum with the power-law index approaching -1.

3. Synchrotron spectrum resulting from both ASA and synchrotron cooling

In the energy range $E < E_{\text{cf}}$, the ASA dominates over the synchrotron cooling and leads to the electron energy spectrum as shown in Eq. (5). At higher energies, the synchrotron cooling plays a dominant role in shaping the electron energy spectrum. Depending on the relation between E_l, E_{cf}, and E_c, where $E_c = 1/\beta t$ is the critical cooling energy [19], the electron spectrum and the corresponding synchrotron spectrum exhibit different forms [17], as illustrated in Figs. 2 and 3. The asymptotic functional forms of the flux in different cases are as follows:

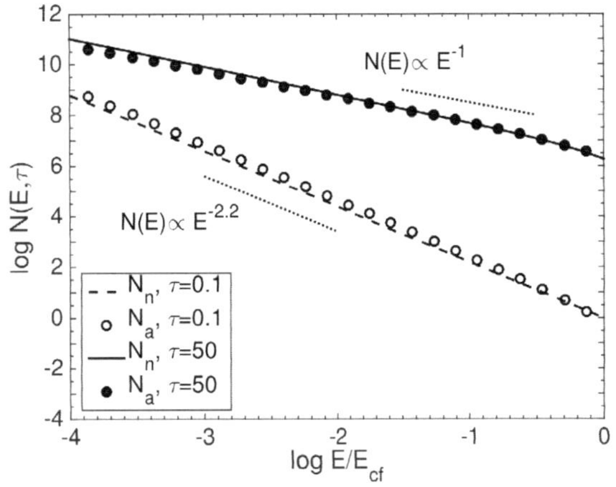

Figure 1. The electron energy spectrum at different times. From [17].

Case (i) $E_l < E_{\rm cf} < E_c$,

$$F_\nu = \begin{cases} F_{\nu,\max}\left(\frac{\nu}{\nu_m}\right)^{\frac{1}{3}}, & \nu < \nu_m, \quad (6a) \\ F_{\nu,\max}, & \nu_m < \nu < \nu_{\rm cf}, \quad (6b) \\ F_{\nu,\max}\left(\frac{\nu}{\nu_{\rm cf}}\right)^{-\frac{p-1}{2}}, & \nu_{\rm cf} < \nu < \nu_c, \quad (6c) \\ F_{\nu,\max}\left(\frac{\nu_c}{\nu_{\rm cf}}\right)^{-\frac{p-1}{2}}\left(\frac{\nu}{\nu_c}\right)^{-\frac{p}{2}}, & \nu_c < \nu. \quad (6d) \end{cases}$$

Case (ii) $E_l < E_c < E_{\rm cf}$,

$$F_\nu = \begin{cases} F_{\nu,\max}\left(\frac{\nu}{\nu_m}\right)^{\frac{1}{3}}, & \nu < \nu_m, \quad (7a) \\ F_{\nu,\max}, & \nu_m < \nu < \nu_{\rm cf}, \quad (7b) \\ F_{\nu,\max}\left(\frac{\nu}{\nu_{\rm cf}}\right)^{-\frac{p}{2}}, & \nu_{\rm cf} < \nu. \quad (7c) \end{cases}$$

Case (iii) $E_c < E_{\rm cf} < E_l$,

$$F_\nu = \begin{cases} F_{\nu,\max}\left(\frac{\nu}{\nu_m}\right)^{\frac{1}{3}}, & \nu < \nu_m, \quad (8a) \\ F_{\nu,\max}, & \nu_m < \nu < \nu_{\rm cf}, \quad (8b) \\ F_{\nu,\max}\left(\frac{\nu}{\nu_{\rm cf}}\right)^{-\frac{1}{2}}, & \nu_{\rm cf} < \nu < \nu_l, \quad (8c) \\ F_{\nu,\max}\left(\frac{\nu_l}{\nu_{\rm cf}}\right)^{-\frac{1}{2}}\left(\frac{\nu}{\nu_l}\right)^{-\frac{p}{2}}, & \nu_l < \nu. \quad (8d) \end{cases}$$

Case (iv) $E_c < E_l < E_{\rm cf}$,

$$F_\nu = \begin{cases} F_{\nu,\max}\left(\frac{\nu}{\nu_m}\right)^{\frac{1}{3}}, & \nu < \nu_m, \quad (9a) \\ F_{\nu,\max}, & \nu_m < \nu < \nu_{\rm cf}, \quad (9b) \\ F_{\nu,\max}\left(\frac{\nu}{\nu_{\rm cf}}\right)^{-\frac{p}{2}}, & \nu_{\rm cf} < \nu. \quad (9c) \end{cases}$$

We note that the low-frequency tail $F_\nu \propto \nu^{1/3}$ comes from the synchrotron single-particle emission spectrum [20, 21].

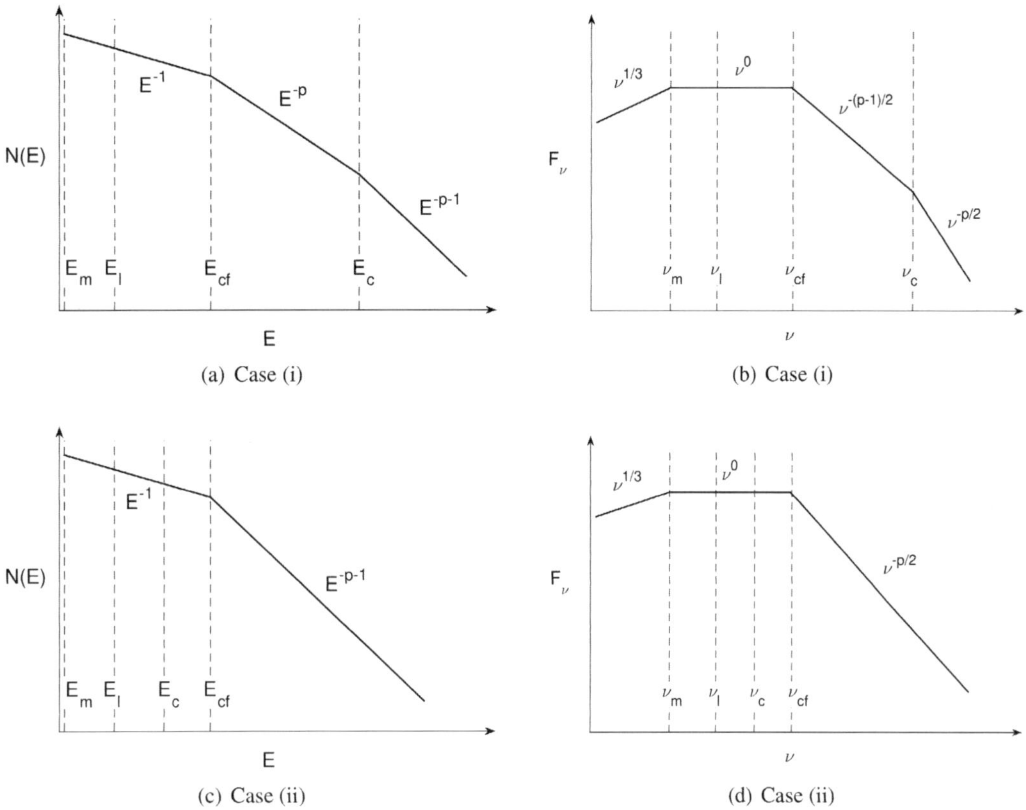

Figure 2. The electron energy spectrum ((a) and (c)) and the corresponding synchrotron spectrum ((b) and (d)) in Case (i) and in Case (ii). From [17].

4. Application to the synchrotron spectra of GRB prompt emission

The ASA can generally take place in MHD turbulence and dominate over other stochastic acceleration mechanisms, e.g., gyroresonance, when the pitch-angle scattering is inefficient. Therefore, we have explored the application of the ASA to the stochastic acceleration process in different astrophysical environments, which are both magnetized and turbulent. In this section and §5, we will discuss the ASA in the context of GRBs and PWNe as examples.

The empirical Band spectrum [15] is usually used to describe the time-averaged synchrotron spectrum of GRB prompt emission. It is characterized by a low-energy spectral index α_s, a break energy E_b, and a high-energy spectral index β_s. The distributions of α_s and β_s are centered around -1 and -2.25, respectively, and E_b is on the order of 100 keV [22]. The above spectral features, especially the hard low-energy spectrum, cannot be well explained by either the thermal model [23, 24] or the synchrotron model [25, 21, 26] of GRBs.

In [16, 17], we investigated the ASA in the magnetized and turbulent GRB outflow [1, 2] and found that it can naturally account for the hard low-energy spectrum of GRB prompt emission. The synchrotron spectrum in both Case (ii) and Case (iv) (§3) agrees with the Band spectrum. As a result of the ASA, there is $N(\nu) \propto F_\nu/\nu \propto \nu^{-1}$ in the frequency range $(\nu_m, \nu_{\rm cf})$. The high-energy synchrotron spectrum is related to the injected electron distribution, and $\beta_s = -2.25$ corresponds to $p = 2.5$, which can

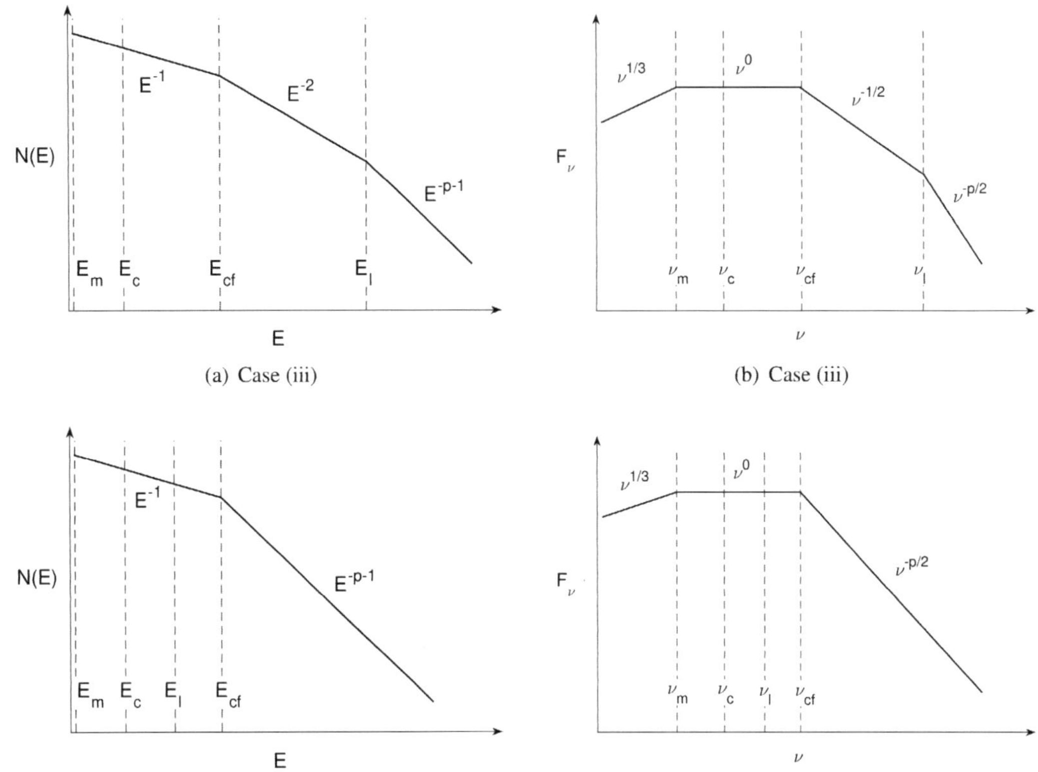

(c) Case (iv) (d) Case (iv)

Figure 3. The electron energy spectrum ((a) and (c)) and the corresponding synchrotron spectrum ((b) and (d)) in Case (iii) and in Case (iv). From [17].

be explained by other first-order Fermi acceleration processes, such as the shock acceleration and the reconnection acceleration [27]. E_b is related to $E_{\rm cf}$. The photon energy corresponding to $E_{\rm cf}$ in the observer frame is [17],

$$E_{s,{\rm cf,obs}} \simeq 385{\rm keV} \left(\frac{1+z}{2}\right)^{-1} \Gamma_2^{-0.2} L_{52}^{0.6} r_{15}^{-1.2}, \tag{10}$$

where z is the redshift, $\Gamma = 100\Gamma_2$ is the bulk Lorentz factor, $L = 10^{52}$ erg s$^{-1} L_{52}$ is the total isotropic luminosity, and $r = 10^{15}$ cmr_{15} is the radius of the emission region [1]. It has the same order of magnitude as E_b that is indicated by observations.

5. Application to the synchrotron spectra of PWNe

In [18], we applied the ASA to explaining the synchrotron spectra of magnetized and turbulent PWNe and found a good agreement between the analytical spectral shapes (§3) and the observed broad-band synchrotron spectra of different PWNe (see Figs. 4 and 5). Moreover, the observed spectral break is related to $E_{\rm cf}$ and can be used to constrain the acceleration timescale of the ASA $\tau_{\rm acc} = 1/a_2$. If the synchrotron spectrum of a PWN falls in Case (ii) or Case (iv), there is

$$\left(\frac{\tau_{\rm acc}}{1\ {\rm kyr}}\right)\left(\frac{B}{100\ \mu{\rm G}}\right)^{\frac{3}{2}} = 26.5\left(\frac{\Gamma}{1}\right)^{\frac{1}{2}}\left(\frac{E_{\rm ph,b}}{10^{-2}\ {\rm eV}}\right)^{-\frac{1}{2}}, \tag{11}$$

where B is the magnetic field strength in the PWN, Γ is the bulk Lorentz factor of the mildly relativistic flow in the PWN, and $E_{\rm ph,b}$ is the observed energy break of synchrotron spectrum. If the synchrotron spectrum of a PWN falls in Case (i) or Case (iii), there is

$$\left(\frac{\tau_{\rm acc}}{1\ {\rm kyr}}\right)\left(\frac{B}{100\ \mu{\rm G}}\right)^{\frac{3}{2}} = 837.0\left(\frac{\Gamma}{1}\right)^{\frac{1}{2}}\left(\frac{E_{\rm ph,b1}}{10^{-5}\ {\rm eV}}\right)^{-\frac{1}{2}}, \tag{12}$$

where $E_{\rm ph,b1}$ is the observed energy break at a lower energy (see Fig. 4(b)). In the case when the spectral shape in the infrared band cannot be well determined, we have

$$26.5\left(\frac{\Gamma}{1}\right)^{\frac{1}{2}}\left(\frac{E_{\rm ph,b}}{10^{-2}\ {\rm eV}}\right)^{-\frac{1}{2}} \leq \left(\frac{\tau_{\rm acc}}{1\ {\rm kyr}}\right)\left(\frac{B}{100\ \mu{\rm G}}\right)^{\frac{3}{2}} \leq 837.0\left(\frac{\Gamma}{1}\right)^{\frac{1}{2}}\left(\frac{E_{\rm ph,b1}}{10^{-5}\ {\rm eV}}\right)^{-\frac{1}{2}}, \tag{13}$$

as illustrated in Fig. 6.

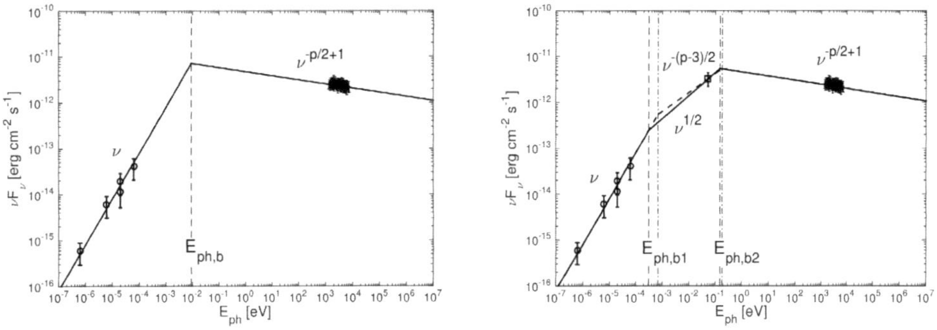

(a) (b)

Figure 4. Comparison between analytical spectral shapes and observational data for the Mouse PWN. Circles, square, and dots represent radio, infrared, and X-ray data, respectively. See the original figure and references in [18].

6. Summary

As a new stochastic acceleration mechanism, the ASA arises from the basic dynamics of MHD turbulence involving both turbulent dynamo and turbulent reconnection of magnetic fields. Different from other stochastic acceleration mechanisms, it is highly efficient as particles undergo the first-order Fermi process within individual turbulent eddies. It is also not subject to the turbulence anisotropy effect, which makes the gyroresonance with Alfvénic turbulence inefficient.

The ASA naturally hardens the injected electron spectrum and results in a hard electron spectrum in the energy range dominated by the ASA. Under the effects of both ASA and radiation cooling, the electron spectrum exhibits a broken power-law shape. The resulting synchrotron spectrum well explains the synchrotron spectrum of GRB prompt emission, as well as the broad-band synchrotron spectrum of a PWN.

The ASA is a general acceleration mechanism in MHD turbulence. Besides GRBs and PWNe, the application of the ASA to other high-energy astrophysical environments, e.g., radio galaxies, blazars, will be investigated in our future work.

Acknowledgement

I acknowledge the support for Program number HST-HF2-51400.001-A provided by NASA through a grant from the Space Telescope Science Institute, which is operated by the Association of Universities for Research in Astronomy, Incorporated, under NASA contract NAS5-26555. I am grateful to Bing Zhang,

Journal of Physics: Conference Series **1332** (2019) 012019 doi:10.1088/1742-6596/1332/1/012019

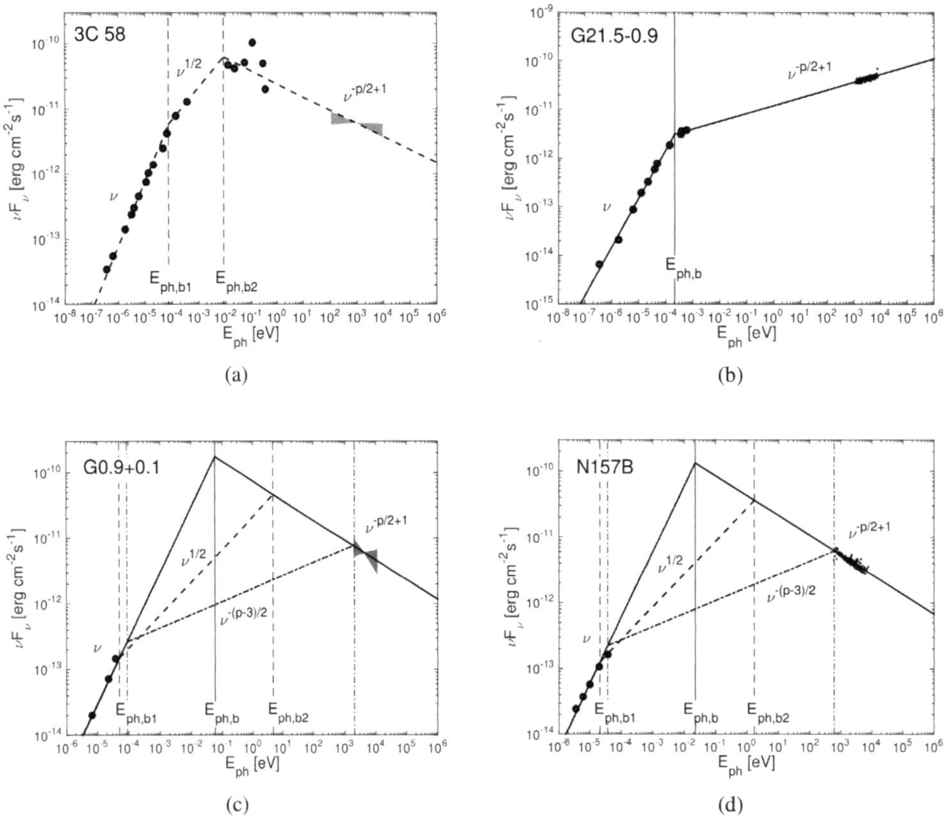

Figure 5. Comparison between analytical spectral shapes and observational data for different PWNe. See the original figures and references in [18].

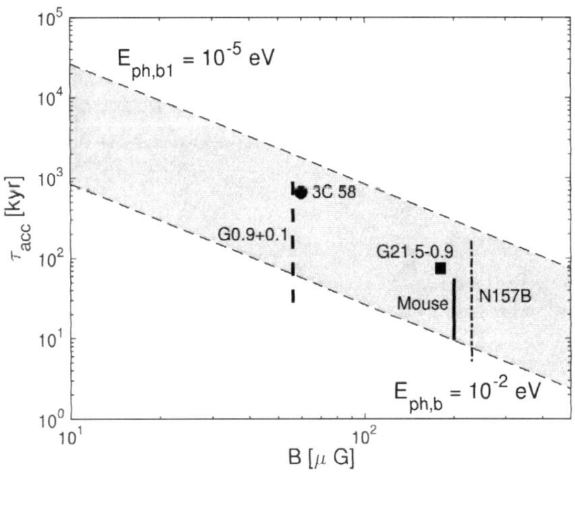

Figure 6. $\tau_{\rm acc}$ vs. $B_{\rm eq}$ of the PWNe, where $B_{\rm eq}$ is the magnetic field strength under the assumption of equipartition between particle and magnetic energies. The shaded region is bounded by Eq. (13). From [18].

Yuan-Pei Yang, Noel Klingler, and Oleg Kargaltsev for their contributions to our studies mentioned here. This paper is based on an invited talk that I have given at the 18th Annual International Astrophysics Conference.

[1] Zhang B and Yan H 2011 *ApJ* **726** 90 (*Preprint* 1011.1197)

[2] Deng W, Li H, Zhang B and Li S 2015 *ApJ* **805** 163 (*Preprint* 1501.07595)
[3] Lazarian A, Zhang B and Xu S 2018 *arXiv: 1801.04061* (*Preprint* 1801.04061)
[4] Xu S and Yan H 2013 *ApJ* **779** 140 (*Preprint* 1307.1346)
[5] Xu S, Yan H and Lazarian A 2016 *ApJ* **826** 166 (*Preprint* 1506.05585)
[6] Xu S and Lazarian A 2018 *ApJ* **868** 36 (*Preprint* 1810.07726)
[7] Goldreich P and Sridhar S 1995 *ApJ* **438** 763–775
[8] Lazarian A and Vishniac E T 1999 *ApJ* **517** 700–718 (*Preprint* arXiv:astro-ph/9811037)
[9] Cho J and Vishniac E T 2000 *ApJ* **539** 273–282 (*Preprint* arXiv:astro-ph/0003403)
[10] Maron J and Goldreich P 2001 *ApJ* **554** 1175–1196 (*Preprint* arXiv:astro-ph/0012491)
[11] Cho J, Lazarian A and Vishniac E T 2002 *ApJ* **564** 291–301 (*Preprint* arXiv:astro-ph/0105235)
[12] Xu S and Lazarian A 2016 *ApJ* **833** 215 (*Preprint* 1608.05161)
[13] Brunetti G and Lazarian A 2016 *MNRAS* **458** 2584–2595 (*Preprint* 1603.00458)
[14] Yan H and Lazarian A 2002 *Physical Review Letters* **89** B1102+ (*Preprint* arXiv:astro-ph/0205285)
[15] Band D, Matteson J, Ford L, Schaefer B, Palmer D, Teegarden B, Cline T, Briggs M, Paciesas W, Pendleton G, Fishman G, Kouveliotou C, Meegan C, Wilson R and Lestrade P 1993 *ApJ* **413** 281–292
[16] Xu S and Zhang B 2017 *ApJL* **846** L28 (*Preprint* 1708.08029)
[17] Xu S, Yang Y P and Zhang B 2018 *ApJ* **853** 43 (*Preprint* 1711.03943)
[18] Xu S, Klingler N, Kargaltsev O and Zhang B 2019 *ApJ* **872** 10 (*Preprint* 1812.10827)
[19] Sari R, Piran T and Narayan R 1998 *ApJL* **497** L17–L20 (*Preprint* astro-ph/9712005)
[20] Meszaros P and Rees M J 1993 *ApJL* **418** L59 (*Preprint* astro-ph/9309011)
[21] Katz J I 1994 *ApJL* **432** L107–L109 (*Preprint* astro-ph/9312034)
[22] Preece R D, Briggs M S, Mallozzi R S, Pendleton G N, Paciesas W S and Band D L 2000 *ApJS* **126** 19–36 (*Preprint* astro-ph/9908119)
[23] Beloborodov A M 2010 *MNRAS* **407** 1033–1047 (*Preprint* 0907.0732)
[24] Deng W and Zhang B 2014 *ApJ* **785** 112 (*Preprint* 1402.5364)
[25] Rees M J and Meszaros P 1992 *MNRAS* **258** 41P–43P
[26] Tavani M 1996 *ApJ* **466** 768
[27] de Gouveia dal Pino E M and Lazarian A 2005 *A&A* **441** 845–853

A possible explanation for the enhancement of energetic particles downstream of the heliospheric termination shock

L L Zhao[1], **G P Zank**[1,2], **L Adhikari**[1]

[1] Center for Space Plasma and Aeronomic Research (CSPAR), University of Alabama in Huntsville, Huntsville, AL 35899, USA

[2] Department of Space Science, University of Alabama in Huntsville, Huntsville, AL 35899, USA

E-mail: `lz0009@uah.edu`

Abstract. *Voyager* 2 observations of the energetic particle "time-intensity" profiles from ~1.8 to ~40 MeV show that the flux peaks downstream of the heliospheric termination shock (HTS), which is inconsistent with the predictions of classical diffusive shock acceleration (DSA). Previous studies suggest that shocks are effective in generating downstream magnetic flux ropes, islands or plasmoids. These dynamically interacting small-scale structures can accelerate charged particles statistically through reconnection-related processes. We present a preliminary study of the magnetic field and plasma properties together with the energetic particle data during the V2 crossing of the HTS. We apply a local stochastic acceleration model associated with solar wind magnetic island dynamics to explain the unusual behavior of energetic particles observed in the vicinity of the HTS. An analytic solution for the particle velocity distribution function derived from the Zank et al. statistical transport theory is used to fit the observed particle flux amplification downstream of the HTS. The results show that stochastic acceleration by interacting magnetic islands can successfully predict the observed (i) peaking of particle intensities behind the HTS, instead of at the shock front; (ii) increasing of the particle flux amplification factor with increasing particle energy; and (iii) increase in distance between the particle intensity peak and the HTS with increasing particle energy; This study illustrates the possibility of local acceleration in the inner heliosheath due to the dynamical interaction of magnetic islands.

1. Introduction

Diffusive shock acceleration (DSA) is usually considered to be responsible for energizing charged particles at shock waves and provides an explanation for abundant gradual solar energetic particle (SEP) events [e.g., 1 - 6]. As a classical mechanism for particle acceleration [7 - 10], DSA, in 1D stationary cases, predicts that the particle intensity peaks at the shock, after which it is a constant. The distribution of diffusively accelerated particles at and downstream of the shock is a power law in momentum p^{-q} with the index $q = 3r/(r-1)$ depending only on the shock compression ratio r [6, 12]. However, steady-state DSA predictions do not always describe observations. Particle energization up to MeV energies that is inconsistent with classical DSA has been observed frequently, both at 1 au [e.g., 14 - 19] and in the outer heliosphere [12, 21, 22]. A notable exception is the *Voyager* spacecrafts observation of the anomalous cosmic ray (ACR)

Content from this work may be used under the terms of the Creative Commons Attribution 3.0 licence. Any further distribution of this work must maintain attribution to the author(s) and the title of the work, journal citation and DOI.
Published under licence by IOP Publishing Ltd

intensity profile that was observed to peak behind the HTS and to posses a spectrum far harder than predicted by classical DSA theory [13, 14]. The accelerated ACR spectrum observed at the HTS crossing in 2007 was hard with a power law index of ∼-4.6 despite the shock compression ratio being ∼2, for which classical DSA predicts an index of -6. Previous studies [23, 24] suggest that the energetic particles were accelerated at an earlier time and trapped behind the shock. They convected more slowly than the propagating shock into the distant heliosphere. Below we present another possible explanation.

The DSA mechanism relies on magnetic turbulence to scatter particles back and forth across the shock, but the contribution of the turbulence to the direct energization of particles may be underestimated. MHD turbulence downstream of the shock dissipates via highly dynamical localized reconnection events, generating magnetic islands, which can further energize charged particles through local stochastic processes. Magnetic reconnection in the solar wind can occur at current sheets of various scales [19]. As a fundamental physical process related to energy conservation, magnetic reconnection has been widely invoked to explain the energization and heating of space plasma through the conversion of magnetic energy into plasma kinetic energy [25, 26]. A statistical acceleration mechanism for charged particles related to magnetic reconnection was first proposed by Matthaeus et al. [27], Goldstein et al. [28], and Ambrosiano et al. [29], based on test-particle simulations. They found that turbulent fluctuations appearing in reconnection regions can trap test particles in the strong electric field for a time sufficiently long to produce significant acceleration. Based on these earlier ideas, several numerical simulations for particle acceleration via turbulent magnetic reconnection associated with multiple interacting magnetic islands have emerged [e.g., 29 - 35]. Simultaneous with these simulation developments, comprehensive transport theories for charged particles traversing numerous contracting and merging small-scale magnetic islands have been proposed by Zank et al. [11, 12] and further developed by le Roux et al. [37, 38, 39]. Reconnection associated with merging and contracting magnetic islands can yield a first-order Fermi process for the direct energization of particles trapped in the islands and also can be combined with other acceleration mechanisms, such as diffusive shock acceleration (DSA), to further accelerate trapped particles [12, 38]. Their theoretical solutions yield a power-law-like spectrum for the particle distribution.

Observational evidence suggests that the heliospheric current sheet (HCS) generates and confines small-scale flux ropes when disturbed. Due to the variability of the solar wind, magnetic reconnection occurs recurrently at the HCS [40], resulting in effective magnetic island merging in the vicinity of the HCS. When the HCS interacts with interplanetary shocks, magnetic island merging and contracting are likely to be more efficient because of the increased reconnection rate and strong compression [17, 18]. After the HTS crossing by V2, the direction of the magnetic field varied in a complicated and irregular manner [42]. Hill et al. [43] suggest that energetic particle intensities are enhanced in sectored inner heliosheath regions, i.e., in the vicinity of the wavy HCS that has been carried out into the inner heliosheath. This may indicate that the stochastic acceleration described by Zank et al. [11] may well be responsible for particle energization in the sectored heliosheath due to the interaction between HTS and the HCS. This provides a possible explanation for the observed strong enhancements in energetic particle flux behind the HTS.

2. Overview of the Observations

Figure 1 displays the variation of magnetic field and plasma properties associated with the V2 crossing of the HTS during the period between 2007 May 1 and 2008 May 1. The panels from top to bottom show, respectively, the interplanetary magnetic field (IMF) strength $|B|$, bulk flow speed V_{sw}, proton density N_p, proton temperature T_p, thermal pressure, magnetic pressure, total plasma pressure, and proton beta β. All data have been averaged over 1 day. During this period, *Voyager* 2 crossed the HTS at ∼84 au on 2007 August 30, as indicated by the dashed vertical line in each panel. The HTS crossing was characterized by a sudden drop in bulk flow speed, from supersonic to subsonic, the latter of which is due to the modification of the sound speed caused by PUIs [12, 45, 46, 47]. It is also accompanied by abrupt increases in the proton density, temperature, and plasma beta. The IMF strength did not show a large increase, but became very turbulent after the HTS crossing. It indicates that the HTS was a weak, quasi-perpendicular shock [45, 46, 47, 48]. The other properties, such as proton density, temperature, total pressure, and plasma beta all increase significantly after the crossing of HTS and exhibit a high level of fluctuations downstream of the shock.

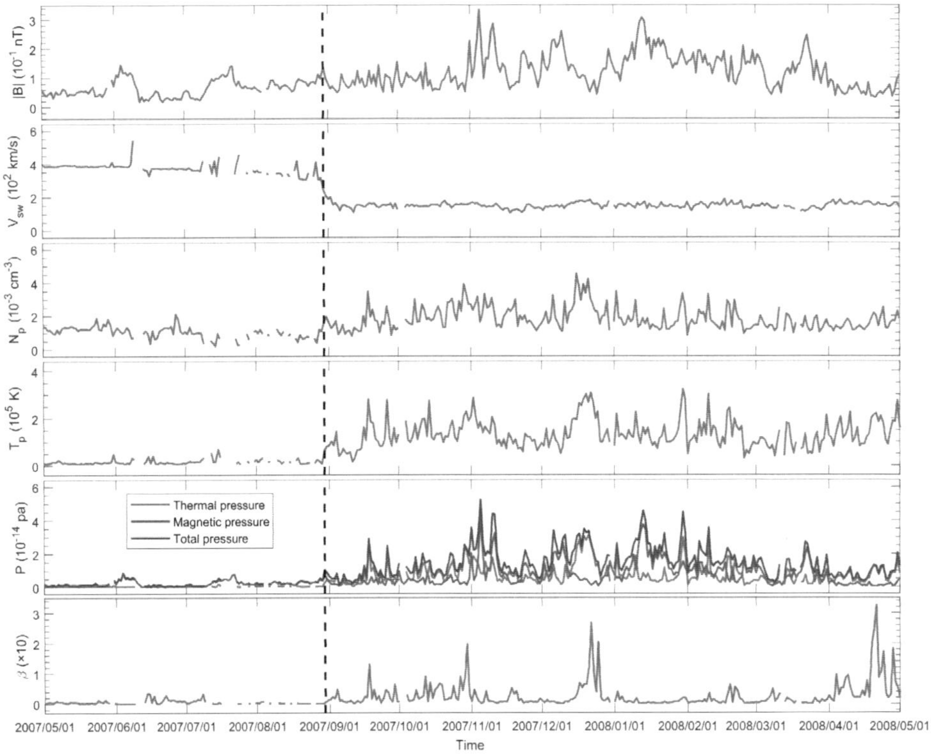

Figure 1. *Voyager* 2 observations from 2007 May 1 to 2008 May 1. From top to bottom, the panels are the daily averaged magnetic field strength B, solar wind speed V_{sw}, proton density N_p, proton temperature T_p, thermal pressure, magnetic pressure, total plasma pressure, and proton beta β. The HTS crossing is taken to be 2007 August 30 (dashed vertical line).

Figure 2 shows the probability distribution of the plasma beta in the upstream, downstream, and further downstream region. The changes of plasma beta in the upstream region were stable and most of them were less than or concentrated around 1, but in the downstream region, beta increased significantly and changed dramatically. About 35% of plasma beta has a value greater than 2 downstream of the HTS. The large plasma beta corresponds to a decrease in the magnetic field magnitude or an increase in plasma density and temperature, which indicates the possible existence of current sheets or discontinuities that bound the small-scale flux ropes. However, this is a curious result since only the thermal plasma is included and PUIs are neglected in the calculation of the plasma beta. Previous studies suggest that the thermal plasma remains relatively cold on transmission through the HTS and it is the PUIs that carry much of the energy [45, 46, 47]. The value of plasma beta calculated here is probably very much a lower bound.

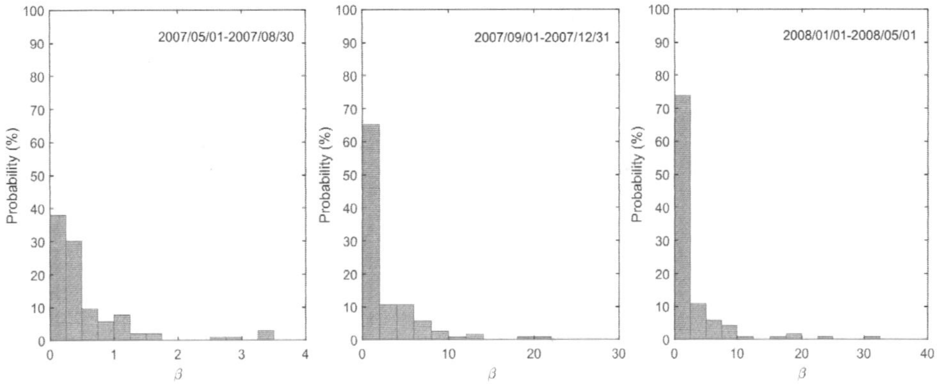

Figure 2. The probability distribution of the plasma beta during the period between 2007 May 1 to 2008 May 1. From left to right, panels correspond to the upstream region, downstream region, and further downstream region.

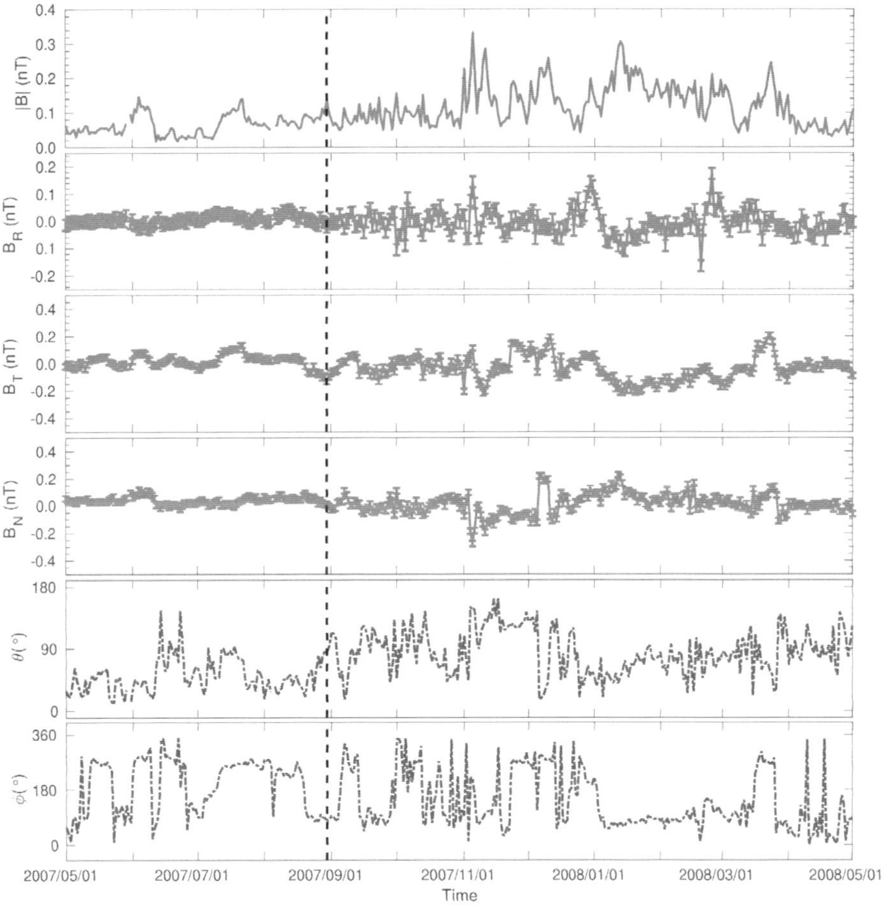

Figure 3. *Voyager* 2 magnetic field observations from 2007 May 1 to 2008 May 1. From top to bottom, the panels are the daily averaged magnetic field strength B, three components B_R, B_T, and B_N in the RTN coordinate system, the elevation θ and azimuthal ϕ angles of the magnetic field direction in the RTN coordinate system. The uncertainty of B_R, B_T, and B_N is also shown, respectively.

Figure 3 shows the evolution of the daily averaged magnetic field strength $|B|$, three magnetic field components B_R, B_T, B_N, and the elevation θ and azimuthal ϕ angles of the magnetic field direction measured by *Voyager* 2 from 2007 May 1 to 2008 May 1. The uncertainties in B_R, B_T, and B_N are also shown by way of reference. In the turbulent downstream region, we can clearly see frequent direction changes of the magnetic field as shown in the IMF elevation θ and azimuthal ϕ panels. It indicates that the HTS may generate high levels of MHD turbulence, waves, vortices, and structures downstream. Furthermore, the HTS may be located in the vicinity of interplanetary sector boundaries, i.e., the wavy HCS, due to the rapid changes in magnetic field directions. The interaction of the HTS and the HCS may trap plasmoids that are formed downstream of the HTS and produce more structures because of an increased reconnection rate. All these factors suggest that plasmoid-related particle acceleration downstream of the HTS is likely to be important.

Figure 4 shows the *Voyager* 2 CRS observations of ACR proton flux immediately upstream and downstream of the HTS. The panels from top to bottom show the daily averaged energetic proton flux for the 7 energy bins corresponding to an energy range of [1.8, 12.8] MeV. Several key points are apparent. The first is that an exponential-like increase in the particle flux ahead of the shock is present, especially for low energies. This is consistent with DSA only and no additional acceleration processes ahead of the HTS are necessary, as discussed in Zank et al [12]. However, the downstream fluxes are completely different from the predictions of conventional DSA theory, which predicts that the particle intensity downstream of the shock should be constant with distance for all energies. Instead, each energy is amplified above its value at the HTS and the amplification is very clearly ordered by increasing energy, i.e., an increasing amplification with increasing particle energy.

Figure 5 shows the ACR proton spectra measured by the *Voyager* 2 CRS detector at 9 representative instances in panels from top left to bottom right, and organized in chronological order. We plot spectra using the 1-day averaged particle flux data. The 10 energy bands used here are 1.8–2.2 MeV, 2.2–3.0 MeV, 3.0–4.6 MeV, 4.6–6.2 MeV, 6.2–7.7 MeV, 7.7–10.3 MeV, 10.3–12.8 MeV, 12.8–15.3 MeV, 15.3–17.9 MeV, 17.9–22.3 MeV. We plot the average energy for each channel as the representative energy. To understand the spectral evolution, a linear fit is applied when a power-law shape is exhibited, and the corresponding power-law index is obtained and listed in the title of each panel. The top three panels show examples of the upstream particle spectra between 2007/7/16 and 2007/8/30. The top right panel shows that at the HTS crossing time, 2007/8/30, the lower end of the particle spectrum is enhanced dramatically and the spectrum looks like a power law. The best-fit power-law index is calculated to be -2.9 using the considered 10 channels (1.8 MeV–22.3 MeV). The number is quite different from the DSA prediction of -2.0, using the shock compression ratio $r \simeq 2.0$. This suggests that DSA may not be fully applicable to explain the energetic particle spectrum.

The observed energetic particle intensities are further enhanced behind the HTS. Here, we illustrate the evolution of the spectrum immediate downstream of the HTS in the middle three panels. The middle left panel shows the spectrum near the HTS crossing at 2007/10/13, where there is a slight increase for all energies compared to the top right panel. After this point, the low-energy particle fluxes do not change much while the medium to high-energy particles fluxes continue increasing until 2007/11/20 (middle panel), and then all the energy channels start to decrease as shown in the middle right panel. From the top right panel to the middle right panel, the spectrum becomes harder in the immediate downstream region and the power-law index varies from -2.9 to -2.0 corresponding to the period between 2007/8/30 to 2007/12/21. The spectra further downstream are plotted in the bottom panels. The spectrum becomes softer as the high energy particle fluxes decrease rapidly. The power-law indices change from -2.07 to -2.85 during this period in the downstream region further from the shock. The observed increase in distance between the particle intensity peak and the shock front with increasing energy was

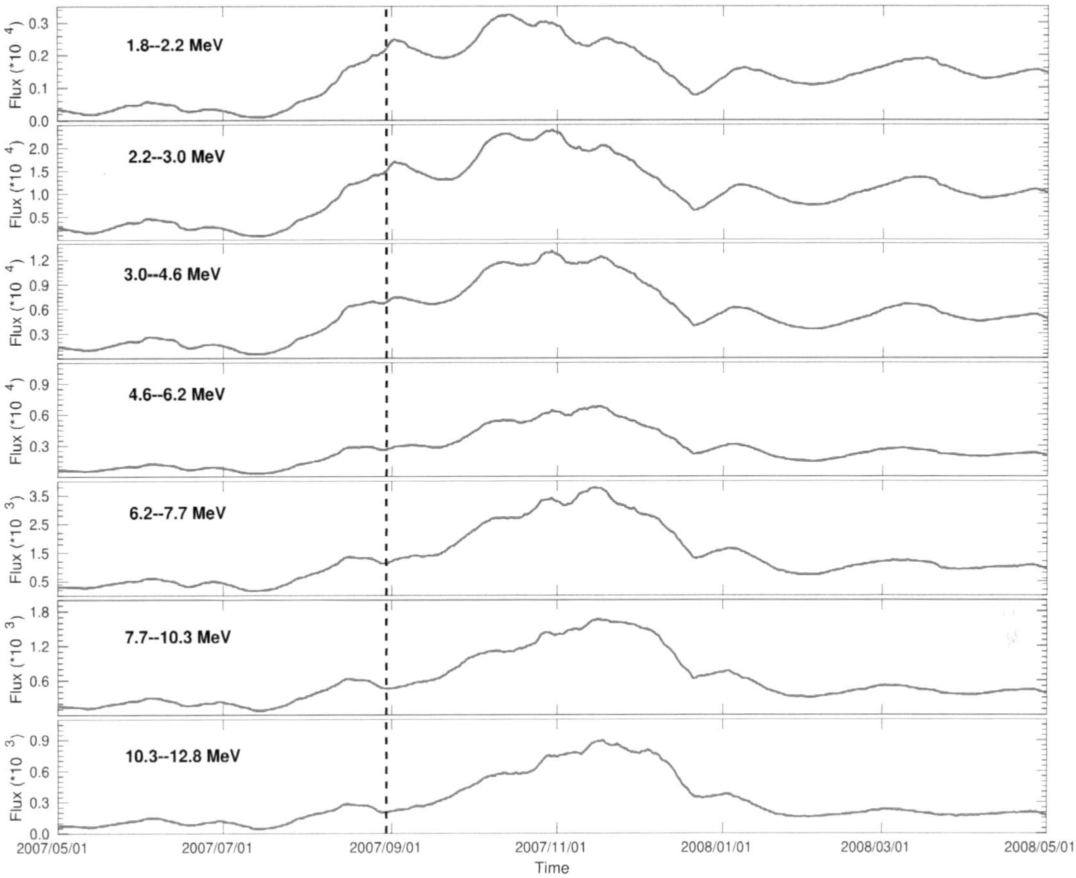

Figure 4. *Voyager* 2 observations of daily averaged energetic proton flux for the 7 energy ranges listed in the panels.

predicted by the Zank et al. [11, 12] statistical transport theory for energetic particle interaction with numerous magnetic islands. The features presented here are similar to previous analyses presented in Zhao et al. [21, 22], and can be reproduced quantitatively using the extended Zank et al. [11] transport model.

3. Particle acceleration by magnetic island dynamics

In the Zank et al. statistical transport theory, three basic processes associated with magnetic islands are thought to be responsible for the energization of ions and electrons. The first is due to magnetic island contraction. Particles trapped within a contracting island experience repeated reflections at the ends of converging mirrors, gaining energy in a first-order Fermi manner [30]. The magnetic island contraction leads to an increase in the particle velocity both parallel and perpendicular to the local magnetic field. A second process for particle energization is due to magnetic islands merging. The magnetic field lines shorten as the merging progresses. By sampling multiple merging magnetic islands, particles can be energized via a second-order Fermi process [44]. In this case, a particle gains in parallel energy and loses in perpendicular energy. A third energization mechanism is due to the reconnection induced electric field generated by island merging. Particles trapped in a merging magnetic island experience an extended period of interaction with the reconnection electric field, and thus gain significant energy. Zank et

Journal of Physics: Conference Series **1332** (2019) 012020 doi:10.1088/1742-6596/1332/1/012020

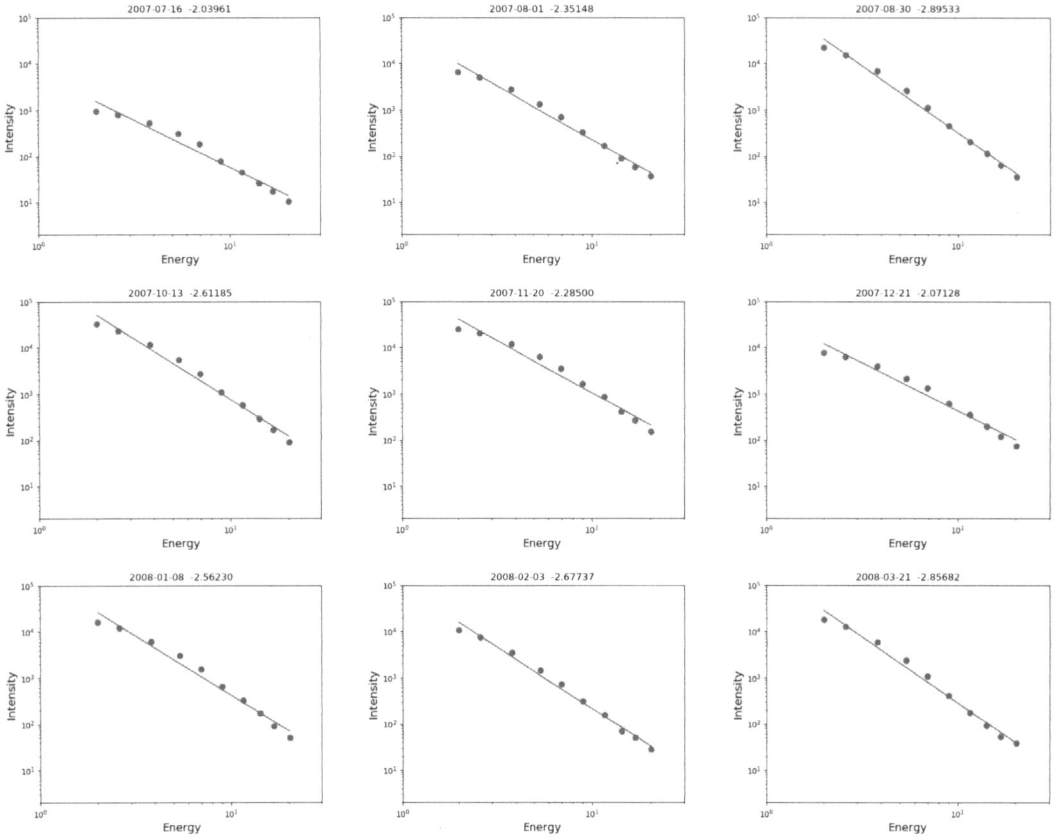

Figure 5. Energetic proton spectra at 9 instances over the energy range 1.8-22.3 MeV. Panels from top left to bottom right show, in chronological order, the spectra in the upstream region (top three panels), immediate downstream region (middle three panels), and further downstream region (bottom three panels). A power-law fitting is applied to the data for each case. Approximate power-law indices are obtained and listed in the title of each panel together with the time of occurrence.

al. combined these three acceleration processes and developed a stochastic transport theory for charged particles propagating in a turbulent region filled with numerous interacting small-scale magnetic islands. They derived a gyrophase-averged transport equation that describes the transport of charged particles in a "sea" of interacting magnetic islands. By supposing that particle scattering over the large-scale is sufficiently fast, one can simplify the gyrophase-averaged transport equation by assuming that the particle distribution is nearly isotropic. The general transport equation for a nearly isotropic particle distribution is given by [11]

$$\begin{aligned}\frac{\partial f}{\partial t} + (U + 3V_E)\frac{\partial f}{\partial x} &- \frac{v}{3}\frac{\partial U}{\partial x}\frac{\partial f}{\partial v} + \frac{1}{v^2}\frac{\partial}{\partial v}\left(\frac{v^3}{3}2\eta_c f\right) \\ &= \frac{\partial}{\partial x}\left(\kappa\frac{\partial f}{\partial x}\right) - v\frac{\partial}{\partial x}\left(V_E\frac{\partial f}{\partial v}\right) - V_E v\frac{\partial^2 f}{\partial x \partial v} \\ &+3\frac{V_E^2}{\tau_s}\frac{1}{v^2}\frac{\partial}{\partial v}\left(v^2\frac{\partial f}{\partial v}\right)\end{aligned}$$

$$+\frac{1}{5}\left(\frac{1}{3}\frac{\partial U}{\partial x}+\frac{\eta_c}{3}+\eta_m\right)^2\frac{1}{v^2}\frac{\partial}{\partial v}\left(\frac{v^4\tau_s}{3}\frac{\partial f}{\partial v}\right), \tag{1}$$

where U is the bulk flow velocity, V_E is the anti-reconnection electric field induced velocity, η_c and η_m are characteristic magnetic island contraction and merging rates, τ_s is the particle scattering time, κ is the spatial diffusion tensor, x and v are particle position and speed, respectively. Two particle energization terms are present, one due to the divergence of the large-scale flow velocity, $\nabla \cdot U$, which is the familiar term responsible for the energization of particles at shocks, and the other energization terms are due to magnetic island contraction (η_c) and merging (η_m). The energization term associated with magnetic island merging enters at the second order.

We consider a steady-state 1D problem with an incompressible large-scale flow, $\nabla \cdot U = 0$. Particles are injected at the origin $x = 0$ with a fixed initial speed v_0. We neglect second-order energy diffusion terms and assume a constant spatial diffusion coefficient. An escape term corresponding to particle loss from a finite acceleration region was introduced in Zhao et al. [21]. Based on these assumptions, Equation (1) can be simplified as follows.

$$\frac{\partial^2 f}{\partial x^2}-2\frac{V_E}{\kappa}\frac{\partial^2 f}{\partial x \partial \xi}-\frac{U+3V_E}{\kappa}\frac{\partial f}{\partial x}-\frac{2\eta_c}{3\kappa}\frac{\partial f}{\partial \xi}-\frac{2\eta_c}{\kappa}f=-\frac{Q}{\kappa}v_0\delta(v-v_0)\delta(x)+\frac{f}{\kappa\tau_e}, \tag{2}$$

where $\xi = \ln(v/v_0)$, τ_e is the particle escape time, and $Q = n_0/(4\pi v_0^2)$ with n_0 particle number density. Equation (2) can be solved analytically by assuming that all parameters (U, V_E, τ_e, κ, and η_c) are homogeneous in time and space, and are energy independent.

Zhao et al. [21] obtained the analytic solution of Equation (2) for the particle velocity distribution function

$$\begin{aligned} f\left(x,\frac{v}{v_0}\right) &= \frac{n_0}{8\pi v_0^2 V_E}\left(\frac{v}{v_0}\right)^{-(3+M_E+2\tau_d/(3\tau_c)M_E^2)/2}\exp\left[-\frac{\tau_d}{3\tau_c}M_E\frac{x}{L_{diff}}\right] \\ &\times I_0(\Phi)H\left(\ln\left(\frac{v}{v_0}\right)\right)H\left[\ln\left(\frac{v}{v_0}\right)+\frac{1}{M_E}\frac{x}{L_{diff}}\right]; \end{aligned} \tag{3}$$

$$\Phi=\sqrt{\frac{\tau_d}{3\tau_c}M_E^2\left(M_E-3+\frac{\tau_d}{3\tau_c}M_E^2\right)-\frac{\tau_d}{\tau_e}M_E}\left[\ln^2\left(\frac{v}{v_0}\right)+\frac{2}{M_E}\ln\left(\frac{v}{v_0}\right)\frac{x}{L_{diff}}\right]^{1/2}, \tag{4}$$

where $\tau_c = 1/\eta_c$ is the characteristic island contraction time, $\tau_d = \kappa/U^2$ is the characteristic particle diffusion time, $L_{diff} = \kappa/U$ is the diffusion length scale, $M_E = U/V_E$ is a dimensionless parameter that characterizes the strength of the anti-reconnection electric field, and I_0 denotes the zeroth order modified Bessel function of the first kind.

Equation (3) is a power-law like solution. The power-law slope depends on M_E, $\frac{\tau_d}{\tau_c}$, and $\frac{\tau_d}{\tau_e}$. Figure 6 shows several examples of the solution (3) for two values of τ_d/τ_c and four values of τ_d/τ_e. The solutions are normalized to $f_0 = f(v/v_0 = 1)$ and evaluated at a fixed location $x/L_{diff} = 1$. For all cases, the spectrum is essentially a power law. Efficient particle escape corresponds to a small escape time τ_e or a large τ_d/τ_e value. Figure 6 shows that a larger τ_d/τ_e value produces a steeper spectrum, which is a natural consequence of particle escape.

Figure 7 illustrates the amplification factor and spatial distribution of particles accelerated via reconnection-associated processes with dynamically evolving and merging magnetic islands. The solution (3) requires the prescription of three parameters, M_E, τ_d/τ_c, and τ_d/τ_e. The colored lines represent solutions for different values of the normalized particle speed v/v_0. Figure 7 shows that the particle intensity amplification for a particular particle energy is predicted to increase with increasing energy.

We use solution (3) to model the observed ACR proton flux enhancement downstream of the HTS. Classical DSA theory predicts that the particle intensity downstream of the shock

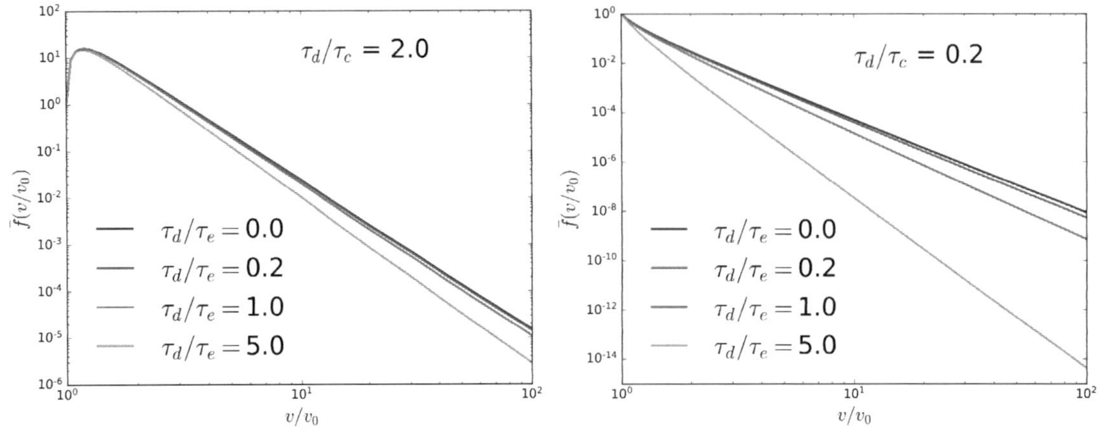

Figure 6. Normalized solution (3) at a fixed spatial location $x/L_{diff} = 1$. The two panels show solutions with $\tau_d/\tau_c = 2.0$ and 0.2, respectively. We choose four different τ_d/τ_e values: 0, 0.2, 1.0 and 5.0. The solutions are normalized to $f_0 = f(v/v_0 = 1)$, and $M_E = 10$ is assumed in all cases.

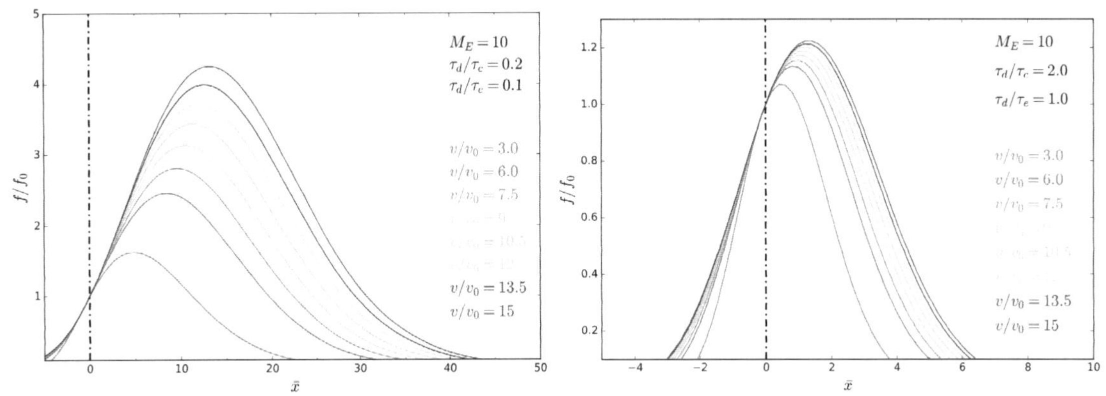

Figure 7. The spatial profile of solution (3) using fixed values of M_E, τ_d/τ_c, and τ_d/τ_e at different energies. The solutions are normalized to $f_0 = f(x = 0, \frac{v}{v_0})$ as measured at a boundary.

is constant with distance. Figure 4 shows that the downstream proton intensities depart significantly from the predictions of DSA since the flux of each energy channel is amplified relative to its value at the shock. High-energy particle (>4.6 MeV) fluxes peak more than 2 months after the HTS crossing. To model the observed flux amplification using solution (3), we need to determine the particle injection point. Although the most natural injection point might appear to be the shock front, low-energy particle fluxes decrease rapidly after the shock passage. To eliminate this DSA-attributed effect, we choose a point ~22 days behind the shock as the injection point (on 2007/9/21) where particle intensities are near a local minimum. This injection point is characterized by abrupt changes in the magnetic field direction (Figure 3), which signals the formation of a current sheet. After the crossing of the injection region, particles interact primarily with magnetic islands and cannot propagate back to the quasi-perpendicular HTS front easily. Therefore, locating the injection point a short distance downstream allows us to examine particle acceleration via flux rope-related processes.

Figure 8(a) shows the theoretical solution (3) with the observed ACR flux amplification overplotted as dashed lines. The boundary of the acceleration region is chosen on 2008/1/28 (corresponding to $x/L_{diff} = 19.5$), where the observed particle fluxes return to quiet levels. Both the analytic solutions and data are normalized to the values at the injection position $x = 0$ (dot-dashed vertical line). The trial-and-error fitting is done by adjusting the dimensionless parameters M_E, τ_d/τ_c, τ_d/τ_e and v_1/v_0. The particle velocity v is calculated as an average velocity in their corresponding energy range, and v_1 is the velocity corresponding to the lowest energy channel. To convert the time series data to the spatial coordinate (x), we assume a constant flow velocity $U_{flow} = 140$ km/s, which is the observed solar wind speed during this period. The spatial coordinate x is normalized by the diffusion length L_{diff}. The set of best fit parameters is: $M_E = 21$, $\tau_d/\tau_c = 0.2$, $\tau_d/\tau_e = 8.0$, $v_1/v_0 = 3$, and $L_{diff} = 8.1 \times 10^{12}$ cm. Although *Voyager* 2 measures the particle differential intensity instead of the velocity distribution function, the amplification factor is the same for both quantities.

Figure 8(a) shows clearly that the particle intensities are strongly enhanced with respect to the injection position for an extended period ($\sim$ 2 months), and the enhancement is more pronounced for high-energy particles. The largest amplification is about 1.6 times for the 1.8–2.2 MeV energy range and 2.6 times for the 10.3–12.8 MeV energy range. The amplification factor is in ascending order with increasing energy. The distance between the particle flux peak and the injection point appears to increase with increasing energy. For example, the flux amplification peaks at $x/L_{diff} \sim 5$ for the 1.8–2.2 MeV range, and at $x/L_{diff} \sim 10$ for the 10.3–12.8 MeV range. These features are quantitatively reproduced by our analytic model.

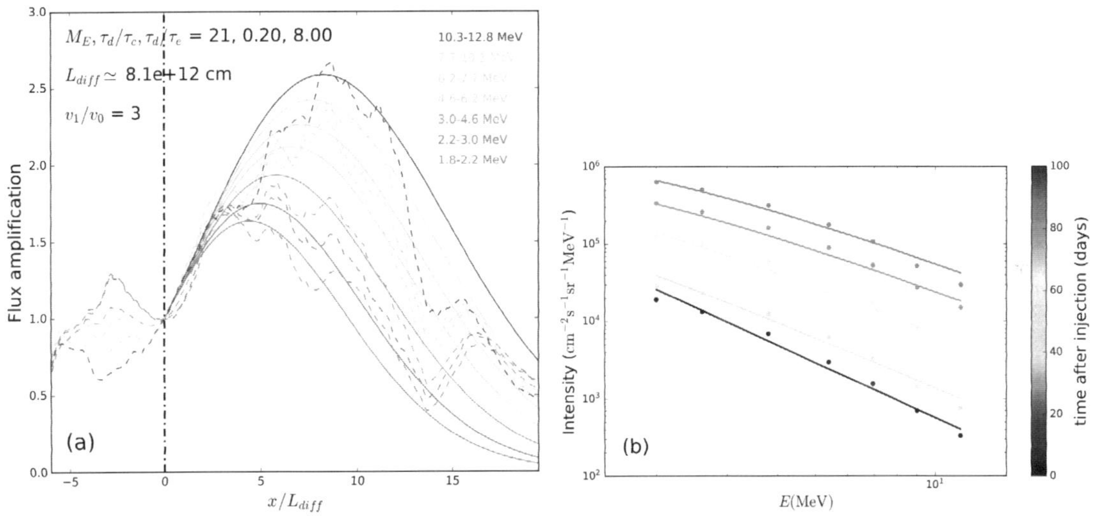

Figure 8. Left panel: particle flux amplification factor as a function of position. The dotted-dashed vertical line identifies the particle injection point at 2007 September 21. Both the observed fluxes and the theoretical distribution functions are normalized to their respective values at the injection point. Dashed lines of different colors denote the observed proton flux amplification factor. Solid lines show the theoretical amplification factor $f(x, v/v_0)/f(0, v/v_0)$ for the corresponding energy channel. Right panel: Evolution of the differential intensity spectrum. Solid dots denote observed particle intensities and solid lines the theoretical solutions. Both observational data and theoretical solutions are color coded according to the time after the injection point.

Consider now the energetic particle spectra downstream of the HTS. To directly compare with observations, Zhao et al. [21] converted the theoretical velocity distribution function f to the particle differential intensity $j(E)$. The differential intensity solution corresponding to Equation (3) is then

$$j\left(x, \frac{E}{E_0}\right) = j_0\left(\sqrt{\frac{E}{E_0}}\right)^{-(3+M_E+2\tau_d/(3\tau_c)M_E^2)/2} \frac{E}{E_0}\exp\left[-\frac{\tau_d}{3\tau_c}M_E\frac{x}{L_{diff}}\right]$$
$$I_0(\Phi)H\left(\ln\left(\sqrt{\frac{E}{E_0}}\right)\right)H\left[\ln\sqrt{\frac{E}{E_0}}+\frac{1}{M_E}\frac{x}{L_{diff}}\right]; \quad (5)$$

$$\Phi = \sqrt{\frac{\tau_d}{3\tau_c}M_E^2\left(M_E - 3 + \frac{\tau_d}{3\tau_c}M_E^2\right) - \frac{\tau_d}{\tau_e}M_E}\left[\left(\ln\sqrt{\frac{E}{E_0}}\right)^2 + \frac{2}{M_E}\ln\sqrt{\frac{E}{E_0}\frac{x}{L_{diff}}}\right]^{1/2}, \quad (6)$$

where j_0 is a normalization factor. The values of M_E, τ_d/τ_c, τ_d/τ_e and L_{diff} have been discussed. The factor j_0 is determined by applying a least squares fit of Equation (5) to the observed intensities at the injection point $x = 0$ (at 2007/9/21), which yields $j_0 = 5 \times 10^7$ $cm^{-2}s^{-1}sr^{-1}MeV^{-1}$.

Figure 8(b) shows the evolution of the particle differential intensity spectra, which are color-coded by time after injection. The observed intensities are shown by solid dots, and the theoretical intensities (5) by solid lines. We choose 5 instances for the spectral analysis: 0, 42, 60, 78, and 84, all in days after the injection time. The three later-time (60, 78, and 84 days) spectra for both observational and theoretical results are multiplied by a factor of 5, 20, 50, respectively, for presentation purposes. The particle spectra harden with increasing distance from the injection point. This feature is well reproduced by our theoretical solution.

4. Summary

We present *Voyager* 2 CRS observations showing the unusual enhancement of ACR proton fluxes downstream of the HTS at ~84 AU. The enhancement is inconsistent with predictions of classical DSA theory but can be explained by local acceleration associated with small-scale magnetic island dynamics. The Zank et al. [11] kinetic transport theory for stochastic particle acceleration via reconnection processes associated with multiple interacting magnetic islands is applied to model the unusual behavior of ACR downstream of HTS. The quasi-perpendicular HTS itself is responsible primarily for generating the downstream islands.

Observational and computational evidence for local particle acceleration associated with magnetically confined magnetic islands is accruing [35, 41]. Here, we employ this local particle acceleration mechanism to explain the puzzling ACR "time-intensity" profile observed at 84 AU. Specifically, we compare the normalized analytical solution (3) and the observed ACR flux amplification in Figure 8(a). The comparison between analytical differential intensities (5) and the observed particle spectra at different locations are shown in Figure 8(b). Our theoretical solutions agree reasonably well with observations. The modeling results show that stochastic acceleration by interacting magnetic islands can successfully predict the observed (i) peaking of ACR intensities downstream of HTS; (ii) increasing in the ACR flux amplification factor with increasing particle energy; and (iii) increase in distance between the particle intensity peak and the HTS with increasing particle energy. This work illustrates that our general theory of stochastic particle acceleration via magnetic reconnection processes, including particle injection, trapping and escape, can explain the observed unusual ACR "time-intensity" profile during the HTS crossing.

Acknowledgments
We acknowledge the partial support of the NSF EPSCoR RII-Track-1 Cooperative Agreement OIA-1655280, an NSF-DOE grant PHY 1707247 and NASA grant/subaward 0000167 (NNX17ABO4G).

References
[1] Reames, D. V. 1999, *Space Science Reviews*, 90, 413
[2] Zank, G. P., Rice, W. K. M., & Wu, C. C. 2000, *J. Geophys. Res.*, 105, 25079
[3] Zank, G. P., Li, G., & Verkhoglyadova, O. 2007, *Space Science Reviews*, 130, 255
[4] Li, G., Zank, G. P., & Rice, W. K. M. 2003, *J. Geophys. Res.*, 108, 1082
[5] Rice, W. K. M., Zank, G. P., & Li, G. 2003, *J. Geophys. Res.*, 108, 1369
[6] Verkhoglyadova, O. P., Zank, G. P., & Li, G. 2015, *Physics Reports*, 557, 1
[7] Axford, W. I., Leer, E., & Skadron, G. 1977, *International Cosmic Ray Conference*, 11, 132
[8] Krymskii, G. F. 1977, *DoSSR*, 234, 1306
[9] Bell, A. R. 1978, *MNRAS*, 182, 147
[10] Blandford, R. D., & Ostriker, J. P. 1978, *ApJ*, 221, L29
[11] Zank, G. P., le Roux, J. A., Webb, G. M., et al. 2014, *ApJ*, 797, 28
[12] Zank, G. P., Hunana, P., Mostafavi, P., et al. 2015, *APJ*, 814, 137
[13] Stone, E. C., Cummings, A. C., & McDonald, F. B., et al. 2005, *Science*, 309, 2017
[14] Decker, R. B., Krimigis, S. M., Roelof, E. C., et al. 2008, *Nature*, 454, 67
[15] Tessein, J. A., Matthaeus, W. H., Wan, M., et al. 2013, *ApJ*, 776, L8
[16] Tessein, J. A., Ruffolo, D., Matthaeus, W. H., et al. 2015, *ApJ*, 812, 68
[17] Khabarova, O., Zank, G. P., Li, G., et al. 2015, *ApJ*, 808, 181
[18] Khabarova, O., Zank, G. P., Li, G., et al. 2016, *ApJ*, 827, 122
[19] Khabarova, O., & Zank, G. P. 2017, *ApJ*, 843, 4
[20] Adhikari, L., Khabarova, O., Zank, G. P., & Zhao, L.-L. 2019, *ApJ*, 873, 72
[21] Zhao, L.-L., Zank, G. P., Khabarova, O., et al. 2018, *ApJ*, 864, L34
[22] Zhao, L.-L., Zank, G. P., Chen, Y., et al. 2019, *ApJ*, 872, 4
[23] Rice, W. K. M., Zank, G. P., Richardson, J. D., & Decker, R. B. 2000, *Geophys. Res. Lett.*, 27, 509
[24] McComas, D. J., & Schwadron, N. A. 2006, *Geophys. Res. Lett.*, 33, L04102
[25] Lu, S., Lu, Q., Huang, C., & Wang, S. 2013, *Physics of Plasmas*, 20, 061203
[26] Wang, H., Lu, Q., Huang, C., & Wang, S. 2016, *ApJ*, 821, 84
[27] Matthaeus, W. H., Ambrosiano, J. J., & Goldstein, M. L. 1984, *Physical Review Letters*, 53, 1449
[28] Goldstein, M. L., Matthaeus, W. H., & Ambrosiano, J. J. 1986, *Geophys. Res. Lett.*, 13, 205
[29] Ambrosiano, J., Matthaeus, W. H., Goldstein, M. L., & Plante, D. 1988, *J. Geophys. Res.*, 93, 14383
[30] Drake, J. F., Swisdak, M., Che, H., & Shay, M. A. 2006, *Nature*, 443, 553
[31] Oka, M., Phan, T.-D., Krucker, S., Fujimoto, M., & Shinohara, I. 2010, *ApJ*, 714, 915
[32] Le, A., Karimabadi, H., Egedal, J., Roytershteyn, V., & Daughton, W. 2012, *Physics of Plasmas*, 19, 072120
[33] Guo, F., Li, H., Daughton, W., & Liu, Y.-H. 2014, *Physical Review Letters*, 113, 155005
[34] Guo, F., Liu, Y.-H., Daughton, W., & Li, H. 2015, *ApJ*, 806, 167
[35] Du, S., Guo, F., Zank, G. P., Li, X., & Stanier, A. 2018, *ApJ*, 867, 16
[36] Lu, Q., Wang, H., Huang, K., Wang, R., & Wang, S. 2018, *Physics of Plasmas*, 25, 072126
[37] le Roux, J. A., Zank, G. P., Webb, G. M., & Khabarova, O. 2015, *ApJ*, 801, 112
[38] le Roux, J. A., Zank, G. P., Webb, G. M., & Khabarova, O. V. 2016, *ApJ*, 827, 47
[39] le Roux, J. A., Zank, G. P., & Khabarova, O. V. 2018, *ApJ*, 864, 158
[40] Gosling, J. T., McComas, D. J., Skoug, R. M., & Smith, C. W. 2006, *Geophys. Res. Lett.*, 33, L17102
[41] Adhikari, L., Khabarova, O., Zank, G. P., & Zhao, L.-L. 2019, *ApJ*, 873, 72
[42] Burlaga, L. F., & Ness, N. F. 2009, *ApJ*, 703, 311
[43] Hill, M. E., Decker, R. B., Brown, L. E., et al. 2014, *ApJ*, 781, 94
[44] Drake, J. F., Swisdak, M., & Fermo, R. 2013, *ApJ*, 763, L5
[45] Mostafavi, P., Zank, G. P., & Webb, G. M. 2017, *ApJ*, 841, 4
[46] Mostafavi, P., Zank, G. P., & Webb, G. M. 2018, *ApJ*, 868, 120
[47] Zank, G. P., Adhikari, L., Zhao, L.-L., et al. 2018, *ApJ*, 869, 23
[48] Florinski, V., Decker, R. B., le Roux, J. A., & Zank, G. P. 2009, *Geophysical Research Letters*, 36, 12

AUTHOR INDEX